Wolfgang Hackbusch (Ed.)

Robust Multi-Grid Methods

Notes on Numerical Fluid Mechanics
Volume 23

Series Editors: Ernst Heinrich Hirschel, München
Kozo Fujii, Tokyo
Keith William Morton, Oxford
Earll M. Murman, M.I.T., Cambridge
Maurizio Pandolfi, Torino
Arthur Rizzi, Stockholm
Bernard Roux, Marseille

(Addresses of the Editors: see last page)

- Volume 1 Boundary Algorithms for Multidimensional Inviscid Hyperbolic Flows (K. Förster, Ed.)
- Volume 2 Proceedings of the Third GAMM-Conference on Numerical Methods in Fluid Mechanics (E. H. Hirschel, Ed.) (out of print)
- Volume 3 Numerical Methods for the Computation of Inviscid Transonic Flows with Shock Waves (A. Rizzi / H. Viviand, Eds.)
- Volume 4 Shear Flow in Surface-Oriented Coordinates (E. H. Hirschel / W. Kordulla)
- Volume 5 Proceedings of the Fourth GAMM-Conference on Numerical Methods in Fluid Mechanics (H. Viviand, Ed.) (out of print)
- Volume 6 Numerical Methods in Laminar Flame Propagation (N. Peters / J. Warnatz, Eds.)
- Volume 7 Proceedings of the Fifth GAMM-Conference on Numerical Methods in Fluid Mechanics (M. Pandolfi / R. Piva, Eds.)
- Volume 8 Vectorization of Computer Programs with Applications to Computational Fluid Dynamics (W. Gentzsch)
- Volume 9 Analysis of Laminar Flow over a Backward Facing Step (K. Morgan / J. Periaux / F. Thomasset, Eds.)
- Volume 10 Efficient Solutions of Elliptic Systems (W. Hackbusch, Ed.)
- Volume 11 Advances in Multi-Grid Methods (D. Braess / W. Hackbusch / U. Trottenberg, Eds.)
- Volume 12 The Efficient Use of Vector Computers with Emphasis on Computational Fluid Dynamics (W. Schönauer / W. Gentzsch, Eds.)
- Volume 13 Proceedings of the Sixth GAMM-Conference on Numerical Methods in Fluid Mechanics (D. Rues / W. Kordulla, Eds.) (out of print)
- Volume 14 Finite Approximations in Fluid Mechanics (E. H. Hirschel, Ed.)
- Volume 15 Direct and Large Eddy Simulation of Turbulence (U. Schumann / R. Friedrich, Eds.)
- Volume 16 Numerical Techniques in Continuum Mechanics (W. Hackbusch / K. Witsch, Eds.)
- Volume 17 Research in Numerical Fluid Dynamics (P. Wesseling, Ed.)
- Volume 18 Numerical Simulation of Compressible Navier-Stokes Flows (M. O. Bristeau / R. Glowinski / J. Periaux / H. Viviand, Eds.)
- Volume 19 Three-Dimensional Turbulent Boundary Layers – Calculations and Experiments (B. van den Berg / D. A. Humphreys / E. Krause / J. P. F. Lindhout)
- Volume 20 Proceedings of the Seventh GAMM-Conference on Numerical Methods in Fluid Mechanics (M. Deville, Ed.)
- Volume 21 Panel Methods in Fluid Mechanics with Emphasis on Aerodynamics (J. Ballmann / R. Eppler / W. Hackbusch, Eds.)
- Volume 22 Numerical Simulation of the Transonic DFVLR-F5 Wing Experiment (W. Kordulla, Ed.)
- Volume 23 Robust Multi-Grid Methods (W. Hackbusch, Ed.)

Wolfgang Hackbusch (Ed.)

Robust Multi-Grid Methods

Proceedings of the Fourth GAMM-Seminar,
Kiel, January 22 to 24, 1988

Friedr. Vieweg & Sohn Braunschweig / Wiesbaden

CIP-Titelaufnahme der Deutschen Bibliothek

Robust multigrid methods: Kiel, January 22 to 24, 1988/Wolfgang Hackbusch (ed.). – Braunschweig; Wiesbaden: Vieweg, 1988
(Proceedings of the ... GAMM seminar; 4)
(Notes on numerical fluid mechanics; Vol. 23)
ISBN 3-528-08097-3

NE: Hackbusch, Wolfgang [Hrsg.]; Gesellschaft für Angewandte Mathematik und Mechanik: Proceedings of the ...; 2. GT

Manuscripts should have well over 100 pages. As they will be reproduced photomechanically they should be typed with utmost care on special stationary which will be supplied on request. In print, the size will be reduced linearly to approximately 75 %. Figures and diagramms should be lettered accordingly so as to produce letters not smaller than 2 mm in print. The same is valid for handwritten formulae. Manuscripts (in English) or proposals should be sent to the general editor Prof. Dr. E. H. Hirschel, Herzog-Heinrich-Weg 6, D-8011 Zorneding.

Vieweg is a subsidiary company of the Bertelsmann Publishing Group.

Produced by W. Langelüddecke, Braunschweig
Printed in Germany

ISSN 0179-9614

ISBN 3-528-08097-3

F O R E W O R D

During the weekend from January 22 to 24, 1988, the "Fourth GAMM-Seminar Kiel" was held at the Christian-Albrechts-Universität in Kiel. The conference was suggested by the GAMM-Committee "Effiziente numerische Verfahren für partielle Differentialgleichungen". We were pleased to have 78 participants from 11 countries.

These proceedings contain twenty contributions to the topic ROBUST MULTI-GRID METHODS. Some of the papers present new algorithmical ideas. Partially, these algorithms are combinations of techniques from conjugate gradient methods and from multi-grid algorithms. Interesting contributions are concerned with the construction and robustness of ILU smoothers (ILU: incomplete LU decompositions). Several papers study the multi-grid application to the Navier-Stokes and Euler equations.

The editor and organiser of the seminar would like to thank the DFG (Deutsche Forschungsgemeinschaft) for their support. We like to express our gratitude to Mr. Burmeister and the other persons involved in the organisation of the seminar and in the preparing of the conference volume.

Kiel, Mai 1988 W. Hackbusch

C O N T E N T S

A robust preconditioner based on algebraic substructuring and two-level grids

O. Axelsson and B. Polman*

Department of Mathematics

University of Nijmegen, Nijmegen, The Netherlands

Abstract

A domain decomposition method is used to construct a new type of block matrix incomplete factorization method. The properties of this method are such that it can be used as an efficient (i.e. with low computational complexity) and robust, corrector on a coarse mesh. Since the cost of it is of optimal order of computational complexity there is no need to use any further levels of grids as it is in a classical multigrid method. Combined with a smoother on the fine mesh the method turns out to perform as well on difficult problems as on model type problems and with a complexity about as low as that for a classical multigrid method on the model problems. The method is well suited for vector- and parallel computers. The smoothing-correction forms a V-cycle step which can be used as a preconditioner for a conjugate gradient method, thus guaranteeing convergence. However, the method is so efficient that there is rarely any need for convergence acceleration.

Keywords: Smoothing-correction, domain-decomposition, algebraic substructuring, robust preconditioner, two-level grids.

AMS Subject classification: 65F10, 65N20.

---

* The second author's research was supported by the Netherlands organization for scientific research (NWO)

1. Introduction

In full multigrid methods for elliptic difference equations one works on a sequence of meshes where a number of pre- and/or postsmoothing steps are performed on each level. As is well known these methods can converge very fast on problems with a smooth solution and a regular mesh, but the rate of convergence can be severely degraded for problems with unisotropy or discontinuous coefficients unless some form of robust smoother is used. Also problems can arise with the increasingly coarser meshes because for some types of discretization methods, coercivity may be lost on coarse meshes and on massively parallel computers the computation cost of transporting information between computer processors devoted to work on various levels of the mesh can dominate the whole computing time. For discussions about some of these problems, see [11].

Here we propose a method that uses only two levels of meshes, the fine and the coarse level, respectively, and where the corrector on the coarse level is equal to a new type of preconditioner which uses an algebraic substructuring of the stiffness matrix. It is based on the block matrix tridiagonal structure one gets when the domain is subdivided into strips. This block-tridiagonal form is used to compute an approximate factorization whereby the Schur complements which arise in the recursive factorization are approximated in an indirect way, i.e. so that its action on some specially chosen vectors is conserved in the incomplete factorization. This construction turns out to give such a favourable eigenvalue distribution that the method gets similar properties as the ideal corrector, i.e. the stiffness matrix itself on the coarse mesh. Therefore it can be used as a corrector on the coarse mesh instead of the stiffness matrix. Since the computational complexity of it is proportional to the number of mesh nodes there is no need to use a full sequence of meshes. Combined with classical smoothers such as Jacobi iteration, red-black block Gauss-Seidel iteration or (block) matrix incomplete factorization iteration methods on the fine mesh (postsmoothing is found to be most efficient), this smoothing-correction method turns out to have excellent convergence properties both for regular problems and for problems with some singular perturbation parameter, such as in discontinuous coefficient type problems.

The computational complexity of the method is so low that it is competitive with classical multigrid methods even on model type problems, where multigrid methods show their best performance. Each iteration step of the method is a two-level V-cycle step and can be used as a preconditioning step. The two-level method can be used as a preconditioner for a conjugate gradient acceleration method, hence guaranteeing convergence for any positive definite problem, or for that matter of any problem with a positive real matrix if a generalized conjugate gradient method such as the GCGLS method in [1] is used as an acceleration method.

Two-level methods have previously been used with conjugate gradient methods in [3] and [5]. For another recent discussion of this, see [14]. In our present case it turns out that the preconditioner is so efficient that there is rarely any need for an acceleration on top of the V-cycles.

Although there is no strict proof of that the present V-cycle method has a computational complexity close to of optimal order, the numerical evidence indicates that the number of iterations increases not faster than $O(\log h)$, when the stepsize parameter $h \to 0$, a result known to hold for the V-cycle classical multigrid method assuming a sufficiently regular problem (see [9]).

The method vectorizes well and can be made to work in parallel between the subdomains if we use an odd-even cyclic reduction ordering method. This latter method has however not been used in the present paper.

Note that if a method which is efficient on a parallel computer with p parallel processors has a computational complexity $\sim C_p h^{-\alpha}$ whereas there exists an algorithm on a serial computer which has a computational complexity $C_1 h^{-\beta}$, of lower order, i.e. with $\beta < \alpha$, then the computing times on the parallel computers will eventually be larger than that on the serial computer, because $\frac{1}{p} C_p h^{-\alpha} > C_1 h^{-\beta}$, if h is sufficiently small. Therefore it is essential to use algorithms of optimal order of computational complexity also on parallel computers.

The present paper is organized as follows. In Section 2 we present the incomplete factorization algorithm and its properties. In Section 3 we describe the smoothing-correction scheme and prove some essential properties for it, while in Section 4 a comparison is made of using the incomplete factorization as a corrector instead of the ideal corrector. Numerical results and conclusions are found in the final section.

2. The incomplete factorization method

Consider an elliptic problem

$$-\underline{\nabla}(A(x,y)\underline{\nabla} u = f(x,y), \quad (x,y) \in \Omega \subset \mathbf{R}^2 \tag{2.1}$$

with some Dirichlet or Neuman type boundary conditions on $\Gamma = \partial\Omega$. Here A, of order 2×2, is uniformly positive definite on Ω. For notational simplicity we assume that the domain is rectangular but this is of no fundamental concern for the method. The problem is discretized with finite differences or finite elements. The domain is divided into strips with k vertical meshlines within each subdomain, where we let the dividing line belong to the domain to the left of it. Using a lexicographic ordering and a matrix partitioning corresponding to this domain decomposition we get a matrix on block tridiagonal form

$$A = \text{tridiag}\,(A_{i,i-1}, A_{i,i}, A_{i,i+1}), \quad i = 1, 2, \ldots, p$$

where p is the number of subdomains.

This block-matrix structure will be used to compute an approximate block-matrix factorization of A. Consider first the structure of the block-matrices. Each diagonal A_{ii} corresponds to the nodes on a subdomain (the ith) and $A_{i,i-1}$ and $A_{i,i+1}$ provides the coupling between the subdomains. Since only the last line of each subdomain, namely the dividing line for that domain, is coupled to the next subdomain, these latter blocks contain just one nonzero block-matrix, whose order is equal to the number of points, m along each vertical line. If we use a lexicographic ordering of the lines the matrices $A_{i,i}$ become themselves block tridiagonal,

$$A_{i,i} = \text{tridiag}(B^{(i)}_{s,s-1}, B^{(i)}_{s,s}, B^{(i)}_{s,s+1}), \quad s = 1, 2, \ldots, k$$

and $A_{i,i-1}$ and $A_{i,i+1}$ contain just one nonzero block, which we denote by $E^{(i-1)}_{1,k}$ and $E^{(i)}_{k,1}$, namely in the upper right and lower left corner, respectively. Finally note that all nonzero blocks, $B^{(i)}_{s,r}$, $r = s-1, s, s+1$, $E^{(i-1)}_{1,k}$ and $E^{(i)}_{k,1}$ are sparse matrices, tridiagonal, bidiagonal or some may even be diagonal, depending on the type of discretization used and on the problem (piecewise constant coefficients or generally variable coefficients).

A block-matrix factorization of A takes the form

$$A = (X - L)(I - X^{-1}U)$$

where

$$L = \text{tridiag}\,(A_{i,i-1}, 0, 0) \tag{2.2a}$$

$$U = \text{tridiag}\,(0, 0, A_{i,i+1}) \tag{2.2b}$$

and

$$X = \text{diag}\,(X_1, X_2, \ldots, X_n).$$

Here the blockdiagonal matrix X must be computed. It satisfies the matrix recursion

$$X_1 = A_{1,1}, \; X_i = A_{i,i} - A_{i,i-1}X^{-1}_{i-1}A_{i-1,i}, \quad i = 2, \ldots, p.$$

Because of the sparsity structure of $A_{i,i-1}$ and $A_{i-1,i}$, an elementary computation shows that the last term has only one nonzero block namely in the upper left corner, so all blocks of X_i are equal to those of $A_{i,i}$, except $(X_i)_{1,1}$, the first matrix block, which satisfies

$$(X_i)_{1,1} = B^{(i)}_{1,1} - E^{(i-1)}_{1,k}(X^{-1}_{i-1})_{k,k}E^{(i-1)}_{k,1}, \quad i = 2, 3, \ldots, p.$$

The matrices $(X_i)_{1,1}$ computed in this way will be full, however, and the computation of the last block of the inverse X^{-1}_{i-1} will be somewhat expensive. (One way to compute it is to factorize the block tridiagonal matrix X_{i-1} as a block-matrix. Then $(X^{-1}_{i-1})_{k,k}$ is equal to the inverse of the final block in the corresponding matrix recursion, cf [4].) Following [7] we compute instead sparse approximations and in such a way that the action of the approximations on two particular vectors $\underset{\sim}{e}_i, i = 1, 2$ is the same as the action of the matrices we are approximating. The choice of vectors we make are the "consistency" vectors,

$$\underset{\sim}{e}_1 = \underset{\sim}{e} = (1, 1, \ldots, 1)^t$$

and

$$\underline{e}_2 = \underline{v} = (1,2,\ldots,n)^t.$$

The structure of $(X_i)_{1,1}$ will be the same as in $B_{1,1}^{(i)}$, namely a symmetric tridiagonal matrix.

As is shown in [7], such a matrix can be computed with little computational effort. We have in fact:

Lemma 2.1. Let $\underline{b}_s = B\underline{e}_s$, $s = 1,2$, where $\underline{e}_s$ is defined above for some, possibly implicitly defined matrix B of order m. Then there exists a unique, symmetric, tridiagonal matrix $G = \text{tridiag}(g_{r,r-1}, g_{r,r}, g_{r,r+1})$, $r = 1,2,\ldots,m$, such that $G\underline{e}_s = B\underline{e}_s$, $s = 1,2$. Here

$$g_{r+1,r} = g_{r,r+1} = g_{r,r-1} + (\underline{b}_2 - r\underline{b}_1)_r, \quad r = 1,2,\ldots,m-1 \quad (g_{1,0} = 0)$$

and

$$g_{r,r} = (\underline{b}_1)_r - g_{r,r-1} - g_{r,r+1}\,. \qquad \diamond$$

Note that the explicit entries of B are not required only its action on vectors must be computable. In our case B corresponds to the matrices $B_{1,1}^{(i)} - E_{1,k}^{(i-1)}(X_{i-1}^{-1})_{k,k}E_{k,1}^{(i-1)}$, which are Schur-complement matrices and are not computed explicitly. Each action means therefore among other things a solution of a linear system with the subdomain block X_{i-1}. We shall comment more about this later on.

Since X_i is equal to $A_{i,i}$ but with the top diagonal block replaced by a matrix G_i of the form defined above, we need to know if G_i is positive definite, for instance.

Theorem 2.1. Let B be s.p.d. with positive rowsums, i.e. with $\underline{b}_1 = B\underline{e} > 0$ and entries $b_{i,j}$ satisfying $b_{i,j} \le 0$, $j \ge i+2$. Then the quadratic form of G defined in Lemma 2.1 is bounded below by the quadratic form of B, i.e.

$$(G\underline{x},\underline{x}) \ge (B\underline{x},\underline{x}) \quad \forall \underline{x} \in \mathbf{R}^m.$$

Proof. It follows from Theorem 3.1 in [7] that $\alpha = 1$, which shows the present special case. $\diamond$

Note that theorem 2.1 shows that there is no "stability problem" because positive definiteness will be preserved. However, when applying Theorem 2.1 to the recursively defined matrices G_i, where G_i is defined by

$$G_i\underline{e}_s = (B_{1,1}^{(i)} - E_{1,k}^{(i-1)}(X_{i-1}^{-1})_{k,k}E_{k,1}^{(i-1)})\underline{e}_s, \quad s = 1,2 \tag{2.3}$$

and

$$X_i = A_{i,i} + \text{diag}(\mathrm{G_i} - \mathrm{B}_{1,1}^{(i)}, 0, \ldots, 0), \quad \mathrm{i} = 2,3,\ldots,\mathrm{p}, \tag{2.4}$$

we must assure that X_i gets no positive entries in positions i,j where $|j-i| \ge 2$. This would be the case if X_{i-1} is a monotone matrix. To examine this we can use the following Lemma (see [7]).

Lemma 2.2. Let B be a symmetric matrix with positive rowsums, i.e. with $B\underline{e} > \underline{0}$. Then G, defined in Lemma 2.1, is a diagonally dominant M-matrix, if and only if

$$\sum_{i=1}^{r} \sum_{\substack{j=r+1 \\ b_{i,j}>0}}^{m} (j-i)b_{i,j} \leq \sum_{i=1}^{r} \sum_{\substack{j=r+1 \\ b_{i,j}<0}}^{m} (j-i)(-b_{i,j}),\ r = 1, \ldots, m\,. \tag{2.5}$$

Clearly, if $b_{i,j} \leq 0,\ i \neq j$, then G is an M-matrix. Also if $b_{i,j} \leq 0,\ |i-j| \geq 2$ and $b_{i,i-1}$ and $b_{i,i+1}$ are sufficiently small compared to the numerical values of the other entries in the same row, then one can show (see [2]) that the LU factorization of B contains factors L, U which are both M-matrices. Hence the inverses L^{-1}, U^{-1} are nonnegative and so is their product. Hence B is monotone. In particular, the first block $A_{1,1}$ is monotone if the entries of $A_{1,1}$ next to the main diagonal are numerically sufficiently small, if they are positive. Next Lemma 2.2 shows that the matrix G_2, defined by the action of $B_{1,1}^{(2)} - E_{1,k}^{(1)}(X_1^{-1})_{kk}E_{k,1}^{(1)}$, is an M-matrix (if 2.5 is valid) so the top diagonal block of X_2 is an M-matrix. Repeating the argument with the factorization of X_2 to show that it is monotone, we can show by recursion that G_i are M-matrices and X_i are monotone. Therefore all matrices $B = B_{1,1}^{(i)} - E_{1,k}^{(i-1)}(X_{i-1}^{-1})_{k,k}E_{k,1}^{(i-1)}$ satisfy the sign requirement $b_{i,j} \leq 0,\ |i-j| \geq 2$ and Theorem 2.1 is applicable for all matrices in the recursion. This shows

Theorem 2.2. Let A be s.p.d. with entries of the diagonal blocks as described above, let X_i be defined by the recursion (2.3), (2.4) and let

$$C = (X-L)(I-X^{-1}U)$$

where L, U are defined by (2.2a,b). Then

a) the spectrum of $C^{-1}A$ is contained in the interval (0,1]

b) at most $(p-1)(n-2)$ eigenvalues are not equal to 1.

Proof. The theorem has been shown in [7]. ◇

As an application consider a regular difference mesh approximation for (2.1). Then the corresponding difference matrix A is an M-matrix and all offdiagonal entries in B are nonpositive. Consider now the same mesh but with piecewise linear finite elements. Then, for variable coefficients, some offdiagonal entries (corresponding to couplings along the diagonals of the mesh) can be positive. However, if we order the points diagonalwise and also use diagonalwise oriented meshlines for the domain decomposition (see figure 2.1), the corresponding matrix will be partitioned into block matrices satisfying the requirements of theorem 2.2.

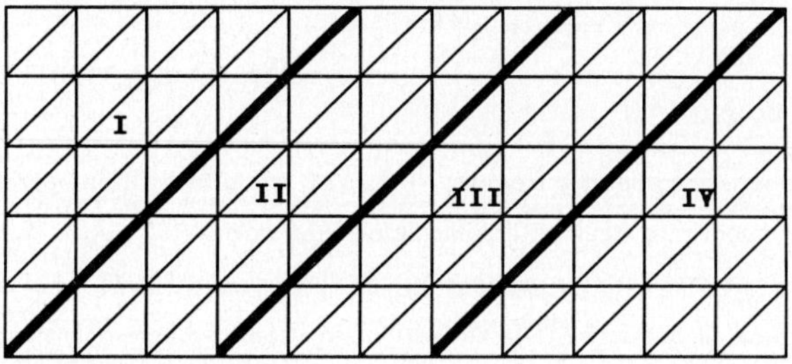

Figure 2.1. Piecewise linear f.e. mesh and subdomain decomposition.

3. The smoothing-correction method

3.1. The algorithm

To solve the system,

$$A_h \underline{u}_h = \underline{f}_h \tag{3.1}$$

on the fine mesh Ω_h, we'll use a standard two level multigrid method (see for instance [9]) except that we shall use the preconditioner defined in the previous section as corrector on Ω_{2h}. To this end we discretize problem (2.1) on Ω_h using a standard finite element or finite difference method to get A_h and on the coarse mesh Ω_H to get A_H and use the latter to construct the corrector C_H as described in the previous section. Note that A_H could also have been derived from A_h using a Galerkin type of method (see [13]) but we haven't considered this here.

Although usually we will take $H = 2h$ other choices are of course possible. On one hand one might take $H = 4h$ or $H = 8h$ for instance, if the problem under consideration is very smooth leading to a very efficient algorithm but on the other hand the choice $H = h$, making the method a one level smoothing-correction scheme, may be the best choice in case of wildly varying coefficients or a solution with very unsmooth behaviour in which case A_{2h} and thus C_{2h} are bad approximations of A_h.

The smoothing-correction scheme takes the following form:

Choose $\underline{x}_0$; set $\underline{r}_0^{(h)} = A_h \underline{x}_0 - \underline{f}_h$

<u>Loop</u> $\quad \underline{r}^{(H)} = R_h^H \underline{r}^{(h)}$

Solve $\quad C_H \underline{\delta}^{(H)} = -\underline{r}^{(H)} \qquad$ (3.2a)

$\underline{\delta}^{(h)} = P_H^h \underline{\delta}^{(H)} \qquad$ (3.2)

$$\underline{x} = \underline{x} + \underline{\delta}^{(h)}$$

$$\underline{r}^{(h)} = \underline{r}^{(h)} + A_h\underline{\delta}^{(h)} \qquad (\text{or } \underline{r}^{(h)} = A\underline{x} - \underline{f}_h)$$

$$\underline{r}^{(h)} = S_\nu(A_h)\underline{r}^{(h)}$$

If ($||\underline{r}^{(h)}|| \geq \varepsilon$) GOTO Loop .

Here R_h^H denotes the restriction operator from $\Omega_h \rightarrow \Omega_H$, P_H^h the prolongation operator from $\Omega_H \rightarrow \Omega_h$, and $S_\nu(A_h)$ denotes ν smoothing steps with some smoothing operator for A_h. Various choices for the smoothing operator are possible, (see for instance [9]), in the numerical tests we have used Jacobi, Red-Black Gauss-Seidel and an incomplete LU-factorization. We consider here only postsmoothing because it turned out to be most efficient in the numerical tests but pre-smoothing or pre- and post-smoothing are also possible.

Note that one iteration of algorithm (3.2) could also be used as a preconditioner in a Preconditioned Conjugate Gradient type of method, in which case for a symmetric positive definite problem convergence is assured.

What makes this algorithm a two level method instead of a full multigrid scheme is of course step (3.2.a) where we solve with C_H on Ω_H instead of using A_H. This choice makes the method much simpler, no nested recursions of the algorithm are needed to solve systems on increasingly coarser grids, and if C_H turns out to have about the same approximation properties as A_H on Ω_H to A_h than the method will have roughly the same convergence behaviour as full multigrid methods. Note that even if C_H doesn't approximate A_h as well as A_H then it is also possible to replace step (3.2.a) by some steps of a P.C.G. method with C_H as preconditioner to solve $A_H\underline{\delta}^{(H)} = -\underline{r}^{(H)}$.

Before returning to the question of the accuracy of C_H as corrector in the next section, we will first give a convergence proof of the two level scheme using A_H instead of C_H as correction (i.e. the ideal case) on Ω_H for the model problem on a square using only an elementary analysis. This will also give a clear insight into the mechanism of this smoothing-correction scheme. To this end we will determine the reduction factor, i.e. the constant factor with which the residual is at least reduced at each iteration. It is easily seen from algorithm (3.2) that the iteration matrix M for the residual is given by:

$$M = S_\nu(A_h)(I - A_h P_H^h C_H^{-1} R_h^H) \tag{3.3}$$

or in case of using the ideal correction

$$M = S_\nu(A_h)(I - A_h P_H^h A_H^{-1} R_h^H) \tag{3.4}$$

and it suffices to estimate $||M||$.

We will start with the one dimensional analogue since the two dimensional case follows mainly from this. Similar analysis are found in [8] and [9],for instance, but the authors have been unable to find all details in previous publications.

3.2. Reduction factor for a one dimensional model problem

Consider

$$\begin{cases} -u'' = f, & 0 < x < 1 \\ u(0) = u(1) = 0 . \end{cases} \tag{3.5}$$

A central difference discretization on Ω_h leads to the following system

$$A_h \underline{x}_h = \underline{f}_h \quad \text{where} \quad A_h = \text{tridiag}\, \frac{1}{h^2}[-1,2,-1] \quad (h = \frac{1}{n+1}) \tag{3.6}$$

where we assume that $n = 2^r - 1$, for some positive integer r.

The eigensolutions of (3.6) are $(\lambda_k^{(h)}, \underline{\phi}_k^{(h)})$, $k = 1,2,\ldots n$ where

$$(\underline{\phi}_k^{(h)})_i = \phi_k(x_i) = \sin k\pi x_i, \quad i = 1,2,\ldots n \quad (x_i = ik)$$

$$\lambda_k^{(h)} = (2h^{-1} \sin \frac{k\pi h}{2})^2.$$

Let $\tilde{\lambda}_k^{(h)} = \frac{h^2}{4}\lambda_k^{(h)}$, then $0 < \tilde{\lambda}_k^{(h)} < 1$.

(Note that $1 - \tilde{\lambda}_k^{(h)} = 1 - \sin^2(\frac{k\pi h}{2}) = \cos^2(\frac{k\pi h}{2}) = \sin^2 \frac{\pi}{2} h(n+1-k) = \tilde{\lambda}_{n+1-k}^{(h)}$).

Initial residual:

$$\underline{r}_h^0 = A_h \underline{x}_h^0 - \underline{f}_h =: \sum_{k=1}^{n} \alpha_k \underline{\phi}_k^{(h)} .$$

Restriction (Full Weighted):

The restriction operator is symbolized by:

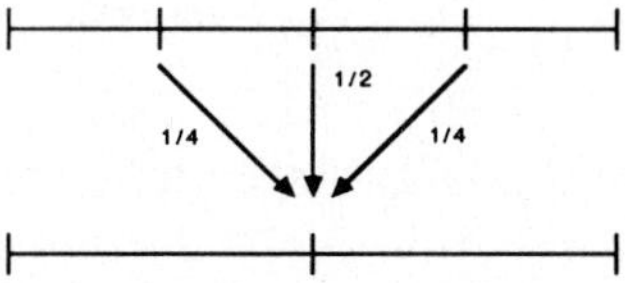

Therefore

$$\begin{aligned} \underline{r}_{2h}(x_i) &= \sum_{k=1}^{n} \alpha_k [\frac{1}{4}\phi_k(x_{i-1}) + \frac{1}{2}\phi_k(x_i) + \frac{1}{4}\phi_k(x_{i+1})] \\ &= \sum_{k=1}^{n} \alpha_k \frac{1}{2} \sin k\pi x_i (\cos k\pi h + 1) = \sum_{k=1}^{n} \alpha_k (1 - \tilde{\lambda}_k^{(h)}) \sin k\pi x_i \\ &= \sum_{k=1}^{n} \tilde{\lambda}_{n+1-k}^{(h)} \alpha_k \sin k\pi x_i \end{aligned}$$

(Note that the restriction acts as a smoothing operator.)

Note that for $k = 1,2,\ldots,\frac{n-1}{2}$: $\sin(n+1-k)\pi x_i = -\sin k\pi x_i$ for $x \in \Omega_{2h}$ i.e. for $i = 2,4,\ldots n-1$.

Therefore:

$$\underline{r}_{2h}(x_i) = \sum_{k=1}^{\frac{n-1}{2}} (\alpha_k \tilde{\lambda}_{n+1-k}^{(h)} - \alpha_{n+1-k}\tilde{\lambda}_k^{(h)}) \sin k\pi x_i =: \sum_{k=1}^{\frac{n-1}{2}} \tilde{\alpha}_k \sin k\pi x_i \quad x_i \in \Omega_{2h}$$

Correction on Ω_{2h}:

In this analysis we assume "ideal" correction i.e.

$$A_{2h}\underline{\delta}_{2h} = -\underline{r}_{2h} \Rightarrow \underline{\delta}_{2h}(x_i) = -\sum_{k=1}^{\frac{n-1}{2}} \frac{1}{\lambda_k^{(2h)}}\tilde{\alpha}_k \sin k\pi x_i$$

Prolongation:

The prolongation operator is symbolized by:

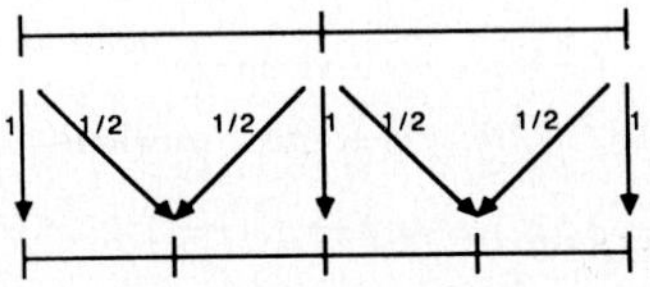

Therefore

$$\underline{\delta}_h(x_i) = \underline{\delta}_{2h}(x_i), \quad i = 2, 4, \ldots, n-1$$

$$\underline{\delta}_h(x_i) = \frac{1}{2}(\underline{\delta}_{2h}(x_{i-1}) + \underline{\delta}_{2h}(x_{i+1})), \quad i = 1, 3, \ldots, n$$

i.e. for $:= 1, 3, \ldots, n$:

$$\begin{aligned}\underline{\delta}_h(x_i) &= -\sum_{k=1}^{\frac{n-1}{2}} \frac{1}{\lambda_k^{(2h)}}\tilde{\alpha}_k \frac{1}{2}(\sin k\pi x_{i-1} + \sin k\pi x_{i+1}) \\ &= -\sum_{k=1}^{\frac{n-1}{2}} \frac{1}{\lambda_k^{(2h)}}\tilde{\alpha}_k(1 - 2\tilde{\lambda}_k^{(h)}) \sin k\pi x_i\end{aligned}$$

$(i.e.)$

$$\begin{aligned}\underline{\delta}_h(x_i) &= -\sum_{k=1}^{n-1} \frac{1}{\lambda_k^{(2h)}}\tilde{\alpha}_k \begin{Bmatrix} 1 - \tilde{\lambda}_k^{(h)} + \tilde{\lambda}_k^{(h)} \\ 1 - \tilde{\lambda}_k^{(h)} - \tilde{\lambda}_k^{(h)} \end{Bmatrix} \sin k\pi x_i; \quad \begin{matrix} i & \text{even} \\ i & \text{odd} \end{matrix} \\ &= -\sum_{k=1}^{\frac{n-1}{2}} \frac{1}{\lambda_k^{(2h)}}\tilde{\alpha}_k[(1 - \tilde{\lambda}_k^{(h)}) \sin k\pi x_i - \tilde{\lambda}_k^{(h)} \sin(n{+}1{-}k)\pi x_i] \\ &= \sum_{k=1}^{\frac{n-1}{2}} \frac{1}{\lambda_k^{(2h)}}\tilde{\alpha}_k(1 - \tilde{\lambda}_k^{(h)})\phi_k(x_i) + \sum_{k=\frac{n+3}{2}}^{n} \frac{1}{\lambda_{n+1-k}^{(2h)}}\tilde{\alpha}_{n+1-k} 4\tilde{\lambda}_{n+1-k}^{(h)}\phi_k(x_i),\end{aligned}$$

where we have used $\sin(n{-}k{+}1)\pi x_i = (-1)^{i-1} \sin k\pi x_i$, $i = 1, 2, \ldots, n$, for $k = 1, 2, \ldots, \frac{n-1}{2}$.

Hence the corrected solution on Ω_h takes the form:

$$\underline{x}_h^* = \underline{x}_h + \underline{\delta}_h \quad \text{with residual}$$

$$\begin{aligned}\underline{r}_h^*(x_i) &= \underline{r}_h(x_i) + A_h\underline{\delta}_h(x_i) \\ &= \sum_{k=1}^{\frac{n-1}{2}}(\alpha_k - \frac{\lambda_k^{(h)}}{\lambda_k^{(2h)}}\tilde{\lambda}_{n+1-k}^{(h)}\tilde{\alpha}_k)\phi_k(x_i) \\ &+ \alpha_{\frac{n+1}{2}}\phi_{\frac{n+1}{2}}(x_i) \\ &+ \sum_{k=\frac{n+3}{2}}^{n} (\alpha_k + \frac{\lambda_k^{(h)}}{\lambda_{n+1-k}^{(2h)}}\tilde{\lambda}_{n+1-k}^{(k)}\tilde{\alpha}_{n+1-k})\phi_k(x_i)\end{aligned}$$

Expanding the $\tilde{\alpha}_k$ we find

$$\underline{r}_h^*(x_i) = \sum_{k=1}^{\frac{n-1}{2}} \left\{ \alpha_k[1 - \frac{\lambda_k^{(h)}}{\lambda_k^{(2h)}}(\tilde{\lambda}_{n+1-k}^{(h)})^2] + \alpha_{n+1-k}\frac{\lambda_k^{(h)}}{\lambda_k^{(2h)}}\tilde{\lambda}_{n+1-k}^{(h)}\tilde{\lambda}_k^{(h)} \right\} \phi_k(x_i)$$

$$+ \alpha_{\frac{n+1}{2}}\phi_{\frac{n+1}{2}}(x_i)$$

$$+ \sum_{k=\frac{n+3}{2}}^{n} \left\{ \alpha_k[1 - \frac{\lambda_k^{(h)}}{\lambda_{n+1-k}^{(2h)}}(\tilde{\lambda}_{n+1-k}^{(h)})^2] + \alpha_{n+1-k}\frac{\lambda_k^{(h)}}{\lambda_{n+1-k}^{(2h)}}\tilde{\lambda}_{n+1-k}^{(h)}\tilde{\lambda}_k^{(h)} \right\} \phi_k(x_i).$$

Elementary computations give:

$$r_h^*(x_i) = \sum_{k=1}^{n} \sin^2\frac{k\pi h}{2}[\alpha_k + \alpha_{n+1-k}]\phi_k(x_i) = \sum_{k=1}^{n} \tilde{\lambda}_k^{(h)}[\alpha_k + \alpha_{n+1-k}]\phi_k(x_i) \tag{3.7}$$

$$=: \sum_{k=1}^{n} \beta_k\phi_k(x_i) \quad ,i = 1, 2, \ldots n.$$

and we see that

$\beta_k \simeq 0(h^2)$ for k small, $k = 1, 2, \ldots$

$\beta_k \simeq \alpha_k$ for $k \simeq \frac{n+1}{2}$

$\beta_k \simeq \alpha_k + \alpha_{n+1-k}$ for k large, $k = n, n-1, \ldots$

Postsmoothing:

For simplicity we will consider here only Jacobi smoothing

$$\underline{x}_h^{(j)} = \underline{x}_h^{(j-1)} - \frac{1}{\tau_j}(A_h\underline{x}_h^{(j-1)} - \underline{f}_h), \quad j = 1, 2, \ldots, \nu;$$

$$\underline{r}_h^{(j)} = \underline{r}_h^{(j-1)} - \frac{1}{\tau_j}A_h\underline{r}_h^{(j-1)} = (I - \frac{1}{\tau_j}A_h)\underline{r}_h^{(j-1)}, \quad j = 1, 2, \ldots, \nu.$$

where τ_j are some positive parameters.

Hence:

$$\underline{r}_h^{(\nu)}(x_i) = \sum_{k=1}^{n} \beta_k S_\nu(\tilde{\lambda}_k^{(h)})\phi_k(x_i) \quad i = 1, 2, \ldots n \tag{3.8}$$

where

$$S_\nu(\tilde{\lambda}_k^{(h)}) = \prod_{j=1}^{\nu}(1 - \frac{\tilde{\lambda}_k^{(h)}}{\tilde{\tau}_j}), \quad \tilde{\tau}_j = \frac{4}{h^2}\tau_j.$$

This brings us to the iteration matrix

$$M = S_\nu(A_h)(I - A_h P_{2h}^h A_{2h}^{-1} R_h^{2h}).$$

Using the $\underline{\phi}_k^{(h)}$ as basis it follows from (3.7) and (3.8) that we may write:

$$M = D_{S_\nu} D_{\tilde{\lambda}} \begin{bmatrix} I & & I \\ & 2 & \\ I & & I \end{bmatrix} \tag{3.9}$$

where D_{S_ν} and $D_{\tilde{\lambda}}$ are diagonal matrices with

$$(D_{S_\nu})_i = S_\nu(\tilde{\lambda}_i^{(h)}) \quad \text{and} \quad (D_{\tilde{\lambda}})_i = \tilde{\lambda}_i^{(h)}, \quad i = 1, 2, \ldots n.$$

Theorem 3.1 The reduction factor for the model problem is bounded by

$$\|M\|_2 \leq 2 \min_{\tilde{\tau}_j \in [0,1]} \max_{x \in [0,1]} |x \prod_{j=1}^{\nu} (1 - \frac{x}{\tilde{\tau}_j})|.$$

Proof. From (3.9) we have immediately:

$$\|M\|_2 \leq 2\|D_{S_\nu} D_{\tilde{\lambda}}\|_2 \leq 2 \max_{1 \leq i \leq n} |S_\nu(\tilde{\lambda}_i^{(h)})\tilde{\lambda}_i^{(h)}| \, .$$

Since $0 < \tilde{\lambda}_i^{(h)} < 1$ we have

$$\|M\|_2 \leq 2 \max_{x \in [0,1]} |S_\nu(x)x|$$ ◇

So for instance with $\nu = 2$ we find as an upper bound for the reduction factor $\frac{1}{3}(2 - \sqrt{3}) \simeq 0.0893$ where $\tilde{\tau}_1 = 2\sqrt{3} - 3$, $\tilde{\tau}_2 = 2\tilde{\tau}_1$ (for more general results on this best approximation problem, see [6]).

3.3 Reduction factor for a two dimensional model problem

$$\begin{cases} -\Delta u = f & \text{on } \Omega = (0,1)^2 \\ u = 0 & \text{on } \Gamma = \partial\Omega \, . \end{cases} \tag{3.10}$$

Central difference discretization on Ω_h leads to

$$A_h \underline{x}_h = \underline{f}_h \quad \text{where} \quad A_h = \text{tridiag}[-1, 2, -1] \otimes I_n + I_n \otimes \text{tridiag}[-1, 2, -1] \quad (h = \tfrac{1}{n+1}) \tag{3.11}$$

where $\otimes$ denotes the tensor product.
The eigensolutions of (3.11) are given by $(\lambda_{p,q}^{(h)}, \underline{\phi}_{p,q}^{(h)})$, $p, q = 1, 2, \ldots n$, where

$$(\underline{\phi}_{p,q}^{(h)})_{i,j} = \phi_{p,q}(x_i, y_j) = \sin p\pi x_i \sin q\pi y_j \, , \; i, j = 1, 2, \ldots, n \quad (x_i = ih, y_j = jh)$$

i.e. $\phi_{p,q}$ is the product of the one dimensional functions in x- and y-direction.

$$\lambda_{p,q}^{(h)} = \lambda_p^{(h)} + \lambda_q^{(h)}$$

where $\lambda_p^{(h)}$, $p = 1, 2, \ldots, n$ are the eigenvalues for the corresponding one dimensional problem.
Let $\tilde{\lambda}_{p,q}^{(h)} = \frac{h^2}{8}\lambda_{p,q}^{(h)} = \frac{1}{2}\tilde{\lambda}_p^{(h)} + \frac{1}{2}\tilde{\lambda}_q^{(h)}$, then $0 < \tilde{\lambda}_{p,q}^{(h)} < 1$.

Initial residual:

$$\underline{r}_h^0 = A_h \underline{x}_h^0 - \underline{f}_h =: \sum_{p,q=1}^{n} \alpha_{p,q} \underline{\phi}_{p,q}^{(h)}$$

Restriction (Full Weighted):

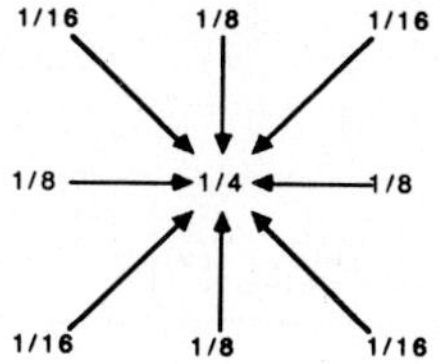

It is readily seen that this is just a tensorproduct of the one dimensional restrictions in x- and y-direction and since $\phi_{p,q}(x,y) = \phi_p(x)\phi_q(y)$ it follows that

$$\underline{r}_{2h}(x_i,y_j) = \sum_{p,q=1}^{n} \alpha_{p,q}\tilde{\lambda}^{(h)}_{n+1-p}\tilde{\lambda}^{(h)}_{n+1-q} \sin p\pi x_i \sin q\pi y_j, \quad i,j = 2,4,6,\ldots,n-1,$$

$$\begin{aligned}\underline{r}_{2h}(x_i,y_j) = \sum_{p,q=1}^{\frac{n-1}{2}} (&\alpha_{p,q}\tilde{\lambda}^{(h)}_{n+1-p}\tilde{\lambda}^{(h)}_{n+1-q} - \alpha_{n+1-p,q}\tilde{\lambda}^{(h)}_{p}\tilde{\lambda}^{(h)}_{n+1-q} - \alpha_{p,n+1-q}\tilde{\lambda}^{(h)}_{n+1-p}\tilde{\lambda}^{(h)}_{q} \\ &+ \alpha_{n+1-p,n+1-q}\tilde{\lambda}^{(h)}_{p}\tilde{\lambda}^{(h)}_{q})\phi_{p,q}(x_i,y_j) \\ =: \sum_{p,q=1}^{\frac{n-1}{2}} &\tilde{\alpha}_{p,q}\phi_{p,q}(x_i,y_j).\end{aligned}$$

Correction on Ω_{2h}:

Using again the "ideal" correction, we have

$$A_{2h}\underline{\delta}_{2h} = -\underline{r}_{2h} \Rightarrow \underline{\delta}_{2h} = -\sum_{p,q=1}^{\frac{n-1}{2}} \frac{1}{\lambda^{(2h)}_{p,q}}\tilde{\alpha}_{p,q}\underline{\phi}^{(2h)}_{p,q}$$

Prolongation

$$\begin{bmatrix} 1/4 & 1/2 & 1/4 \\ 1/2 & 1 & 1/2 \\ 1/4 & 1/2 & 1/4 \end{bmatrix} = [1/2 \;\; 1 \;\; 1/2] \otimes \begin{bmatrix} 1/2 \\ 1 \\ 1/2 \end{bmatrix}.$$

Since this is again a tensorproduct of the linear interpolation in x- and y-direction, we get following the previous analysis:

$$\underline{\delta}_h(x_i,y_j) = -\sum_{p,q=1}^{\frac{n-1}{2}} \frac{1}{\lambda^{(2h)}_{p,q}}\tilde{\alpha}_{p,q} \begin{Bmatrix} 1-\tilde{\lambda}^{(h)}_p+\tilde{\lambda}^{(h)}_p \\ 1-\tilde{\lambda}^{(h)}_p-\tilde{\lambda}^{(h)}_p \end{Bmatrix} \sin p\pi x_i \begin{Bmatrix} 1-\tilde{\lambda}^{(h)}_q+\tilde{\lambda}^{(h)}_q \\ 1-\tilde{\lambda}^{(h)}_q-\tilde{\lambda}^{(h)}_q \end{Bmatrix} \sin q\pi y_j \quad \begin{cases} i,j & \text{even} \\ i,j & \text{odd} \end{cases}$$

$$\begin{aligned} = -\sum_{p,q=1}^{\frac{n-1}{2}} \frac{1}{\lambda^{(2h)}_{p,q}}\tilde{\alpha}_{p,q}&[(1-\tilde{\lambda}^{(h)}_p)\sin p\pi x_i - \tilde{\lambda}^{(h)}_p \sin(n+1-p)\pi x_i]\cdot \\ &[(1-\tilde{\lambda}^{(h)}_q)\sin q\pi y_j - \tilde{\lambda}^{(h)}_q \sin(n+1-q)\pi y_j]. \end{aligned}$$

Expanding the $\tilde{\alpha}_{p,q}$, using the eigenvectors as basis and by a suitable reordering of the nodes we may write:

$$\hat{\underline{\delta}}_{p,q} = D_{p,q}\hat{\underline{\alpha}}_{p,q} \quad p,q = 1,2,\ldots \tfrac{n-1}{2}$$

where $\hat{\underline{\delta}}_{p,q} = \begin{bmatrix} (\underline{\delta}_h)_{p,q} \\ (\underline{\delta}_h)_{n+1-p,n+1-q} \\ (\underline{\delta}_h)_{n+1-p,q} \\ (\underline{\delta}_h)_{p,n+1-q} \end{bmatrix}$, $\hat{\underline{\alpha}}_{p,q} = \begin{bmatrix} \alpha_{p,q} \\ \alpha_{n+1-p,n+1-q} \\ \alpha_{n+1-p,q} \\ \alpha_{p,n+1-q} \end{bmatrix}$

and

$$D_{p,q} = \frac{-1}{\lambda_{p,q}^{(2h)}} \begin{bmatrix} \tilde{\lambda}_{n+1-p}^{(h)}\tilde{\lambda}_{n+1-q}^{(h)} \\ -\tilde{\lambda}_p^{(h)}\lambda_q^{(h)} \\ -\tilde{\lambda}_p^{(h)}\tilde{\lambda}_{n+1-q}^{(h)} \\ \tilde{\lambda}_{n+1-p}^{(h)}\tilde{\lambda}_q^{(h)} \end{bmatrix} \cdot \left[\tilde{\lambda}_{n+1-p}^{(h)}\tilde{\lambda}_{n+1-q}^{(h)}, -\tilde{\lambda}_p^{(h)}\tilde{\lambda}_q^{(h)}, -\tilde{\lambda}_p^{(h)}\tilde{\lambda}_{n+1-q}^{(h)}, \tilde{\lambda}_{n+1-p}^{(h)}\tilde{\lambda}_q^{(h)}\right] .$$

For p and/or q equal to $\frac{n+1}{2}$ we define:

$$\hat{\underline{\delta}}_{p,q} = (\underline{\delta}_h)_{p,q} \qquad (\text{i.e. } \hat{\underline{\delta}}_{p,q} \in \mathbf{R}^1)$$

$$\hat{\underline{\alpha}}_{p,q} = \alpha_{p,q}$$

$$D_{p,q} = 0$$

and therefore we may write: $\hat{\underline{\delta}} = \text{blockdiag}\,(D_{p,q})\hat{\underline{\alpha}}$.

Correction on Ω_h:

$$\underline{x}_h^* = \underline{x}_h + \underline{\delta}_h \quad \text{with residual}$$

$$\underline{r}_h^* = \underline{r}_h + A_h\underline{\delta}_h = (I - A_h P_{2h}^h A_{2h}^{-1} R_h^{2h)})\underline{r}_h .$$

Using the above ordering and basis we find

$$\hat{\underline{r}}_h^* = \text{blockdiag}\,(I + \Lambda_{p,q} D_{p,q})\hat{\underline{r}}_h$$

where
$$\Lambda_{p,q} = \text{diag}(\lambda_{\text{p,q}}^{(\text{h})}, \lambda_{\text{n+1-p,n+1-q}}^{(\text{h})}, \lambda_{\text{n+1-p,q}}^{(\text{h})}, \lambda_{\text{p,n+1-q}}^{(\text{h})}),\ \text{p, q} = 1, 2, \ldots \tfrac{\text{n}-1}{2}$$
$$= \lambda_{p,q}^{(h)} \quad p \text{ and/or } q \text{ equal to } \tfrac{n+1}{2} .$$

Straightforward computation gives that

$$\frac{1}{\lambda_{p,q}^{(2h)}} \cdot \Lambda_{p,q} = \frac{1}{\tilde{\lambda}_p^{(h)}\tilde{\lambda}_{n+1-p}^{(h)} + \tilde{\lambda}_q^{(h)}\tilde{\lambda}_{n+1-q}^{(h)}}$$
$$\text{diag}(\tilde{\lambda}_{\text{p}}^{(\text{h})} + \tilde{\lambda}_{\text{q}}^{(\text{h})}, \tilde{\lambda}_{\text{n+1-p}}^{(\text{h})} + \tilde{\lambda}_{\text{n+1-q}}^{(\text{h})}, \tilde{\lambda}_{\text{n+1-p}}^{(\text{h})} + \tilde{\lambda}_{\text{q}}^{(\text{h})}, \tilde{\lambda}_{\text{q}}^{(\text{h})} + \tilde{\lambda}_{\text{n+1-q}}^{(\text{h})})$$
$$=: \frac{1}{\tilde{\lambda}_p^{(h)}\tilde{\lambda}_{n+1-p}^{(h)} + \tilde{\lambda}_q^{(h)}\tilde{\lambda}_{n+1-q}^{(h)}}\tilde{\Lambda}_{p,q} \quad p, q = 1, 2, \ldots, \tfrac{n-1}{2}$$
$$\tilde{\Lambda}_{p,q} := \tilde{\lambda}_p^{(h)} + \tilde{\lambda}_q^{(h)} \quad \text{for } p \text{ and/or } q \text{ equal to } \tfrac{n-1}{2} .$$

Finally defining

$$\tilde{D}_{p,q} := \begin{cases} \dfrac{\lambda_{p,q}^{(h)}}{\tilde{\lambda}_p^{(h)}\tilde{\lambda}_{n+1-p}^{(h)} + \tilde{\lambda}_q^{(h)}\tilde{\lambda}_{n+1-q}^{(h)}} D_{p,q} & p, q = 1, 2, \ldots \tfrac{n-1}{2} \\ 0 & \text{otherwise} \end{cases}$$

leads to:

$$\hat{\underline{r}}_h^* = \text{blockdiag}\,(\tilde{\Lambda}_{p,q}(\tilde{\Lambda}_{p,q}^{-1} + \tilde{D}_{p,q}))\hat{\underline{r}}_h .$$

Postsmoothing:

We use again Jacobi smoothing and analogously we find

$$\underline{r}_h^{(\nu)} = S_\nu(A_h)\underline{r}_h^* = S_\nu(A_h)(I - A_h P_{2h}^h A_{2h}^{-1} R_h^{2h})\underline{r}_h = M\underline{r}_h$$

$$S_\nu(A_h) = \text{blockdiag}\,(S_{p,q}^{(\nu)})$$

where

$$S_{pq}^{(\nu)} = \begin{cases} \text{diag}(S_\nu(\tilde\lambda_{p,q}^{(h)}), S_\nu(\tilde\lambda_{n+1-p,n+1-q}^{(h)}), S_\nu(\tilde\lambda_{n+1-p,q}^{(h)}), S_\nu(\tilde\lambda_{p,n+1-q}^{(h)})), & p,q \le \frac{n-1}{2} \\ S_\nu(\tilde\lambda_{p,q}^{(h)}) \text{ for } p \text{ and/or } q \text{ equal to } \frac{n+1}{2} \end{cases}$$

where we have used again the above ordering and basis.

Writing M as

$$\begin{aligned} M &= S_\nu(A_h)\text{blockdiag}(\tilde\Lambda_{p,q})\text{blockdiag}(\tilde\Lambda_{p,q}^{-1} + \tilde D_{p,q}) \\ &= \text{blockdiag}(S_{p,q}^\nu \tilde\Lambda_{p,q}) \cdot \text{blockdiag}(\tilde\Lambda_{p,q}^{-1} + \tilde D_{p,q}) \end{aligned}$$

we can bring the estimation of $||M||_2$ back to estimates for 4×4 matrices.

Lemma 3.1

A. $||S_{p,q}^{(\nu)}\tilde\Lambda_{p,q}||_2 \le 2 \min\limits_{\tilde\tau_j \in (0,1)} \max\limits_{x\in[0,1]} x \prod\limits_{j=1}^{\nu} (1 - \frac{x}{\tilde\tau_j})$

B. $||\tilde\Lambda_{p,q}^{-1} + \tilde D_{p,q}||_2 \le 2$

Proof. A. This follows immediately from

$$\begin{aligned} S_{p,q}^{(\nu)}\tilde\Lambda_{p,q} = 2\text{diag}(&S_\nu(\tilde\lambda_{p,q}^{(h)})\tilde\lambda_{p,q}^{(h)}, S_\nu(\tilde\lambda_{n+1-p,n+1-q}^{(h)})\tilde\lambda_{n+1-p,n+1-q}^{(h)}, S_\nu(\tilde\lambda_{n+1-p,q}^{(h)})\tilde\lambda_{n+1-p,q}^{(h)}, \\ &S_\nu(\tilde\lambda_{p,n+1-q}^{(h)})\tilde\lambda_{p,n+1-q}^{(h)}) \end{aligned}$$

so

$$||S_{p,q}^{(\nu)}\tilde\Lambda_{p,q}||_2 \le 2 \max_{x\in[0,1]} x \prod_{j=1}^{\nu}(1 - \frac{x}{\tilde\tau_j}) \quad \forall p,q\,.$$

B. There are two cases to consider

1. p and/or q equals $\frac{n+1}{2}$ then $\tilde D_{p,q} = 0$ and $\tilde\Lambda_{p,q}^{-1} = \frac{1}{\tilde\lambda_p^{(h)} + \tilde\lambda_q^{(h)}} \le \frac{1}{\sin^2\frac{\pi}{4}} = 2$
2. $p,q \le \frac{n-1}{2}$. In this case, it suffices to show that $||D_{\lambda,\mu}||_2 \le 2 \;\; \lambda,\mu \in (0,1)$ where

$$D_{\lambda,\mu} = \begin{bmatrix} \frac{1}{\mu+\lambda} & & & \\ & \frac{1}{2-\mu-\lambda} & & \\ & & \frac{1}{1-\mu+\lambda} & \\ & & & \frac{1}{\mu+1-\lambda} \end{bmatrix} +$$

$$\frac{1}{(1-\mu)\mu+\lambda(1-\lambda)} \begin{bmatrix} -(1-\mu)^2(1-\lambda)^2 & -\mu(1-\mu)\lambda(1-\lambda) & \mu(1-\mu)(1-\lambda)^2 & (1-\mu)^2\lambda(1-\lambda) \\ -\mu(1-\mu)\lambda(1-\lambda) & -\mu^2\lambda^2 & \mu^2\lambda(1-\lambda) & \mu(1-\mu)\lambda^2 \\ (1-\mu)^2\lambda(1-\lambda) & \mu^2\lambda(1-\lambda) & -\mu^2(1-\lambda^2 & -\mu(1-\mu)\lambda(1-\lambda) \\ (1-\mu)^2\lambda(1-\lambda) & \mu(1-\mu)\lambda^2 & -\mu(1-\mu)\lambda(1-\lambda) & -(1-\mu)^2\lambda^2 \end{bmatrix}.$$

Straightforward computations give that $D_{\lambda,\mu}$ is symmetric, diagonally dominant so that $D_{\lambda,\mu}$ is positive definite. We will show that

i) Trace $(D_{\lambda,\mu}) \le 4 \quad \forall\lambda,\mu \in (0,1)$

ii) $\text{Det}(D_{\lambda,\mu} - 2I) > 0 \quad \forall\lambda,\mu \in (0,1)$

From i) it follows that at most one eigenvalue of $D_{\lambda,\mu}$ is larger than 2 from ii) it follows that 0, 2, or 4 eigenvalues of $D_{\lambda,\mu}$ are larger than 2, so all eigenvalues of $D_{\lambda,\mu}$ are bounded by 2, i.e. $||D_{\lambda,\mu}||_2 \le 2$.

Proof of i). We split the trace into two pieces and show that both are bounded by two

(ia) $$\frac{1}{\mu+\lambda}+\frac{1}{2-\mu-\lambda}-\frac{(1-\mu)^2(1-\lambda)^2+\mu^2\lambda^2}{\mu(1-\mu)+\lambda(1-\lambda)}\le 2 \iff$$

$$\frac{f}{g}:=\frac{2[\mu(1-\mu)+\lambda(1-\lambda)]-(\mu+\lambda)(2-\mu-\lambda)[(1-\mu)^2(1-\lambda)^2+\mu^2\lambda^2]}{(\mu+\lambda)(2-\mu-\lambda)[\mu(1-\mu)+\lambda(1-\lambda)]}\le 2$$

$$\iff 2g-f\ge 0$$

$$2g-f=-2\mu\lambda(\mu+\lambda-1)^2(1-\mu)(1-\lambda)+\mu^2(1-\mu)^2+\lambda^2(1-\lambda)^2$$

$$\ge -2\mu\lambda(1-\mu)(1-\lambda)+\mu^2(1-\mu)^2+\lambda^2(1-\lambda)^2=[\mu(1-\mu)-\lambda(1-\lambda)]^2\ge 0$$

(ib) $$\frac{1}{1-\mu+\lambda}+\frac{1}{\mu+1-\lambda}-\frac{(1-\mu)^2\lambda^2+\mu^2(1-\lambda)^2}{\mu(1-\mu)+\lambda(1-\lambda)}\le 2 \iff$$

$$\frac{f}{g}:=\frac{2[\mu(1-\mu)+\lambda(1-\lambda)]-(1-\mu+\lambda)(\mu+1-\lambda)[(1-\mu)^2\lambda^2+\mu^2(1-\lambda)^2]}{(1-\mu+\lambda)(\mu+1-\lambda)[\mu(1-\mu)+\lambda(1-\lambda)]}\le 2$$

$$\iff 2g-f\ge 0$$

$$2g-f=-2\mu\lambda(1-\mu)(1-\lambda)(\mu-\lambda)^2+\mu^2(1-\mu)^2+\lambda^2(1-\lambda)^2$$

$$\ge -2\mu\lambda(1-\mu)(1-\lambda)+\mu^2(1-\mu)^2+\lambda^2(1-\lambda)^2=[\mu(1-\mu)-\lambda(1-\lambda)]^2\ge 0$$

Proof of ii).

$\det(D_{\lambda,\mu}-2I)=$

$$=\frac{8\mu\lambda(1-\mu)(1-\lambda)[4\mu^4-8\mu^3-8\mu^2\lambda^2+8\mu^2\lambda+\mu^2+8\mu\lambda^2-8\mu\lambda+3\mu+4\lambda^4-8\lambda^3+\lambda^2+3\lambda]}{(\mu+\lambda)(2-\mu-\lambda)(1-\mu+\lambda)(\mu+1-\lambda)[\mu(1-\mu)+\lambda(1-\lambda)]}$$

But since

$$4\mu^4-8\mu^3-8\mu^2\lambda^2+8\mu^2\lambda+\mu^2+8\mu\lambda^2-8\mu\lambda+3\mu+4\lambda^4-8\lambda^3+\lambda^2+3\lambda$$

$$=-8\mu(1-\mu)\lambda(1-\lambda)+\mu(1-\mu)(3-2\mu)(1+2\mu)+\lambda(1-\lambda)(3-2\lambda)(1+2\lambda)$$

$$\ge -4\mu^2(1-\mu)^2-4\lambda^2(1-\lambda)^2+\mu(1-\mu)(3-2\mu)(1+2\mu)+\lambda(1-\lambda)(3-2\lambda)(1+2\lambda)$$

$$=3\mu(1-\mu)+3\lambda(1-\lambda)>0 \quad \forall\mu,\lambda\in(0,1),$$

we have $\det(D_{\lambda,\mu}-2I)>0 \quad \forall\mu,\lambda\in(0,1)$. ◇

4. Comparisons with ideal corrector

Unfortunately we haven't been able to get an estimate of the reduction factor if we use C_{2h} as corrector. This is due to the fact that as with all incomplete factorizations C_{2h} doesn't preserve the eigenmodes of A_{2h} so that it is not possible to carry out an analysis based on fourier expansions as in the previous section.

In this section we will show for some numerical test problems that the action of C_{2h} is very close to the action of A_{2h} and with a somewhat rough analysis we will show that this is especially the case for smooth vectors.

Consider therefore $\|(I - C_H^{-1}A_H)\underline{\phi}_{p,q}^{(H)}\|$ where $\underline{\phi}_{p,q}^{(H)}$ is an eigenvector of A_H. We will use the discrete L_2-norm in the following, i.e.

$$\|\underline{u}\|^2 = \frac{1}{n^2}\sum_{i,j=1}^{n} u_{i,j}^2 \, .$$

Note that $\|\underline{\phi}_{p,q}^{(H)}\|^2 \underset{n\to\infty}{\longrightarrow} \int_0^1\int_0^1 \phi_{p,q}^2(x,y)dxdy = \frac{1}{4}$

Now recall that if we write $C_H = A_H + R_H$, then R_H is a zero matrix except for the diagonal blocks corresponding to the first lines of the substructures $2, 3, \ldots, \hat{p}$ ($n = \hat{p}.k$, k the number of lines per structure). So for any vector $\underline{u}$ it follows that if we split it in $\underline{u} = \underline{u}_1 + \underline{u}_2$ where $\underline{u}_1$ is zero on these lines and $\underline{u}_2$ is zero elsewhere then:

$$\|(I - C_H^{-1}A_H)\underline{u}\| = \|(I - C_H^{-1}A_H)\underline{u}_2\| = \|C_H^{-1}R_H\underline{u}_2\|.$$

Since $R_H\underline{e} = R_H\underline{v} = \underline{0}$ we may substract a vector with linearly growing components from $\underline{u}_2$. This may be done for each line separately and since $\|C_H^{-1}R_H\| \leq 1$ we find:

$$\|(I - C_H^{-1}A_H)\underline{u}\|^2 \leq \frac{1}{n}\sum_{i=1}^{\hat{p}-1}\|\underline{u}_{ik+1} - (a_i\underline{e} + b_i\underline{v})\|^2$$

where $\underline{u}_{ik+1} \in \mathbf{R}^n$ is the part of $\underline{u}$ corresponding to the $ik+1$ line (i.e. the first line of structure $i+1$) in the mesh and a_i and b_i are free to choose constants.

Note that if $\underline{u}$ is linear on each one of these $\hat{p}-1$ lines in the y-direction then $C_H\underline{u} = A_H\underline{u}$. Setting $\underline{u} = \underline{\phi}_{p,q}^{(H)}$ we find:

$$\begin{aligned}
\|(I - C_H^{-1}A_H)\underline{\phi}_{p,q}^{(H)}\|^2 &\leq \frac{1}{n}\sum_{i=1}^{\hat{p}-1}\sin^2 p\pi x_i[\frac{1}{n}\sum_{j=1}^{n}(\sin q\pi y_j - (a_i + b_i y_j))^2] \\
&= \frac{1}{k}\frac{1}{\hat{p}}\sum_{i=1}^{\hat{p}-1}\sin^2 p\pi x_i[\frac{1}{n}\sum_{j=1}^{n}(\sin q\pi y_j - (a_i + b_i y_j))^2] \\
&\underset{n\to\infty}{\longrightarrow} \frac{1}{k}\cdot\frac{1}{2}\cdot\begin{Bmatrix} \frac{1}{2} - \frac{4}{q^2\pi^2} \\ \frac{1}{2} - \frac{12}{q^2\pi^2} \end{Bmatrix} \quad \begin{matrix} q \ \text{odd} & (a_i = \frac{2}{q\pi},\ b_i = 0) \\ q \ \text{even} & (a_i = \frac{6}{q\pi},\ b_i = \frac{-12}{q\pi}) \, . \end{matrix}
\end{aligned}$$

This indicates that C_H will approximate A_H best for the smooth eigenmodes and of course C_H approximates A_H better if we increase k, i.e. the number of lines per structures (for $k = n$ we have $C_H = A_H$). This gives only a very rough indication of the accuracy of C_H as approximation of A_H. In figure 4.1 we have connected the points

$$(\tilde{\lambda}_{p,q}^{(H)}, \|(I - C_H^{-1}A_H)\underline{\phi}_{p,q}^{(H)}\|) \quad \text{for } H = \frac{1}{17}, \ \hat{p} = k = 4.$$

Here we have reordered the eigenvectors so that the corresponding eigenvalues are in increasing order.

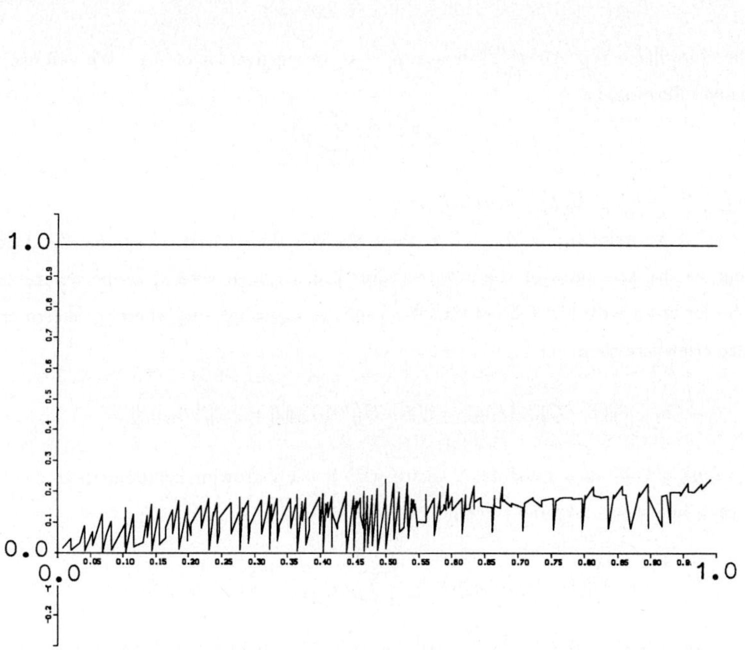

fig. 4.1. $(\tilde{\lambda}^{(H)}_{p,q}, \|(I - C_H^{-1}A_H)\underline{\phi}^{(H)}_{p,q}\|)$, $H = \frac{1}{17}$, $\hat{p} = k = 4$

We see indeed that especially for the small eigenvalues (i.e. the smooth eigenvectors) C_H is very close to A_H.

In figure 4.2.a (resp. 4.2.b) we have plotted the actions of the correction matrices $M_0 = (I - A_h P^h_{2h} A^{-1}_{2h} R^{2h}_h)$ (respectively $\tilde{M}_0 = I - A_h P^h_{2h} C^{-1}_{2h} R^{2h}_h$) on the eigenvectors. To this end we have connected the points $(\tilde{\lambda}^{(h)}_{p,q}, \|M_0 \underline{\phi}^{(h)}_{p,q}\|)$ and $(\tilde{\lambda}^{(h)}_{p,q}, \|\tilde{M}_0 \underline{\phi}^{(h)}_{p,q}\|)$ again with the eigenvalues in increasing order. The dotted line, i.e. the graph corresponding to the quadratic polynomial, illustrates the results of two steps of a Jacobi smoother (with smoothing parameters $\tilde{\tau}_1^{-1} = 0.3599$, $\tilde{\tau}_2^{-1} = 0.8901$) on the eigenvalues. Also the product of these two lines is displayed. This shows the qualitative behaviour of the iteration matrix $M = S_\nu(A_h)M_0$ on the eigenmodes of A_h.

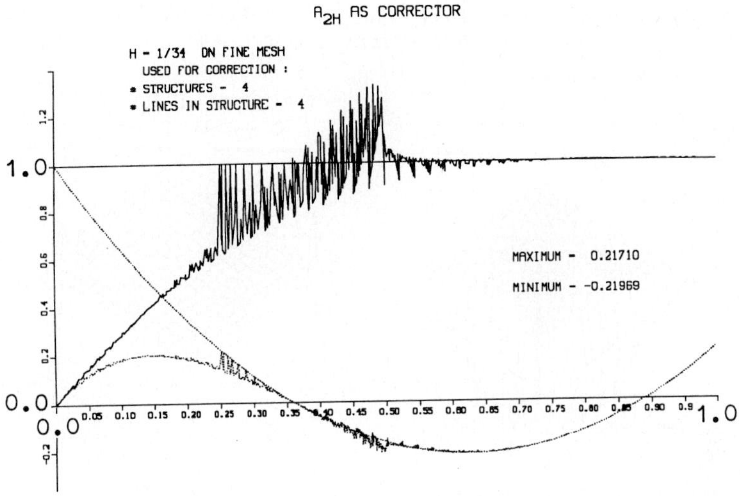

figure 4.2.a. $(\tilde{\lambda}^{(h)}_{p,q}, ||(I - A_h P^h_{2h} A^{-1}_{2h} R^{2h}_h)\underline{\phi}^{(h)}_{p,q}||)$, $h = \frac{1}{33}$.

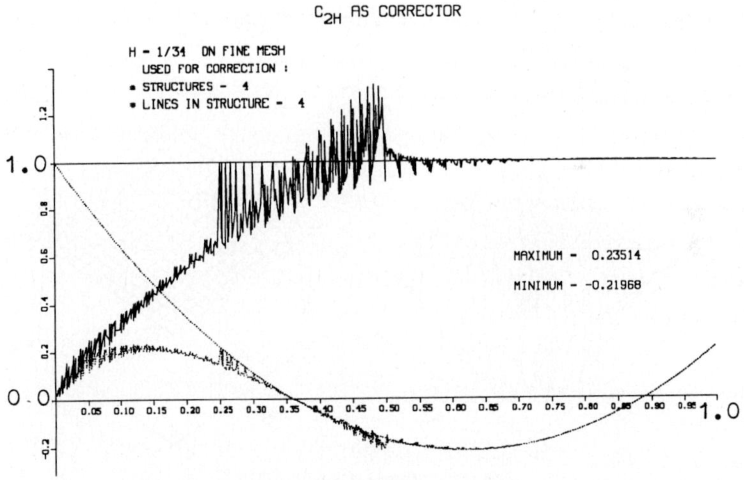

figure 4.2.b. $(\tilde{\lambda}^{(h)}_{p,q}, ||(I - A_h P^h_{2h} C^{-1}_{2h} R^{2h}_h)\underline{\phi}^{(h)}_{p,q}||)$, $h = \frac{1}{33}$.

We see that there is a remarkable close resemblance between these two graphs, indicating that replacing A_{2h} by C_{2h} in the smoothing-correction process will hardly influence the convergence of the process. Finally in figures 4.3 and 4.4 we show intermediate residual vectors in the smoothing-correction process for the model problem with Jacobi smoothing (see section 5). In figure 4.3 we show the residual vector after the first correction step, i.e. $\underline{r}_0 = A_h \underline{x}_o - \underline{f}_h = \underline{f}_h$ (we have chosen $\underline{x}_0 = \underline{0}$, so that $\underline{r}_0$ is a smooth vector). The left picture shows the residual vector on the domain whereas the right picture shows the vector expanded in the basis of eigenvectors of A_h. In 4.3.a we have used A_{2h} as corrector on Ω_{2h} whereas in 4.3.b we used C_{2h}. In 4.4.a,b we show the residual after postsmoothing with two steps Jacobi.

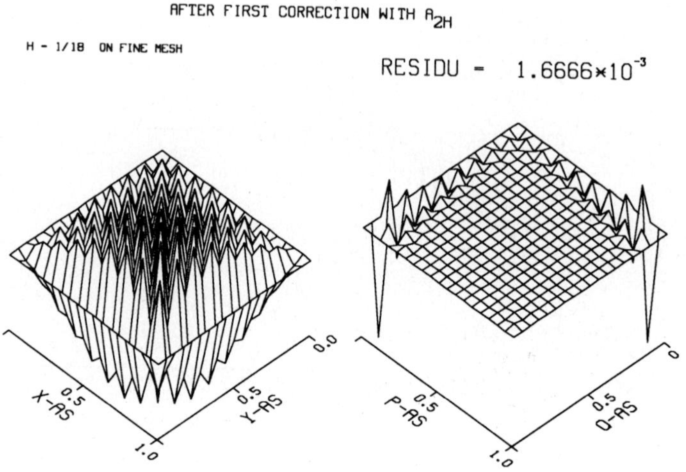

figure 4.3.a.

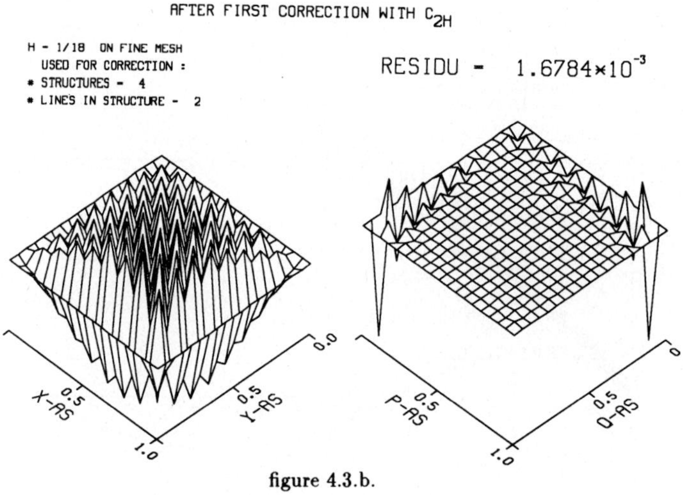

figure 4.3.b.

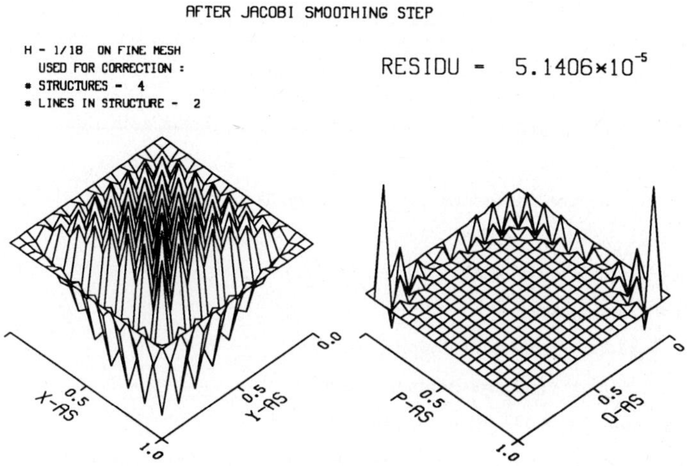

figure 4.4.a.

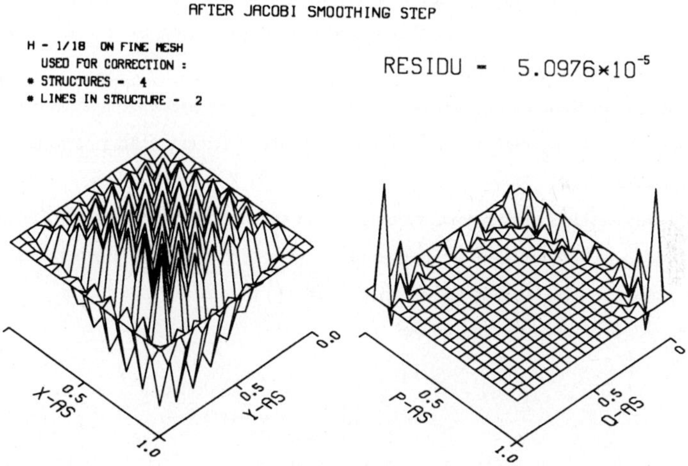

figure 4.4.b.

Again we see that the resemblance is striking; from these figures we expect that using C_{2h} instead of A_{2h} as correction operator on Ω_{2h} will give approximately the same convergence behaviour. The numerical tests in section 5 show that this is indeed the case, not only for the model problem on which our analysis is based but also for some problems with varying or discontinuous coefficients.

5. Numerical Results

We consider first the model problem

$$-\Delta u = f \quad \text{on} \quad \Omega = (0,1)^2 \tag{5.1}$$
$$u = 0 \quad \text{on} \quad \Gamma = \partial\Omega$$

discretized with central differences on a uniform mesh. f is chosen such that the solution becomes $u(x,y) = e^{xy}(1-x)x(1-y)y$.

As initial vector for the smoothing-correction method we have chosen:

1) $\underline{x}_h^0 = \underline{0}$
2) $\underline{x}_h^0$, a randomly filled vector with $(\underline{x}_h^0, \underline{x}_h^0) = (\underline{f}_h, \underline{f}_h)$.

The iterations were stopped when

a) $(\underline{r}^k, \underline{r}^k) \leq \frac{1}{100}h^4$ (absolute stopping criterion).
b) $(\underline{r}^k, \underline{r}^k) \leq 10^{-10}(\underline{r}^0, \underline{r}^0)$ (relative stopping criterion).

The results are found in table Ai, Bi and Ci, $i = 1,2$.

In all tables we have used $H = 2h$. In the tables n denotes the number of unknowns in a vertical meshline on Ω_h, k is the number of substructures on Ω_{2h} as used in the construction of C_{2h}. The tables show the number of iterations of the smoothing-correction scheme for various values of n and k. The number above is with stopcriterion a) and the number below with stopcriterion b). For the model problem we have used three choices for the smoothing operator:

A: Jacobi; it turned out to be most effective to use only one or two Jacobi steps between the corrections. For the results in table A1 and A2 we have used two steps with smoothing parameters $\tilde{\tau}_1^{-1} = 0.8901$ and $\tilde{\tau}_2^{-1} = 0.3599$.

B: Red Black Gauss-Seidel; in this case we only use one smoothing step in every iteration. The Red Black step looks as follows: Reorder the points to get $A_h = \begin{bmatrix} D_1 & L^T \\ L & D_2 \end{bmatrix}$ where D_1 (resp. D_2) is a diagonal matrix corresponding to the Red (resp. Black) points.

Let $\underline{x} = (\underline{x}_1, \underline{x}_2)$ be split accordingly then

$$\underline{x}_1^{(i+1)} = D_1^{-1}(\underline{f}_1 - L^T \underline{x}_2^{(i)})$$
$$\underline{x}_2^{(i+1)} = D_2^{-1}(\underline{f}_2 - L\underline{x}_1^{(i+1)})$$

C: Block incomplete LU factorization, in this case we also use only one smoothing step per iteration. (For the approximate inverses that occur in the block factorization we have used halfbandwidth 1, for a full description of these methods see [7a]).

Note that on a serial computer in all three choices the computational complexity of one step of the smoother is equal to that of a matrix-vector product. But of course A and B are preferable on a vector or parallel computer.

table A1 Jacobi $\underline{x}_0 = 0$

	$k \backslash n$	9	17	33	65
a)		3	3	4	6
b)	1	6	7	11	15
a)		3	3	4	5
b)	2	6	7	8	11
a)		3	3	4	4
b)	4	6	7	7	9
a)			3	4	4
b)	8		7	7	7

table A2 Jacobi $\underline{x}_0$ random

$k \backslash n$	9	17	33	65
	3	4	5	7
1	7	8	10	15
	3	4	4	5
2	7	7	8	10
	3	4	4	5
4	7	7	8	9
		4	4	5
8		7	7	9

table B1 Red Black $\underline{x}_0 = 0$

$k \backslash n$	9	17	33	65
	3	3	4	6
1	7	8	11	16
	3	3	4	5
2	7	8	8	12
	3	3	4	4
4	7	8	7	9
		3	4	4
8		7	7	7

table B2 Red Black $\underline{x}_0$ random

$k \backslash n$	9	17	33	65
	3	4	5	8
1	7	7	11	17
	3	4	5	6
2	7	7	8	11
	3	4	4	5
4	7	7	8	10
		4	4	5
8		7	8	10

table C1 *LU* $\underline{x}_0 = 0$

$k \backslash n$	9	17	33	65
	2	2	3	5
1	4	4	6	10
	2	2	3	4
2	4	4	5	8
	2	2	2	3
4	4	4	4	6
		2	2	3
8		4	4	5

table C2 *LU* $\underline{x}_0$ random

$k \backslash n$	9	17	33	65
	2	2	3	5
1	4	4	3	9
	2	2	2	4
2	4	4	5	8
	2	2	2	3
4	3	3	4	6
		2	2	3
8		3	4	5

Remark 5.1 In all cases the distance of the iterative approximation to the exact solution of the differential equation when using the absolute stopping criterion was almost equal to the distance when using the relative stopping criterion. This means that for the relative stopping criterion, we iterate longer than is needed to reach the level of the truncation error.

Remark 5.2 The results in the previous tables for $k = 4$ or 8 are equal, up to a difference of at most one iteration, to the results we obtain if we replace the correction operator C_{2h} by a PCG algorithm to solve $A_{2h}\underline{\delta}^{(2h)} = -\underline{r}^{(2h)}$, i.e. the ideal corrector. Therefore we didn't include these tables. It means that the two level method presented in the present paper performs equally well as a full multigrid V-cycle method for the model problem, but the computational cost of our method is less and the method is much simpler to implement.

Finally to test the robustness of our method we have also tested it on a some problems with anisotropy or discontinuous coefficients. We have only used the LU-smoother since it is well known that for such problems Jacobi and Red Black are bad smoothers. We consider the following problems;

I $-\varepsilon u_{xx} - u_{yy} = f$

with $u(x, y) = x(1-x)y(1-y)e^{xy}$ as exact solution.

$\varepsilon = 10^{-5}$

II $-u_{xx} - \varepsilon u_{yy} = f$

u, ε as by I.

III $-(a(x,y)u_x)_x - (a(x,y)u_y)_y = f$

where $a(x,y) = 1 + 0.65atan(x - \frac{1}{2}) + 0.35atan(10(y - \frac{1}{2}))$ (see [10]).

IV Insulator problem

$-a\Delta u = f$ in $\Omega = (0,1)^2$

where $a = \begin{cases} 0.01 & \text{in the shaded area} \\ 1 & \text{elsewhere} \end{cases}$

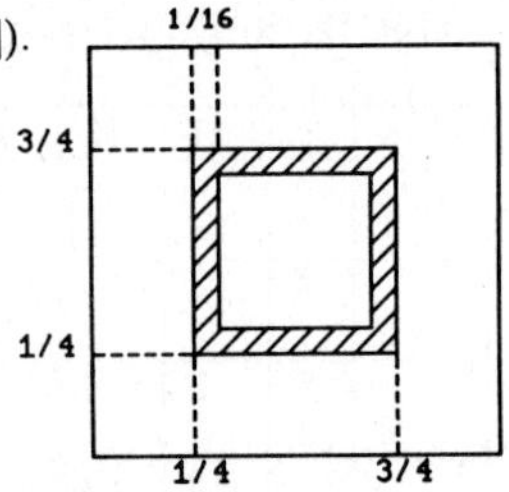

For problems I and II we can be short since for both problems the method needs only one iteration for all values of n and k. This is mostly due to the incomplete block factorization we use as smoother (see [12]).

table CIII.1 LU $\underline{x}_0 = 0$

$k \backslash n$	9	17	33	65
	2	2	3	5
1	4	4	6	10
	2	2	3	4
2	4	4	5	8
	2	2	2	3
4	4	4	4	6
		2	2	3
8		4	4	5

table CIII.2 LU $\underline{x}_0$ random

$k \backslash n$	9	17	33	65
	2	2	3	5
1	3	4	5	9
	2	2	3	4
2	3	3	4	6
	2	2	2	3
4	3	3	4	6
		2	2	3
8		3	4	5

table CIV.1 LU $\underline{x}_0 = 0$

$k\backslash n$	17	33	65
	2	3	5
1	5	6	10
	2	3	4
2	5	6	8
	2	3	3
4	4	6	6
	2	3	3
8	4	6	6

table CIV.2 LU $\underline{x}_0$ random

$k\backslash n$	17	33	65
	2	3	5
1	4	5	9
	2	3	4
2	4	5	7
	2	3	3
4	4	5	6
	2	3	3
8	4	5	5

We see from these tables that the method works equally well for these latter problems as for the model problem.

Furthermore we see, from the results in all tables, that the number of iterations is already so small, especially for $k \geq 4$, that the use of an outer acceleration procedure can hardly reduce the overall cost.

References

1 Axelsson, O. A generalized conjugate gradient, least square method, Numer. Math. 51 (1987), 209-227.

2 Axelsson, O., 1988 in preparation.

3 Axelsson, O., On multigrid methods of two-level type. Multigrid methods, Proceedings, Köln-Porz (1981), ed. Hackbusch W., Trottenberg U., Lect. notes in Math. 960, Springer, Berling, Heidelberg, New York.

4 Axelsson, O., Incomplete block matrix factorization preconditioning methods. The ultimate answer? J. Comp. Appl. Math. 12/13 (1985), 3-18.

5 Axelsson, O., Gustafsson, I., Preconditioning and two-level multigrid methods of arbitrary degree of approximation. Math. Comp. 40 (1983), 219-242.

6 Axelsson, O., Lindskog, G., On the rate of convergence of the preconditioned conjugate gradient method, Numer. Math. 48 (1986), 499-523.

7 Axelsson, O., Polman, B., Block preconditioning and domain decomposition method I. Report 8735 (1987), Department of Mathematics, University of Nijmegen.

8 Fedorenko, R.P., The speed of convergence of one iterative process. USSR Comput. Math. and Math. Phys. 4, 3(1964), 227-235.

9 Hackbusch, W., Multigrid Methods and Applications, Springer series in Comp. Math. 4(1985), Springer, Berlin, Heidelberg, New York.

10 Keyes, D.E., Gropp, W.D., A comparison of domain decomposition technique for elliptic partial differential equations and their parallel implementation, SIAM J. Sci. Statist. Comput. 8 (1987),

166-202.

11 Multigrid Methods II, proceedings Köln 1984, ed. Hackbusch, W., Trottenberg, U., Lecture notes in Math. 1228, Springer, Berlin, Heidelberg, New York.

12 Polman, B., Preconditioning matrices based on incomplete blockwise factorizations. M.Sc. thesis, Nijmegen, 1984.

13 Wesseling, P., A robust and efficient Multigrid Method. Multigrid methods, Proceedings, Köln-Porz (1981), ed. Hackbusch, W., Trottenber, U., Lect. notes in Math. 960, Springer, Berlin, Heidelberg, New York.

14 Wachpress, E.L., Split-level iteration, Comp. & Maths. with Applic., 10 (1984), 453-456.

ADAPTIVE MULTIGRID SOLUTION OF THE CONVECTION-DIFFUSION EQUATION ON THE DIRMU MULTIPROCESSOR

P. Bastian, J. H. Ferziger*, G. Horton, J. Volkert

Institut für Mathematische Maschinen und Datenverarbeitung III, Universität Erlangen-Nürnberg, Martensstr. 3, D-8520 Erlangen

* Department of Mechanical Engineering, Stanford University, Stanford, CA 94305

SUMMARY

The two-dimensional convection-diffusion equation is solved using a nested iteration multigrid scheme with local grid refinement on a parallel computer with distributed shared memory. The adaptive grid refinement is based on estimates of local truncation errors. An array of processors is used; the grids on all levels are subdivided equally amongst all processors. The parallel efficiencies obtained range from 57% to 81%.

INTRODUCTION

The discretization of partial differential equations leads to large systems of equations when a high accuracy is required. On the other hand, the convergence rate of classical iterative solution algorithms deteriorates as the number of unknowns increases. The solution of such systems can be accelerated in the following three ways:

1) The use of multigrid methods, whose rate of convergence is independent of the number of unknowns.

2) The introduction of local grid refinement to reduce computation time and storage requirements.

3) The use of parallel processors for faster computation.

Methods 1) and 2) were combined in Thompson and Ferziger [1], and 1) and 3) in Geus [2]. In this paper all three methods are combined to solve the two-dimensional convection-diffusion equation on a memory coupled multiprocessor. The adaptive grid refinement is based on the work of Berger [3], who used local truncation errors as refinement criterion.

In the next section the model problem, boundary conditions and discretization are presented. This is followed by a discussion of the local refinement strategy, error estimation and the nested iteration scheme. The DIRMU multiprocessor [4] and the parallelization are described in the third section. Finally results for parallel efficiency are presented, together with conclusions and recommendations for further work.

MODEL PROBLEM AND DISCRETIZATION

Model Problem. The model problem is the two-dimensional convection-diffusion equation on the unit square

$$U\frac{\partial \phi}{\partial x} + V\frac{\partial \phi}{\partial y} = \epsilon(\frac{\partial^2 \phi}{\partial x^2} + \frac{\partial^2 \phi}{\partial y^2}) \tag{1}$$

with the mixed boundary conditions shown in Fig. 1. U and V are the constant velocities in x and y directions, respectively, and $\alpha = \arctan(V/U)$ is the flow angle. The diffusivity is denoted by ε. The continuous transition from one to zero at x = 0 is obtained from the error function.

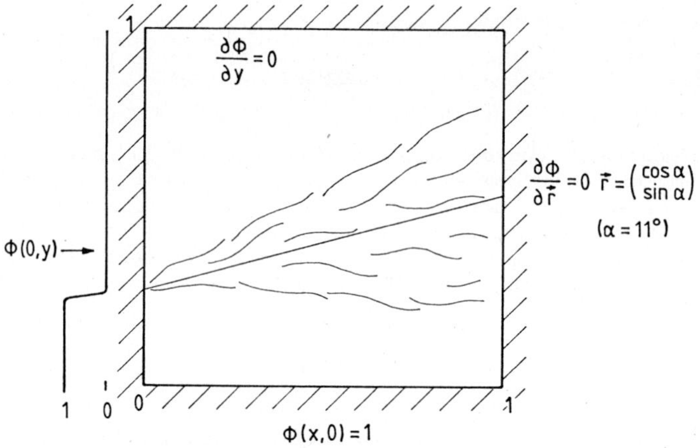

Fig. 1: Model Problem

Discretization. Upwind differences are used for the discretization of the convective terms, and central differences for the diffusive terms. This yields the following discretization formula for the interior points:

$$U\frac{u_{i,j}-u_{i-1,j}}{h}+V\frac{u_{i,j}-u_{i-1,j}}{h} = \frac{\epsilon}{h^2}(u_{i-1,j}+u_{i,j-1}-4u_{i,j}+u_{i+1,j}+u_{i,j+1}). \tag{2}$$

The boundary condition at x = 1 is approximated as

$$u_{i,j} - (1-\tan\alpha)u_{i-1,j} - \tan\alpha\, u_{i-1,j-1} = 0. \tag{3}$$

Equations (2) and (3) are written as a system of linear equations:

$$Lu = f. \tag{4}$$

ERROR ESTIMATION AND LOCAL GRID REFINEMENT

The aim of the local grid refinement is to reduce the solution error, defined as the difference between the exact solution of the differential equation (1) and of the discretized system (4), to a specified limit with a minimum number of grid points. As will be shown below, solution errors are caused by local truncation errors. The method therefore performs local grid refinements in regions with large truncation errors.

<u>Error Estimation</u>. In the following the grids are denoted by Ω_0, Ω_1, Ω_2, ... with corresponding mesh sizes h_0, h_1, h_2, ... , obtained by standard coarsening $h_{l-1} = 2h_l$. Grid boundaries are denoted by Γ_l. It is assumed that the solution errors e_l and e_{l-1} on grids Ω_l and Ω_{l-1} can be written as Taylor series. This yields

$$e_l = h_l^p F(x,y) + h_l^q G(x,y) + O(h_l^r) \tag{5}$$

$$e_{l-1} = 2^p h_l^p F(x,y) + 2^q h_l^q G(x,y) + O(2^r h_l^r) \tag{6}$$

where F(x,y) and G(x,y) are the p-th and q-th derivatives of e, respectively. Subtracting Eqn. (6) from (5) gives an estimate for the solution error:

$$\tilde{e}_l = \frac{u_l - u_{l-1}}{2^p - 1}. \tag{7}$$

This estimate can only be computed on the grid points of Ω_l that coincide with points on Ω_{l-1}. The error estimates of the remaining points are obtained by bilinear interpolation.

The truncation error on grid Ω_l is defined as

$$\tau_l = L_l \phi - f_l. \tag{8}$$

As L_l is a linear operator, this can be written as

$$\tau_l = L_l e_l \tag{9}$$

and approximated by

$$\tilde{\tau}_l = L_l \tilde{e}_l \,. \tag{10}$$

Thus the local truncation error can be interpreted as a source term for the solution error, which is then convected and diffused. The additional cost of the truncation error estimate is equal to one Gauß-Seidel type relaxation step on the finest grid.

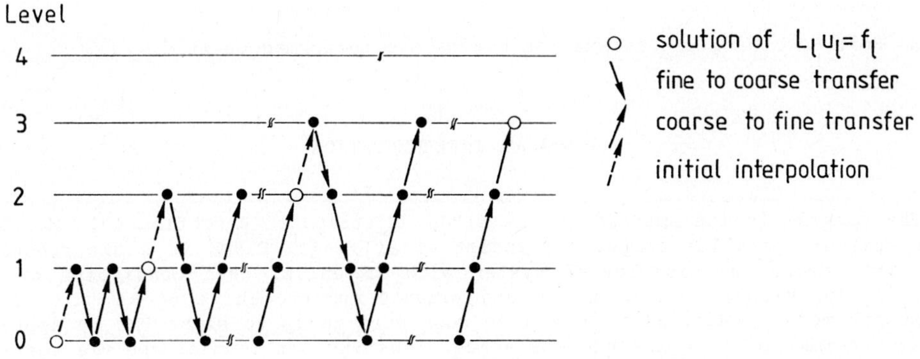

Fig. 2: Adaptive Multigrid Method

Multigrid Method. The multigrid method proceeds as shown in Fig. 2. It is a nested iteration scheme with Full Approximation Storage (FAS) and V-cycles, see Hackbusch [5]. In the following it is assumed that a solution on grid level l has been obtained. Using the solutions on grids Ω_l and Ω_{l-1}, a solution error estimate is obtained according to Eqn. (7). If this exceeds a prescribed limit at any point, the method continues by calculating the truncation errors (Eqn. 10). Grid nodes whose errors exceed a given limit are flagged, and the bounding rectangle defines the new local grid Ω_{l+1}. Interior boundary conditions for the solution on the refined grid Ω_{l+1} are obtained by interpolating solution values from grid Ω_l. At this point adaptive FAS cycles are started to obtain a converged solution on grid Ω_{l+1}. This FAS method is given by the following procedure:

PROCEDURE AFAS(l)

1) Pre-Smoothing: ν_1 relaxation sweeps are performed on grid Ω_l.

2) Restriction: The restriction operator I_l^{l-1} is applied to yield the defect d_{l-1} and the grid function u_{l-1} which is used as starting value for the coarse grid iteration

$$d_{l-1} = I_l^{l-1}(f_l - L_l u_l) \tag{11a}$$

$$u_{l-1} = I_l^{l-1} u_l \tag{11b}$$

3) Coarse grid solution: Eqn. (12) is solved by γ calls to AFAS(l-1),

$$L_{l-1}x_{l-1} = \begin{cases} L_{l-1}u_{l-1} + d_{l-1} & \text{on } \Omega_{l-1} \cap (\Omega_l \setminus \Gamma_l) \\ f_{l-1} & \text{otherwise} \end{cases} \tag{12}$$

where x_{l-1} is the full approximation $x_{l-1} = u_{l-1} + v_{l-1}$.

4) Compute Correction v_{l-1}:

$$v_{l-1} = x_{l-1} - u_{l-1} \tag{13}$$

5) Interpolation:

$$u_l = u_l + I_{l-1}^l v_{l-1} \tag{14}$$

6) Post-Smoothing: ν_2 relaxation sweeps are performed on grid Ω_l.

PARALLEL IMPLEMENTATION

The program is implemented on the DIRMU (Distributed Reconfigurable Multiprocessor) parallel computer (Händler et al. [4]). DIRMU is a distributed shared memory multiprocessor system with 26 units, each consisting of a processor module (PM) with CPU and private memory, and a 64 kB multiport memory module (MM). Each MM can be connected to up to eight PMs to appear as a normal part of its address space. Thus PMs can communicate via shared variables stored in the MMs, see Fig. 3.

Various processor configurations can be achieved by utilizing different combinations of processor to multiport links. For the present work, a processor

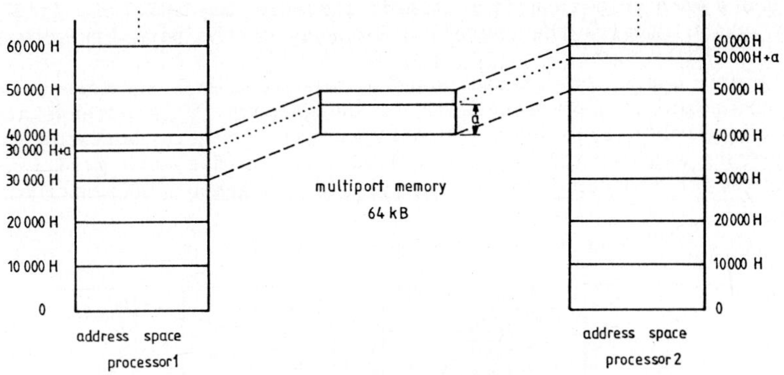

Fig. 3: Shared Memory Processor Configuration

array is used with a minimum of four and a maximum of twenty processors. The parallelization of the method can be divided into two parts:

- the parallelization of the data structure, and
- the parallelization of the algorithm.

Parallelization of the Data Structure. The object of the parallelization of the data structure is to achieve an optimum load balance between the processors whilst retaining a manageable data structure. Here the grid points are distributed among the MM's by dividing the rows and columns of the numerical grid as equally as possible among the processors of the array. This is illustrated in Fig. 4.

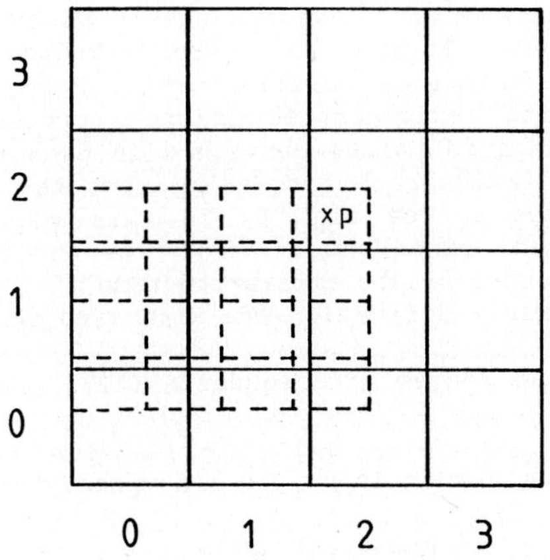

Fig. 4: Grid Distribution

The restriction of the point P requires the transportation of data from MM (3,3) (see Fig. 4) in the fine local grid to MM (2,2) in the global coarse grid. Hence each grid-to-grid multigrid operation may involve a grid redistribution, which impairs the overall efficiency of the parallel program.

The multigrid operations coarse-to-fine, fine-to-coarse and relaxation require each grid point to have access to its neighbours. This necessitates a data structure which allows each processor access to grid portions located in neighbouring MMs. To this end, one MODULA record for each grid is stored in each MM, which contains the necessary grid management information, see Fig. 5.

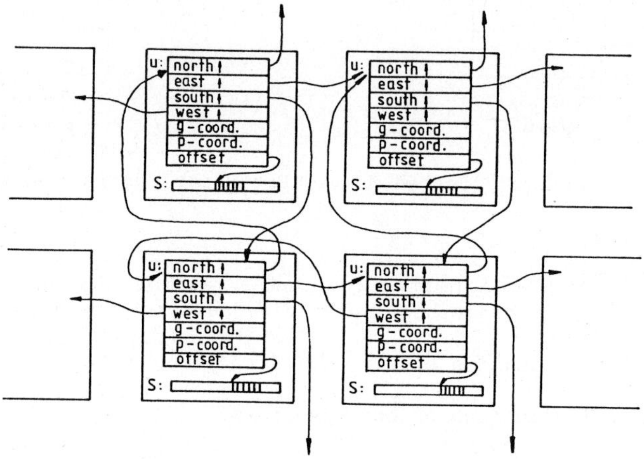

Fig. 5: Grid Function Management

<u>Parallelization of the Algorithm</u>. In this section the parallelization of the three main elements of the multigrid scheme, namely relaxation, restriction and interpolation, is described.

Multigrid theory requires the relaxation scheme to be a fast solver of the convective problem ($\varepsilon = 0$ in Eqn. (1)), see Hackbusch [5]. In our case the smoother should also be highly parallelizable. Here a modified pointwise Gauß-Seidel method with lexicographic ordering corresponding to the flow direction is used. The grid points contained in each row of processors are relaxed in a genuine lexicographic ordering, but the processor rows begin relaxation simultaneously, see Fig. 6. In this manner a good degree of parallelism is achieved, since the waiting time for the processors at the end of each row is bounded by the relaxation time of one row of grid points. The implications of this modification are discussed in the conclusions.

In order to prepare the coarse grid equation (12), the following steps are performed:

1) Restrict starting values $I_l^{l-1} u_l$ to a temporary grid function t_{l-1}

2) Redistribute: $u_{l-1} := \text{REDISTRIBUTE}(t_{l-1})$ on $\Omega_{l-1} \cap (\Omega_l \setminus \Gamma_l)$

3) Calculate and Restrict Defect: $t_{l-1} := I_l^{l-1}(L_l u_l - f_l)$

4) Redistribute: $f_{l-1} := \text{REDISTRIBUTE}(t_{l-1})$

5) Correct: $f_{l-1} := f_{l-1} + L_{l-1}u_{l-1}$ on $\Omega_{l-1} \cap (\Omega_l \setminus \Gamma_l)$.

The temporary grid function t_{l-1} corresponds to the coarse grid points below the finer grid. However, like the fine grid, it is also stored in all MMs, in order to provide temporary storage for restricted variables between computation and redistribution, cf. Fig 4.

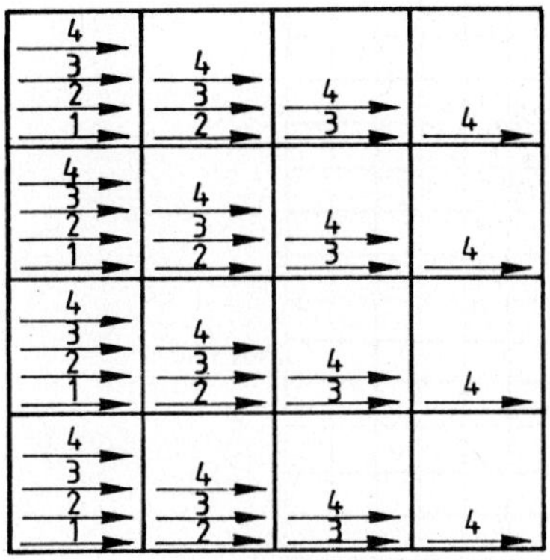

Fig. 6: Relaxation Ordering

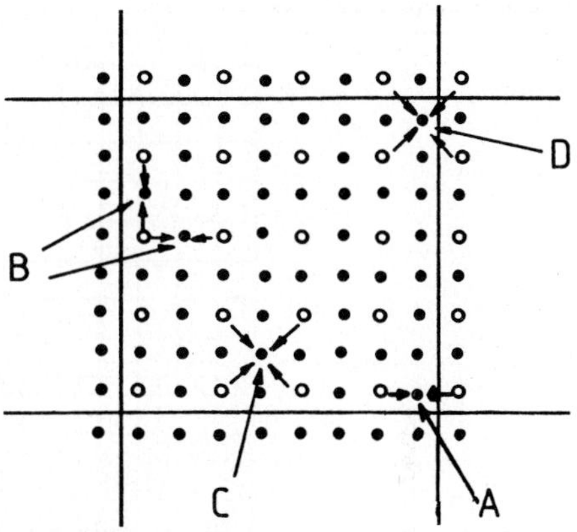

Fig. 7: Bilinear Interpolation Points of type A and D need values from neighbouring processors.

The introduction of temporary grid functions leads to a reduction in efficiency, as does the redistribution in steps 2) and 4). Neither of these is necessary in a monoprocessor implementation. A further loss stems from the fact that step 5) can only be computed by those processors with access to coarse grid points directly below the fine grid, while the remaining processors are idle.

A coarse-to-fine transfer comprises the following four steps:

1) Calculate Correction: $v_{l-1} := x_{l-1} - u_{l-1}$ on $\Omega_{l-1} \cap \Omega_l$

2) Redistribute: $t_{l-1} := \text{REDISTRIBUTE}(v_{l-1})$

3) Interpolate Correction: $t_l := I_{l-1}^{l} t_{l-1}$ on $\Omega_{l-1} \cap \Omega_l$

4) Add Correction: $u_l := u_l + t_l$ on $\Omega_{l-1} \cap \Omega_l$.

Here both fine and coarse temporary grid functions are necessary. Step 1) can only be performed by those processors with access to coarse grid points beneath the fine grid. Local synchronization between neighbouring processors is necessary, as the computation of certain fine grid points at MM boundaries requires that values in adjacent MMs are already available, see Fig. 7.

RESULTS

Fig. 8 shows an example of a refined grid. The computing time on each grid level is proportional to the number of unknowns. The savings resulting from adaptive grid refinement are thus proportional to the areas not covered by the corresponding local grids. In addition to the algorithmic savings, the computation time is further reduced by parallel processing.

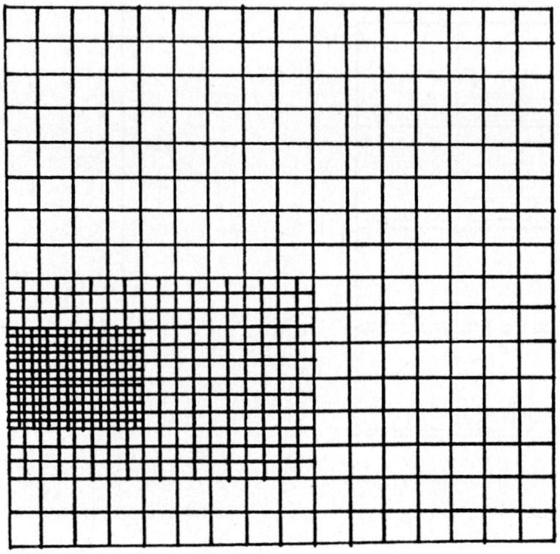

Fig. 8: Locally Refined Grids

The efficiency of a parallel program is defined by

$$E(p) = \frac{T_m}{p T_p} \tag{15}$$

with T_m the monoprocessor, and T_p the parallel computation time, and p the number of processors used. The efficiencies for three test problems are shown in Fig. 9.

The full domain calculations are performed more efficiently than the problem with local refinement. This is due to the grid redistributions which become necessary during interpolation and restriction. The parallel computer becomes more efficient as the number of grid points per MM increases, as can be seen by comparing the results of the two full domain calculations. This is caused by the decreasing ratio of interprocessor communication to parallelizable grid point operations.

For a fixed number of processors, the efficiency depends on the sidelength of the processor array. Arrays with a larger number of processors in the x-direction perform less well. This can be explained by the increased row-wise local synchronization during relaxation.

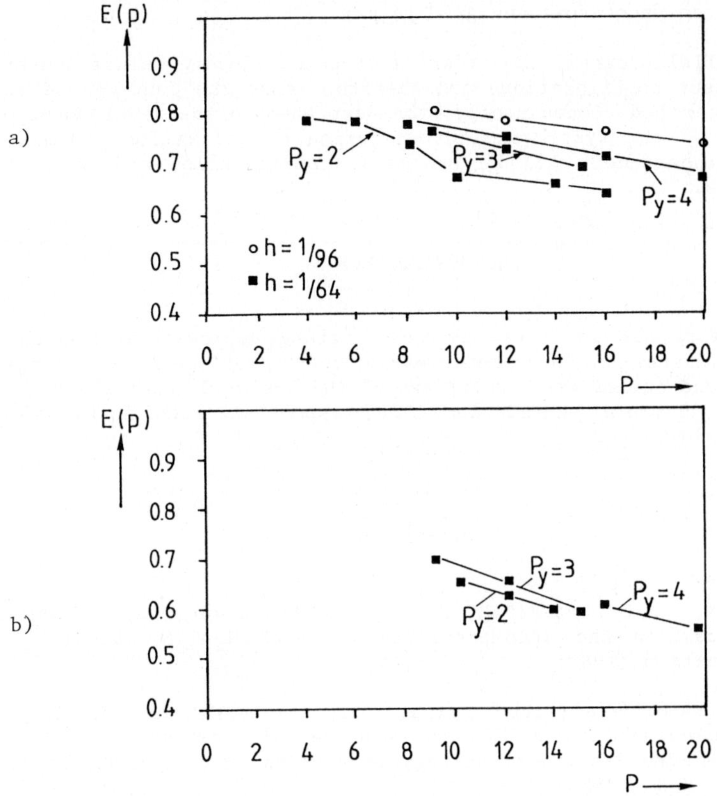

Fig. 9: Parallel Efficiencies; a) Full Domain Calculations, b) Calculations with Local Refinement

CONCLUSIONS AND OUTLOOK

In the present work the two-dimensional convection-diffusion equation is solved using a multigrid scheme with local grid refinement on a parallel MIMD-computer. The parallel efficiencies for full domain problems range from 65% to 81%, and from 57% to 70% for a problem with local refinement. However, these losses in efficiency are far outweighed by the gains due to the reduction in the number of grid points, making local refinement also attractive for parallel computers.

The efficiency for a given problem decreases as the number of processors is increased, and hence the number of grid points per MM decreases. Measurements show, however, that synchronization times are short enough to expect little loss in efficiency if the problem size were to be increased with the number of processors.

The main loss in efficiency is due to unequal processor load, since it is in general not possible to divide NxN grid points equally among PxP MMs using the present strategy. However, this effect becomes negligible for large numbers of processors and grid points.

With the parallel program, the ordering of points during relaxation depends on the processor configuration, and therefore also the convergence rate of the multigrid method. Consequently, the next step in the development of the program will be the systematic investigation of relaxation schemes which are robust, independent of the processor configuration and still highly parallelizable.

ACKNOWLEDGEMENTS

The present work was sponsored by the Stiftung Volkswagenwerk within its program "Entwicklung von Berechnungsverfahren für Probleme der Strömungstechnik". The second author was a recipient of the Senior Scientist Award of the Alexander von Humboldt Foundation. The support of both institutions is gratefully acknowledged.

REFERENCES

[1] THOMPSON, M. C., FERZIGER, J. H.: "An Efficient Adaptive Multigrid Technique for the Incompressible Navier-Stokes Equations" private communication (1987).

[2] GEUS, L.: "Parallelisierung eines Mehrgitterverfahrens für die Navier-Stokes-Gleichungen auf EGPA-Systemen", Arbeitsber. des Inst. für Math. Maschinen und Datenverarbeitung, Universität Erlangen-Nürnberg, Band 18, Nummer 3 (1985).

[3] BERGER, M., "Adaptive Mesh Refinement for Hyperbolic Partial Differential Equations", Ph.D. Thesis, Dept. Computer Sc., Stanford University (1982)

[4] HÄNDLER, W., MAEHLE, E., WIRL, K., "The DIRMU Testbed for High-Performance Multiprocessor Configurations", Proc. of the first Intern. Conf. on Supercomputing Systems, St. Petersburg (1985).

[5] HACKBUSCH, W., "Multi-Grid Methods and Applications", Springer Verlag, 1985.

FINITE VOLUME MULTIGRID SOLUTIONS OF THE TWO-DIMENSIONAL INCOMPRESSIBLE NAVIER-STOKES-EQUATIONS

C. Becker, J. H. Ferziger*, M. Perić, G. Scheuerer

Lehrstuhl für Strömungsmechanik, Universität Erlangen-Nürnberg, Egerlandstr. 13, D-8520 Erlangen

* Department of Mechanical Engineering, Stanford University, Stanford, CA 94305

SUMMARY

A multigrid scheme for solving the two-dimensional, laminar, incompressible Navier-Stokes equations is presented. It uses V-cycles in a nested iteration scheme and full approximation storage. The iterative SIMPLE algorithm is employed to relax the coupled momentum and continuity equations. Within the SIMPLE algorithm, linear systems are solved with an ILU factorization method. The multigrid scheme is extended to handle local grid refinement, with the refinement based on estimates of local truncation errors. Results are reported for backward facing step flows at various Reynolds numbers.

INTRODUCTION

Recently, Barcus et al. [1] presented a cell-centered multigrid method for solving the two-dimensional, incompressible, laminar Navier-Stokes equations; it is based on the iterative SIMPLE algorithm of Patankar and Spalding [2]. This procedure employs the so-called correction scheme (CS), see Hackbusch [3]. The present paper describes an extension of the method to a full approximation scheme (FAS) in which approximate solutions are used as coarse grid variables instead of corrections as in the CS-method. FAS is better suited to coupled systems of equations with strongly non-linear source terms. Such systems occur in the solution of the time-averaged Navier-Stokes equations with two-equation turbulence models.

Another important feature of FAS is its straightforward extension to local grid refinement methods. Local grid refinement is a technique that allows the concentration of grid points where required, while retaining the advantage of structured grids: the resulting difference equations can be solved with efficient iterative schemes. In this paper, a local grid refinement method embedded in a FAS multigrid scheme for the Navier-Stokes equations is presented. The refinement philosophy is adopted from the work of Berger [4], Caruso et al. [5] and Thompson and Ferziger [6]. It is based on estimation of the local truncation error, as this is the source of solution error.

In the following section, the two-dimensional Navier-Stokes equations are presented. Next, the finite volume method and the relaxation scheme used as the base for the multigrid cycle are described. The nested iteration and local refinement method are discussed in the third and fourth sections. Finally, results for two-dimensional flow over a backward facing step at various Reynolds numbers are shown.

MODEL PROBLEM

The model problem consists of the conservation equations for mass and momentum for laminar, two-dimensional flow of an incompressible, Newtonian fluid. In cartesian coordinates, these equations read:

$$\frac{\partial(\rho U)}{\partial x} + \frac{\partial(\rho V)}{\partial y} = 0 \tag{1}$$

$$\frac{\partial}{\partial x}(\rho U^2 - \mu\frac{\partial U}{\partial x}) + \frac{\partial}{\partial y}(\rho UV - \mu\frac{\partial U}{\partial y}) = -\frac{\partial P}{\partial x} \tag{2}$$

$$\frac{\partial}{\partial x}(\rho UV - \mu\frac{\partial V}{\partial x}) + \frac{\partial}{\partial y}(\rho V^2 - \mu\frac{\partial V}{\partial y}) = -\frac{\partial P}{\partial y}\,. \tag{3}$$

The velocities in the x- and y-direction are denoted by U and V; P is the static pressure. ρ and μ stand for the constant density and dynamic viscosity of the fluid.

The test case is the flow over a backward facing step. Figure 1 shows the geometry and the mixed Dirichlet/von Neumann boundary conditions.

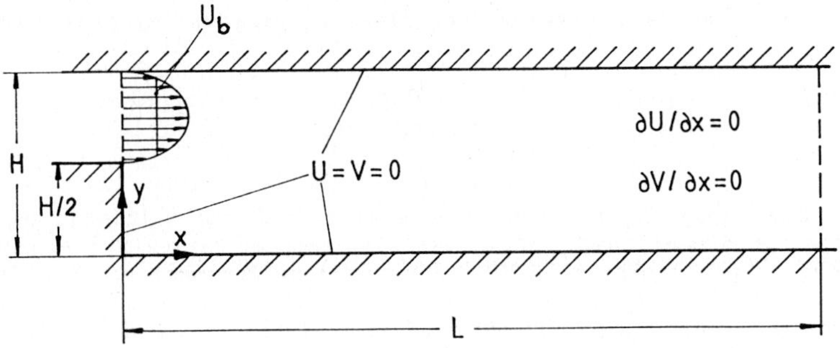

Fig. 1: Model Problem

RELAXATION SCHEME

A combination of conservative finite volume differencing and the iterative SIMPLE algorithm of Patankar and Spalding [2] is used for relaxations within the multigrid cycle. Uniformly spaced grids with colocated (non-staggered) variable arrangement, cf. Fig. 2, are employed. In connection with multigrid methods, the colocated arrangement has advantages over the traditional staggered grids as only one set of control volumes needs to be refined and coarsened.

The Navier-Stokes equations (2) and (3) are discretized by first integrating them over a control volume and then approximating the fluxes across the cell faces in terms of nodal values. In the present work, first order upwind differences are used for the convective fluxes and second order central differences for the diffusive fluxes. The resulting difference equations are

$$\sum_m A_m U_m = -\delta y \, \delta P_x \tag{4}$$

$$\sum_m A_m V_m = -\delta x \, \delta P_y \tag{5}$$

where the index m runs over the nodal points P, W, E, N, S and δP_x, δP_y are the pressure differences across a control volume in the coordinate directions. For an east cell face the coefficients A_m have the form

$$A_m = \rho U_e \delta y + \frac{\mu \, \delta y}{\delta x} \tag{6}$$

Linearization is performed by evaluating the mass fluxes (eg. $\rho U_e \delta y$) using velocities from the previous iteration.

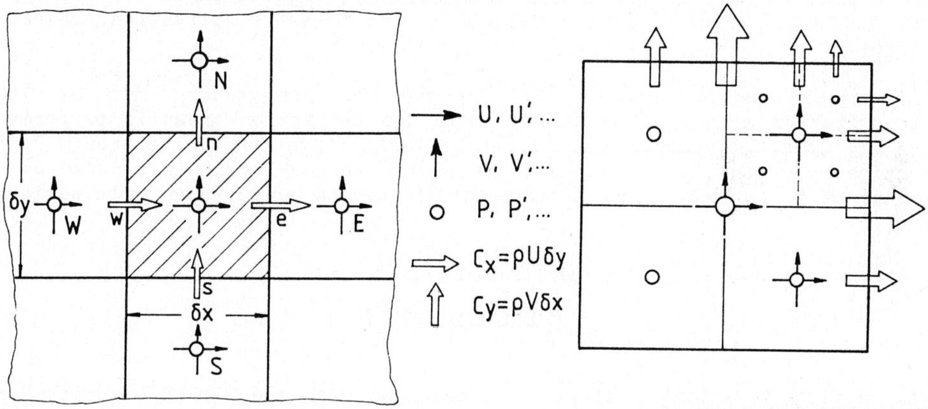

Fig. 2: a) Colocated Variable Arrangement; b) Location of Coarse and Fine Grids

In the SIMPLE-algorithm the discretized momentum equations are initially solved using a guessed pressure field. The resulting velocities are introduced into the discretized continuity equation, leading to an effective mass source $\dot{m}$:

$$(\rho U_e - \rho U_w)\,\delta y + (\rho V_n - \rho V_s)\,\delta x = \dot{m} \; . \tag{7}$$

In the colocated arrangement, the cell face velocities appearing in the mass fluxes are calculated from the discretized momentum equations using a special interpolation practice, see Perić et al. [7]. The velocity at the east control volume face is, for instance,

$$U_e = \overline{\left(\frac{\sum_m A_m U_m}{A_P}\right)}_e - \overline{\left(\frac{1}{A_P}\right)}_e \delta y (P_E - P_P) \tag{8}$$

where the overbar denotes linear interpolation of the corresponding terms in the difference equations for points P and E. Analogous expressions are used for the remaining velocities. The mass source $\dot{m}$ is eliminated by applying corrections to the velocity and pressure fields. The link between the velocity and pressure corrections is derived from a truncated form of Eqn. (8):

$$\tilde{U}_e = -\delta y \overline{\left(\frac{1}{A_P}\right)}_e (\tilde{P}_E - \tilde{P}_P). \tag{9}$$

Substitution of Eqn. (9) into the discretized continuity equation yields the pressure correction equation

$$\sum_m B_m \tilde{P}_m = -\dot{m}. \tag{10}$$

Equation (10) has zero gradient boundary conditions on all boundaries. The correction is used to update the pressure, velocities and mass fluxes, and, in turn, the coefficients A_m for the next iteration. The efficiency of the SIMPLE algorithm depends considerably on the underrelaxation factors applied to the momentum equations and the pressure correction, see the detailed discussion in [7]. In the present calculations, underrelaxation factors of 0.7 and 0.3 are used for the velocities and pressure correction, respectively.

The linear equation systems (4), (5) and (10) are relaxed with the ILU factorization method of Stone [9]. Only one relaxation sweep is performed for the momentum equations within each SIMPLE iteration. The pressure correction equation (10) is iterated until the residual norm is reduced by a factor of 5. In most cases, no more than 10 sweeps were necessary to achieve this.

MULTIGRID METHOD

Coarse Grid Equations. After ν_1 iterations with the SIMPLE algorithm on the fine grid, the momentum equations are satisfied within residuals R^u and R^v, respectively:

$$\sum_m A^1_m U^1_m = -\delta y\, \delta P^1_x + R^u \tag{11}$$

$$\sum_m A^1_m V^1_m = -\delta x\, \delta P^1_y + R^v. \tag{12}$$

The notation A_m indicates that the coefficients are assembled using the velocities U^1 and V^1. To reduce the residuals to zero, corrections must be added, yielding

$$\sum_m (A^1_m + A'_m)(U^1_m + U'_m) = -\delta y\, \delta(P^1_x + P'_x) \tag{13}$$

and a corresponding equation for the the V-velocity. Addition of Eqns. (11) and (13) and restriction to a coarse grid leads to the coarse grid equation for the U-velocity:

$$\sum_m \hat{A}_m \hat{U}_m = -\delta y\, \delta P'_x + R^u + \sum_m \hat{A}_m (I^{2h}_h U^1_m) \tag{14}$$

where the "full approximations" are defined as $\hat{U}_m = U^1_m + U'_m$ and $\hat{A}_m = A^1_m + A'_m$. I^{2h}_h is a bilinear restriction operator. Since the pressure term is linear, it is sufficient to define only pressure errors on the coarse grid. The coarse grid equation for the V-velocity is obtained in an analogous way.

The SIMPLE algorithm is also employed on the coarse grid. Restricted values

$I_h^{2h} U_m^1$ and $I_h^{2h} V_m^1$ are used as starting guesses for the velocities; the pressure errors P'are initially set to zero. After relaxing the momentum equations with Stone's method, a pressure correction equation, identical to the one on the base grid, is solved to provide corrections for $\hat{U}$, $\hat{V}$ and P'. The entire process on the coarse grid is repeated for a specified number of iterations. Before returning to the fine grid, velocity errors are calculated from the full approximations via:

$$U'_m = \hat{U}_m - U_m^1, \; V'_m = \hat{V}_m - V_m^1. \tag{15}$$

<u>Restriction and Prolongation</u>. In order to solve Eqn. (14) on the coarse grid, the finite volume coefficients A_m, the residuals R^u and the dependent variable U_m need to be restricted. Since a cell-centered multigrid scheme is employed, the coefficients and residuals can be restricted without any interpolation. The mass fluxes appearing in the coarse grid coefficients A_m are evaluated by adding the corresponding fine grid mass fluxes, see Fig. 2. The diffusion fluxes are recalculated with central differences. The residuals are restricted by summing up the contributions of the four related fine grid control volumes. Bilinear interpolation is used to obtain the cell center values of the dependent variables on the coarse grid. The prolongation of the errors back to the fine grid is also performed with bilinear interpolation.

<u>Full Multigrid Scheme</u>. The two-grid V-cycle just described is repeated until the solution on the fine grid has converged. The velocity and presssure fields are then prolongated to the next finer grid. The V-cycle is restarted, but now with two coarser grids. The procedure can be extended to a specified number of grids. Fig. 3. shows the method for a five-grid system. The numbers in Fig. 3 indicate how many smoothing iterations are performed on each grid level. More relaxations are made on coarser grids. This strategy is nearly optimum in terms of computing time.

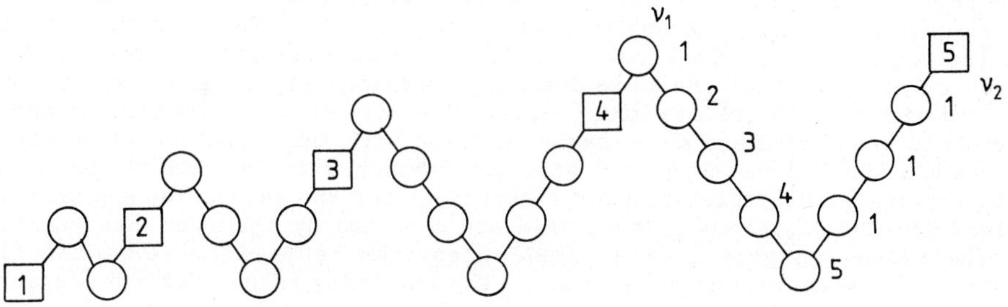

Fig. 3: Nested Iteration Scheme

LOCAL GRID REFINEMENT

The aim of local grid refinement is to reduce the solution errors below a specified level with minimum computational effort. Solution error is defined as difference between the exact solution and a discrete solution on a grid with spacing h:

$$\varepsilon_h = \phi - \phi_h \,. \tag{16}$$

Solution errors originate from truncation errors defined by

$$\tau_h = \sum_m A_m \phi_m - \sum_m A_m (\phi_h)_m \,. \tag{17}$$

For linear problems the solution and truncation errors are related by

$$\tau_h = \sum_m A_m (\varepsilon_h)_m \,. \tag{18}$$

Eqn. (18) shows that truncation errors act as source terms for solution errors. Hence, the local grid refinement is based on estimated truncation errors.

In order to compute truncation errors from Eqn. (18), values of ε_h are required. These are obtained using a technique based on Taylor series described by Caruso et al. (1985). For grid distances h and 2h, respectively, and a first order discretization scheme, the following relations emerge:

$$\phi \simeq \phi_h + h\phi' + \frac{h^2}{2}\phi'' + \dots \tag{19a}$$

$$\phi \simeq \phi_{2h} + 2h\phi' + \frac{(2h)^2}{2}\phi'' + \dots \tag{19b}$$

Subtracting Eqn. (19a) from (19b) yields a second order estimate for the solution error

$$\varepsilon_h \simeq \phi_h - \phi_{2h} \tag{20}$$

which can then be used together with Eqn. (18) to approximate the truncation error.

The local refinement scheme can be summarized as follows. The nested iteration described in the previous section is used until a prescribed grid level l is reached, see Fig. 4. A solution error estimate is then based on solutions on the successive grids l and l-1. If the solution errors exceed a given value at any point in the flow domain, truncation errors are computed with the aid of Eqn. (18). Points with truncation errors above an empirically chosen threshold are marked and a rectangle, aligned with the coordinate directions, is drawn around them. A refined grid (of spacing h/2) is constructed within this rectangle. Dirichlet boundary conditions for the velocities and starting values for the iteration on this grid are obtained by bilinear interpolation of the values on grid l. The SIMPLE algorithm is used to relax the flow equations ν_1 times on the refined grid. The resulting velocities and pressures are then restricted and the following equations are solved on grid level l:

$$\sum_m \hat{A}_m \hat{U}_m = -\delta y\, \delta P_x' + R^u + \sum \hat{A}_m (I_h^{2h} U_m^1) \qquad \text{on grid } l \cap l+1 \tag{21a}$$

$$\sum_m \hat{A}_m \hat{U}_m = -\delta y\, \delta P_x' + \hat{A}_m (U_m)_l \qquad \text{otherwise.} \tag{21b}$$

Here, $(U_m)_l$ denotes existing solution values at grid level l. The last term is added in order to have pressure *errors* on the right hand sides of both Eqs. (21). After performing smoothing relaxations on this grid level, the V-cycle as described above is continued.

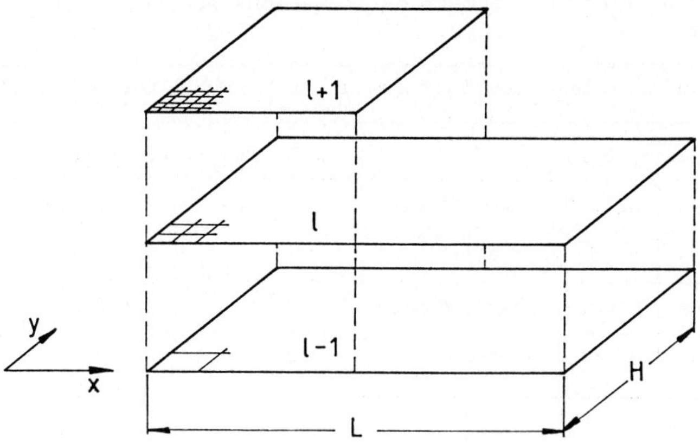

Fig. 4: Local Grid Refinement

RESULTS

Laminar flow over a backward facing step at various Reynolds numbers is taken as the test case. The Reynolds number is based on the entrance bulk velocity and the channel height. As the reattachment length of a laminar backward facing step flow increases with Reynolds number, the length of the computation domain is also increased. In the present calculations, L is equal to 5, 9 and 16 step heights for Re = 100, 250 and 500. Uniform, rectangular grids with cell aspect ratios of 2.5 are used in all cases. The residuals of the momentum and continuity equations are normalized with the inlet momentum and mass fluxes. Convergence is assumed when all three normalized residual norms are reduced below 10^{-4}.

Table 1 shows the computing times and numbers of fine grid iterations for the nested iteration scheme (FAS) without local refinement. The abbreviation SGM denotes a conventional single grid method and CS stands for the correction scheme of Barcus et al. [1]. For Re = 100, the convergence rates of both multigrid methods are almost independent of the number of control volumes, while the computing times increase approximately linearly. In contrast, computing times of the SGM increase quadratically. Although CS demands less computational work per multigrid cycle, its overall efficiency is slightly worse. This is probably due to terms neglected in the CS coarse grid equations, see Barcus et al. [1]. The picture is similar at higher Reynolds numbers. The relative efficiency of the FAS method decreases as the Reynolds number increases. Similar behaviour was already noted for the flow in a lid driven cavity [1]. The effect can be caused by a grid inadequate to resolve the flow. However, this point needs further investigations. Multigrid calculations with fewer control volumes in the x-direction and higher cell aspect ratios were tried at Reynolds numbers greater than 250. These calculations did not converge on the finer grids with any combination of smoothing relaxations on the various grid levels, see Becker [9].

First results of the local refinement scheme are now presented. In these calculations at Re = 100, only one level of grid refinement is used. Figure 5

Table 1: Computing Times and Number of Iterations for Single and Multigrid Schemes

Re = 100; under-relaxation factors α^u = 0.7 (0.8 for SGM), α^p = 0.3										
	Computing Times					Number of Iterations				
CV	20x10	40x20	80x40	160x80	320x160	20x10	40x20	80x40	160x80	320x160
SGM	0.50	4.10	88.30	1257.70	19.000*	25	71	249	883	
CS	0.65	3.09	13.72	62.28	265.36	31	9	8	9	9
FAS	0.66	2.62	10.28	51.35		31	9	8	9	

Re = 250; under-relaxation factors α^u = 0.7, α^p = 0.3								
	Computing Times				Number of Iterations			
CV	36x10	72x20	144x40	288x80	36x10	72x20	144x40	288x80
SGM	1.63	28.16	425.12		51	156	490	
FAS	1.63	6.93	32.81	218.04	51	15	15	16

Re = 500; under-relaxation factors α^u = 0.7, α^p = 0.3								
	Computing Times				Number of Iterations			
CV	64x10	128x20	256x40	512x80	64x10	128x20	256x40	512x80
SGM	5.48				95			
FAS	5.48	40.23	146.54	891.69	95	54	33	38

shows the estimated solution and truncation errors for the U-velocity for a 40x20 control volume grid. Based upon this estimate, the region $0 < x < L/2$ is locally refined. The computing times and number of fine grid iterations for this calculation are summarized in Table 2. For this case, a factor of eight in computing time is saved relative to the conventional SGM. However, the calculation is somewhat more expensive than a FAS calculation on a 80x40 grid. This difference comes from the reduced convergence rate of the problem on the refined grid caused by applying Dirichlet boundary conditions at $x = L/2$. The additional cost for performing the solution and truncation error estimates is almost negligible. Figure 5 shows the differences between solutions on the 80x40 and the locally refined 40x40 control volume grid. As expected, the differences are confined to the left half of the flow domain. They are nowhere larger than 3% of the bulk velocity U_b. The differences in the V-velocity, although numerically much smaller than those of U, are concentrated at the left boundary of the local grid. They can be further reduced by using more stringent convergence criteria.

Table 2: Computing Times and Number of Iterations for the Local Refinement Method

		Number of Iter.	Computing Time
FAS with local refinement	20x10 CV	25	0.52
	40x20 CV	8	2.21
	40x40 CV	12	9.76
SGM	80x40 CV	249	88.30

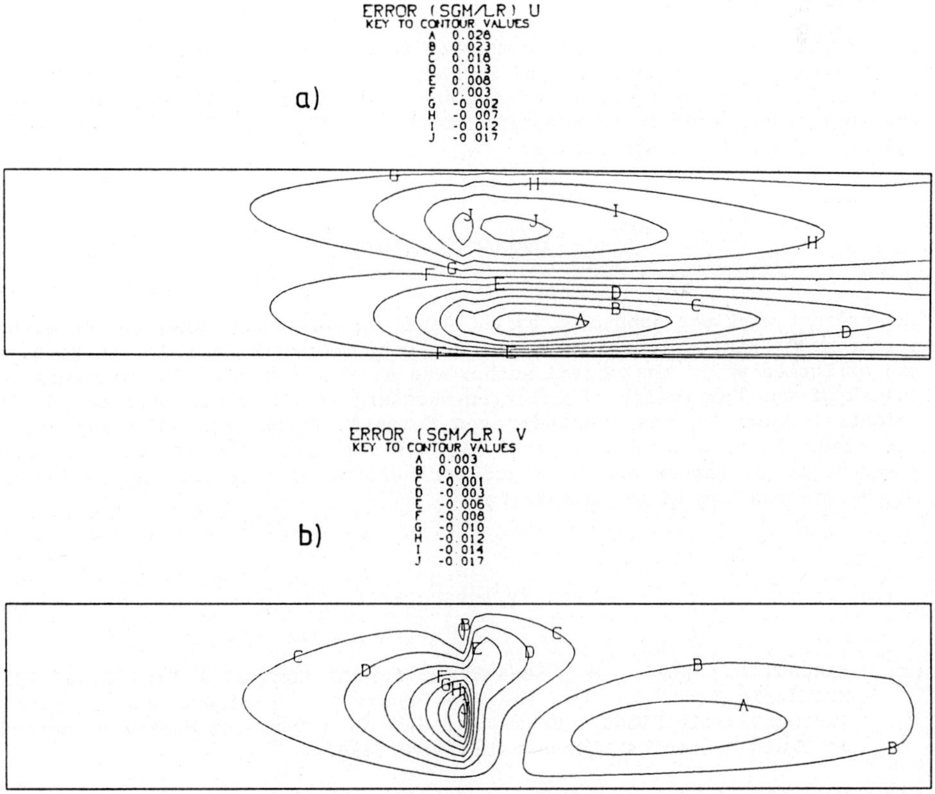

Fig. 5: Differences between solutions on global and locally refined grids: a) U-velocity, b) V-velocity

CONCLUSIONS

In this paper, a full approximation scheme for the solution of the two-dimensional, incompressible Navier-Stokes equations is presented. It uses V-cycles and a nested iteration scheme. The iterative SIMPLE algorithm of Patankar and Spalding [2] is employed as the relaxation scheme. The strongly implicit method of Stone [8] is used to solve the linear equation systems. The FAS method is extended to a scheme with local grid refinement based on estimates of local truncation errors.

Results are obtained for laminar backward facing step flows at several Reynolds numbers. They show the expected multigrid characteristics, i.e. the number of iterations is independent of the number of grid cells, and the computing times increase linearly. The newly developed FAS method is slightly more efficient than the CS scheme of Barcus et al. [1]. For fine grids, the convergence rates of both multigrid schemes are sensitive to the aspect ratio of the grid control volumes. Initial experience with the combination of FAS and local grid refinement shows marked improvements in convergence rates and computing times relative to the single grid method. It is expected that the advantage of local refinement will increase when more levels of refinement are used.

The sensitivity of the multigrid methods to the aspect ratios of the control volumes suggests increased use of local refinement and/or the generalization of a multigrid method suited to non-equidistant grids. We plan to perform the local refinement in an adaptive manner, similar to the work of Thompson and Ferziger [6], in the future.

ACKNOWLEDGEMENTS

The present work was sponsored by the Deutsche Forschungsgemeinschaft within its program "Finite Approximationen in der Strömungsmechanik". The research was performed while the second author was at the Lehrstuhl für Strömungsmechanik of the University of Erlangen-Nürnberg as a recipient of the Senior Scientist Award of the Alexander von Humboldt Foundation. The support of all three institutions is gratefully acknowledged. The authors are also grateful to M. Barcus and M. Rüger for helpful discussions and assistance in the preparation of the manuscript.

REFERENCES

[1] BARCUS, M., PERIĆ, M., SCHEUERER, G.: "A Control Volume Based Full Multigrid Procedure for the Prediction of Two-Dimensional, Laminar, Incompressible Flows", Proc., 7th GAMM Conference on Numerical Methods in Fluid Mechanics, Louvain-la-Neuve (1987).

[2] PATANKAR, S. V., SPALDING, D. B.: "A Calculation Procedure for Heat, Mass and Momentum Transfer in Three-Dimensional Parabolic Flows" Int. J. Heat Mass Transfer, 15 (1972), pp. 1787-1806.

[3] HACKBUSCH, W.: "Multigrid Methods and Applications", Springer Verlag, Berlin, 1985.

[4] BERGER, M.: "Adaptive Mesh Refinement for Hyperbolic Partial Differential Equations", Ph.D. Thesis, Dept. Computer Sc., Stanford Univ., 1982.

[5] CARUSO, S. C., FERZIGER, J. H., OLIGER, J.: "Adaptive Grid Techniques for Elliptic Problems", Rept.-Nr. TF-23, Thermosc. Div., Stanford Univ., 1985.

[6] THOMPSON, B., FERZIGER, J. H.: "An Efficient Adaptive Multigrid Technique for the Incompressible Navier-Stokes Equations", priv. communication, 1987.

[7] PERIĆ, M., KESSLER, R., SCHEUERER, G.: "Comparison of Finite-Volume Numerical Methods with Staggered and Colocated Grids", Rept.-Nr. 163/T/87, Lehrst. für Strömumgsmechanik, Univ. Erlangen-Nbg., 1987 (acc. for publ. in Computers and Fluids).

[8] STONE, H. L.: "Iterative Solution of Implicit Approximations of Multi-Dimensional Partial Differential Equations" SIAM J. Num. Anal., 5 (1968) pp. 530-558.

[9] BECKER, C.: "Berechnung zweidimensionaler, elliptischer Strömungen mit Mehrgitterverfahren unter Verwendung lokaler Gitterverfeinerung", Diploma Thesis, Lehrst. für Strömungsmechanik, Univ. Erlangen-Nbg., 1988.

CONCEPTS FOR A DIMENSION INDEPENDENT APPLICATION OF MULTIGRID ALGORITHMS TO SEMICONDUCTOR DEVICE SIMULATION

P. Conradi, D. Schröder

Technische Universität Hamburg-Harburg
- Technische Elektronik -
Eißendorfer Straße 38
2100 Hamburg 90

Summary

The solution of semiconductor device equations requires fast and flexible numerical algorithms. It is shown that a program construction based on multigrid methods combined with structured programming techniques helps to overcome the computational bottleneck apparent in 3 dimensional problems. In our approach, modified classical multigrid methods have been incorporated into a program using a tree structured data organization. Thus, a space dimension independent formulation of grid transfer operations and adaptive grid refinement techniques can be introduced. The program code is independent of the domain configuration as well as the problem dimension. This flexibility is inherent in the underlying data structure which is operated and manipulated by the program.

1. Introduction

An important means in the design of integrated electronic circuits is the device simulation. Hereby, the intrinsic behaviour of the integrated devices (e.g. MOS transistor) is calculated with the aid of a computer analysis. Based on a physical and technological model, an evaluation and optimization of the circuit is gained before a time and cost consuming fabrication.
Since the initializing work of van Roosbroeck [1] and Gummel [2], many one and two dimensional simulators have been developed. Mostly, these simulators are based on a finite difference or finite element discretization of the basic coupled equation system, and a subsequent solution by relaxation or Newton-Raphson methods. Due to the advance of the semiconductor fabrication technology and the decreasing size of the devices of the next generation, boundary effects become more and more important, and a simulation in three dimensions will be necessary.

The conventional numerical device simulation methods prove to have a too severe consumption of computational power to be applied to 3D problems on today's computers. Thus, it is necessary to exploit every suitable idea to decrease the requirements of the numerical work. In this context, the superiour efficiency of the multigrid method in the solution of partial differential equations seemed to justify an attempt to apply this method to the problem of device simulation.
In the design of the program, the following requirements have been considered.

Since the physics of submicron devices becomes more advanced, emphasis has been put on the possibility to change the basic equation system easily, because in parallel ongoing work new models of electron transport are developed and will be incorporated into the program.

The amount of computational work in 3D problems is always an order of magnitude higher than in 2D calculations. For a rough estimation of the properties of a certain device family, it is often sufficient to utilize an approximate two dimensional model, and do a quick 2D calculation. Once the most promising candidates have been identified, the model can be refined, and a detailed 3D simulation can be performed. Since the data structure and algorithms in our program have been designed independent of problem dimension, an easy replacement of 1D or 2D models by 3D models is supported.

Finally, the program was intended to be flexible with regard to problem adapted grid generation and local grid refinements. The tree-like data structure has been especially designed to facilitate this possibility.

In this report, we present the basic concept and preliminary results of our prototype program. In chapter 2, the basic semiconductor device equations are recalled, and the discretization schemes and the geometry of the sample problem are given. Chapter 3 explains in detail the special data structure according to the above mentioned requirements. The employed multigrid strategy together with the used prolongation, restriction and relaxation operators is presented. In chapter 4, preliminary numerical results and experiences with the program are reported.

2. Problem description

The basic equations describing electron transport in semiconductor devices have been known for many years, and can be found in a lot of textbooks [3, 4].

They are referred to as the "semiconductor device equations". These equations consist of a system of coupled partial differential equations describing the behaviour of the electric field, the response to the field and the continuity of the electrically charged particles:

$$\Delta\varphi = -\frac{q}{\varepsilon}(p - n + N_D - N_A), \qquad (2.1)$$

$$\frac{\partial n}{\partial t} = \frac{1}{q}\,\mathrm{div}\,\vec{J}_n - R + G, \qquad (2.2)$$

$$\frac{\partial p}{\partial t} = -\frac{1}{q}\,\mathrm{div}\,\vec{J}_p - R + G. \qquad (2.3)$$

The main variables of the system are the electric potential φ, the electron concentration n, and the hole concentration p. The electron and hole current densities are given by the auxiliary equations

$$\vec{J}_n = -q\mu_n\,\mathrm{grad}\,\varphi + qD_n\,\mathrm{grad}\,n, \qquad (2.4)$$

$$\vec{J}_p = -q\mu_p\,\mathrm{grad}\,\varphi - qD_n\,\mathrm{grad}\,p. \qquad (2.5)$$

The quantity q is the electric charge, ε is the dielectric permittivity, μ_n and μ_p the electron and hole mobilities, D_n and D_p are the electron and hole diffusivities. N_D and N_A are the donator and acceptor doping concentrations. R and G are the recombination and generation rates; in our case Shockley-Read- Hall recombination and no generation is assumed [3]:

$$R = \frac{np - n_i^2}{\tau_p(n+n_1) + \tau_n(p+p_1)} , \qquad G = 0 . \qquad (2.6)$$

The permittivity, the mobilities and the diffusivities as well as the intrinsic concentration n_i are treated as material specific constants.

The mathematical properties of the equation system (2.1 - 2.6) have been studied by several authors [5, 6, 7] and shall not be recalled here.

Our first computational experiments using the multigrid method were performed on a simplified regime of the above equations, namely the "Thermodynamic Equilibrium". In this case, no electric currents flow, and (2.1-2.3) can be reduced to

$$\Delta\varphi = -\frac{q}{\varepsilon}\left[n_i\, e^{-\varphi/v_T} - n_i\, e^{\varphi/v_T} + N_D - N_A\right], \qquad (2.7)$$

$$n = n_i\, e^{\varphi/v_T} , \qquad (2.8)$$

$$p = n_i\, e^{-\varphi/v_T} . \qquad (2.9)$$

where v_T is the thermal voltage according to the temperature of the device [8], and N means

$$N = N_D - N_A . \qquad (2.10)$$

The geometry of the sample problem consisted of a simplified model of a semiconductor diode (Fig. 2.1). A box-shaped piece of semiconductor bears metallic contacts at both ends. The constant dopings of acceptor material $N_o = N_A$ on the left half and donor material $N_o = N_D$ on the right produce an abrupt pn-junction in the middle of the device [3]. The values of the physical parameters chosen for the sample problem were

$$\left.\begin{aligned} N_o &= 10^{16}\ \mathrm{cm}^{-3} \\ \ell &= 1\ \mu\mathrm{m} \\ d &= 1\ \mu\mathrm{m} \\ w &= 1\ \mu\mathrm{m} \end{aligned}\right\} \qquad (2.11)$$

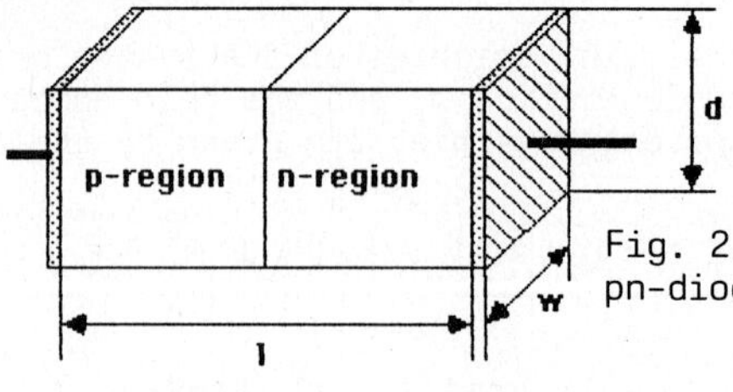

Fig. 2.1
pn-diode

The material specific constants have been selected as the standard values for silicon [3, 4].

For the boundary at the metallic contacts the flat band case has been assumed. This means electric neutrality at the boundary:

$$p - n + N = 0, \tag{2.12}$$

which leads to

$$\varphi = v_T \ln\left(\frac{N}{2n_i} + \sqrt{\left(\frac{N}{2n_i}\right)^2 + 1} \right). \tag{2.13}$$

On the free surfaces, homogeneous Neumann conditions have been imposed [5].
The problem domain is discretized by subdividing the volume into small cuboids, where the point in the center of the box represents the value of the potential in the subdomain. In fig. 2.2, two possible discretizations of the diode are shown; fig. 2.2a shows a one-dimensional discretization which consists of slices parallel to the contact planes, while in fig. 2.2b a two-dimensional discretization is sketched. The right-hand contact plane has been drawn somewhat lifted to make the discretization better visible.

The discretized form of eq. (2.7) is obtained by the well known box integration method [9, 10, 11], which consists of an integration of (2.7) with respect to the volume of the sub- domain. Using Gauss' law for the LHS term, and approximating the normal derivatives on the surfaces by finite differences yields the discretization formula

$$\sum_{\{r\}} \frac{2h_{r-}(\varphi_{r+} - \varphi_z) + 2h_{r+}(\varphi_{r-} - \varphi_z)}{h_{r+}h_{r-}(h_{r+} + h_{r-})}$$

$$= -\frac{q}{\varepsilon}\left[n_i\, e^{-\varphi_z/v_T} - n_i\, e^{\varphi_z/v_T} + N \right]. \tag{2.14}$$

which is illustrated for the case of two dimensions in fig. 2.3. The summation index r in (2.14) varies over the set of coordinate directions used in the specific discretization. In fig. 2.2a, r equals one, and no summation is needed, while in fig. 2.2b, r takes on the values 1 and 3. This kind of formulation is part of our above mentioned dimension independent concept. The quantities $h_{r\pm}$ in (2.14) represent the respective distances of the grid points (see fig. 2.3).

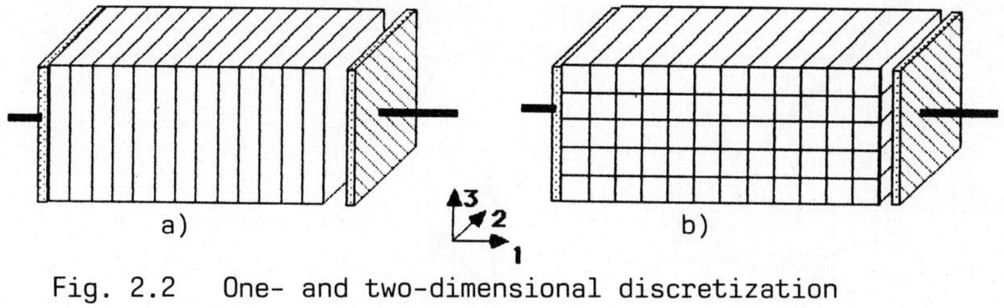

Fig. 2.2 One- and two-dimensional discretization

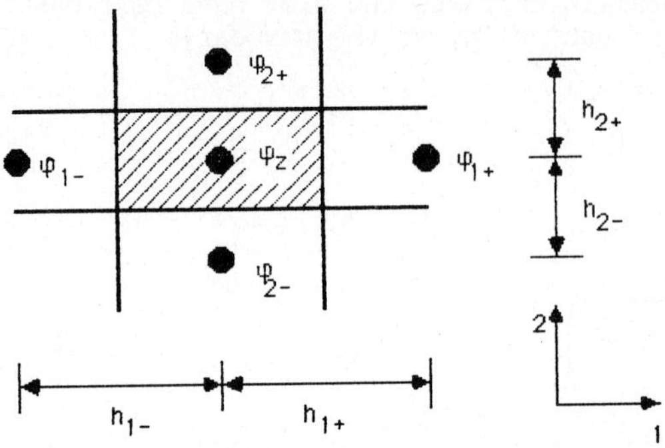

Fig. 2.3

Discretization of Poisson's equation

3. Data structure and muliigrid operators

To meet the demands, a special data structure for the representation of the problem domain has been designed. At first, the problem domain is transformed into a cube with unit lengths.

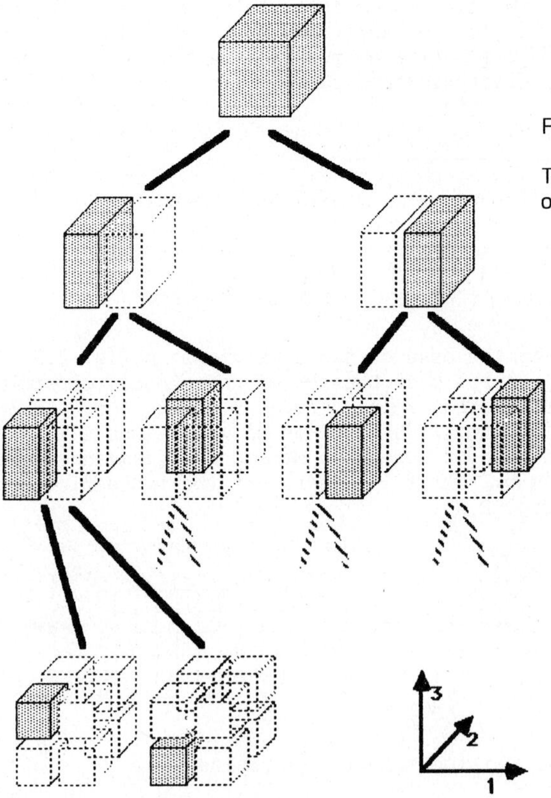

Fig. 3.1

Tree-structured data organization

The administration of the various multigrid levels inside the program is done in a tree structure. The "coarsest" level consists just of the unit cube itself and constitutes the root of the tree (see fig. 3.1). The first level of refinement is obtained by subdivision of the root in a certain direction, say 1. The so created subboxes are connected with the "father" box by respective pointers, and vice versa. The next level of refinement is created for instance by half dividing each of the boxes in level one in the direction 2, as explained in fig 3.1. By subsequent subdivisions, a discretization of the problem domain is obtained, which can in principle be made as fine as necessary for the given problem.

The advantages of this structure are:

a) By omission of respective division processes in building up the tree, it is possible to generate a specific one- or two-dimensional representation (cf. fig. 2.2).

b) If the process of box division is limited to certain regions in space, a local refinement can easily be performed, thus forming subtrees on those nodes representing the regions to be refined.

c) As shown below, this data structure allows to formulate the multigrid operators in a manner which is independent of the dimension.

The data structure of the tree is represented in the computer by records of data for each node in the tree. The record contains pointers to the father and the 2 son nodes as well as to the neighbouring nodes in space. Furthermore, space is allocated in each record for the storage of the unknown quantities, e.g. the potential, the defect, and other values necessary for the numerical algorithm. The neighbour pointers allow a quick reference of the values in neigbouring boxes, which is needed for the computation of the finite differences. The use of pointers in a device simulation program has been reported by Barton [12] and Franz et al. [9], whereas the concept of neighbour pointers in a tree structure for three dimensions is new in this context.

The mesh size h is now merely a function of tree position, which makes the numerical operators very simple.

The essential operation in the multigrid method is the transition between the different grids. In our approach, this is done by a simple up or down climbing on the tree along the respective pointers. As can be seen in fig. 3.1, one level in the tree corresponds to a coarsening or refinement of the grid in a single direction. Thus, a full three dimensional grid transition is assembled by doing three of these simple operations. This "partial coarsening" is a 3D generalization of the "semi-coarsening" mentioned in [13], where a two dimensional grid coarsening is done by successive grid transitions in both of the two directions. The partial coarsening allows the program code of the grid transition operators to be independent of the actual dimension of the problem, since it is always done only in one dimension at once as determined by the actual tree levels.

Fig. 3.2a illustrates the restriction operation of our partial coarsening. The characters y and z represent the values on the fine mesh, while x is the value on the coarse mesh; a, b, and c are the respective distances of the box centres. The restriction always involves the two sons of a tree node (cf. fig. 3.1), giving the value of the father node by the linear interpolation formula

$$x = \frac{a-b}{c-b} z + \frac{c-a}{c-b} y . \qquad (3.1)$$

The prolongation is illustrated by fig. 3.2b. The value of the son node is computed by linear interpolation of the value of the father node and of the father's neighbour node nearest to the son. If the distances are marked as shown in fig. 3.2b, the prolongation formula is identical to (3.1).

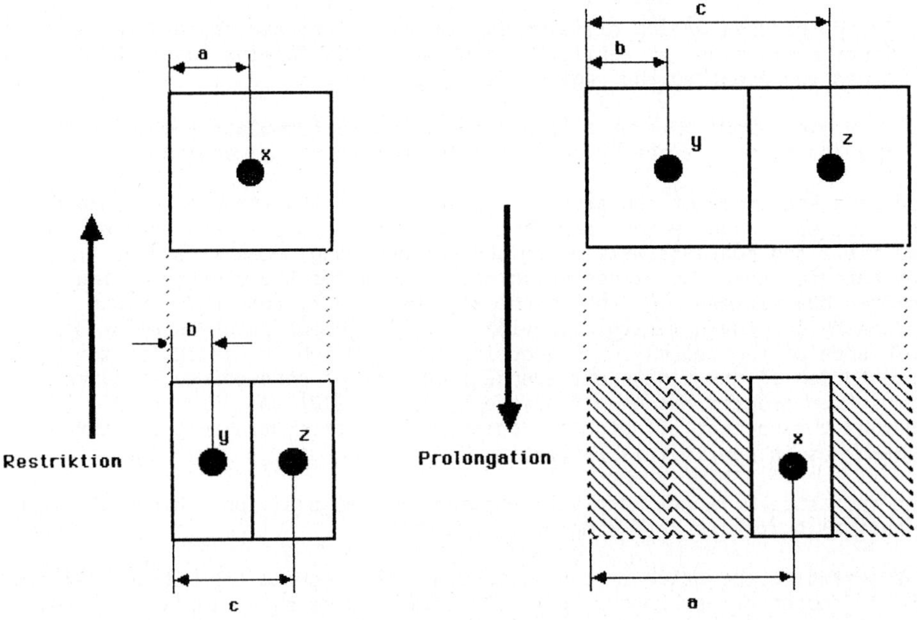

a) b)

Fig. 3.2 Grid transfer operations

The quantities a, b, and c are fully determined by the position of the involved nodes in the tree. Thus, the restriction and prolongation operators can be formulated independently of the problem dimension or an globally prescribed mesh size.

As the relaxation operator on the various multigrid levels we chose the Jacobi-Picard-ω-relaxation [13], which has the advantage that no differentiations need to be computed as in Newton-like algorithms. Using ω=1, the (i+1)st iteration of a certain value φ_z (cf. fig. 2.3) is calculated from the ith iteration by

$$\varphi_z^{(i+1)} = \frac{1}{\sum\limits_{\{r\}} \frac{1}{h_{r+}h_{r-}}} \left\{ \sum_{\{r\}} \frac{h_{r-}\varphi_{r+}^{(i)} + h_{r+}\varphi_{r-}^{(i)}}{h_{r+}h_{r-}(h_{r+}+h_{r-})} + \frac{1}{2}\frac{q}{\varepsilon}\left[n_i\, e^{-\varphi_z^{(i)}/v_T} - n_i\, e^{\varphi_z^{(i)}/v_T} + N \right] \right\}. \tag{3.2}$$

This relaxation formula is obtained by inserting (2.14) into eq. (5.17) of [13]. The mesh distances h_{r+} and h_{r-} are fully determined by the position of the corresponding node in the tree. Thus, the relaxation formula is independent of any actual mesh size or refinement. Hence, in the case of local grid refinement it is not necessary to change the relaxation operator as well as the restriction and prolongation operators in any way.

4. Preliminary results and conclusion

We performed the first numerical experiments on the semiconductor diode presented in chapter 2 in a one-dimensional as well as a two-dimensional discretization. Since the problems to be solved are nonlinear, the Full Approximation Scheme [13] has been adopted. The multigrid operations in both calculations were done in a V-cycle [13], doing 3 relaxations while going from fine to coarse, 1 relaxation going from coarse to fine, and 4 relaxations on the coarsest level. The initial distribution has been selected in a way such that the charge neutrality (2.12) is satisfied.

The finest one-dimensional grid had 64 points. Relaxations were performed on the 64-, 32-, 16-, and 8-point levels. The resulting potential distribution is shown in fig. 4.1. It can be clearly seen how the smooth potential step at the junction of the p- and n-regions arises [3]. Since the potential behaves in the expected way [3], it can be concluded that the program works correctly.

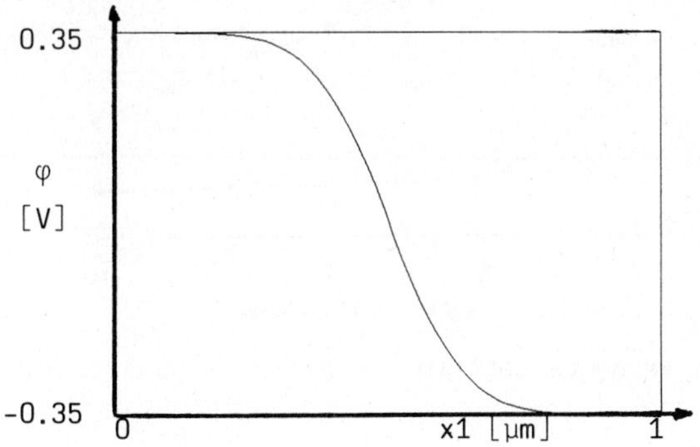

Fig. 4.1

One-dimensional potential distribution

The finest mesh in the 2D configuration was 64x64. Relaxations have been done on the levels 64x64, 64x32, 32x32, 32x16, 16x16, 16x8, and 8x8. The resulting two-dimensional distribution is shown in fig. 4.2. Since the problem is in principle one-dimensional (cf. fig. 2.1), the solution must resemble the 1D result, as can be seen in fig 4.1..

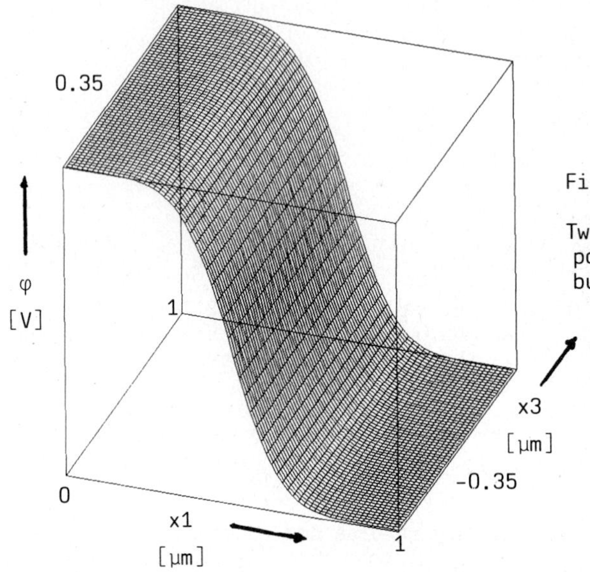

Fig. 4.2

Two-dimensional potential distribution

Fig. 4.3 shows the decrease of the residual in the one- as well as the two-dimensional calculation. The logarithm of the L2-norm of the residual as a function of multigrid cycle has been plotted. A satisfying decrease of about one decade per cycle can be stated.

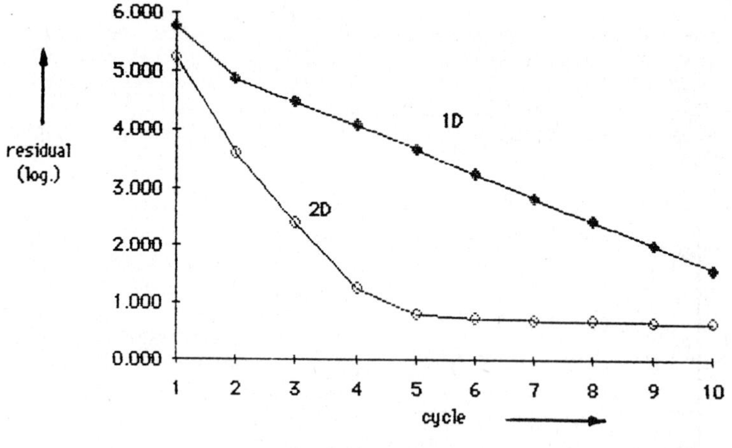

Fig. 4.3 Convergence pattern

In conclusion, the following items can be noted:

The concept of a data structure has been presented which allows the flexible multigrid treatment of one-, two- and three- dimensional problems, and supports an adaptive local grid refinement (yet not implemented). Multigrid operations have been formulated in a structure independent way; all informations concerning mesh distances and problem dimensions needed by the operators are determined by the position of the actual node in the tree structure. After implementation in a prototype program, the concept has been tested and could be validated by the simulation of a standard semiconductor device where the results are known.

In forthcoming work, further numerical experiments according to the various multigrid parameters as cycle type, relaxation counts, relaxation parameters etc. have to be done. Computations of three dimensional problems will be performed and compared to the 1D and 2D calculations. For device simulations of practical interest, the program must be extended to treat the semiconductor device equations (2.1-2.3). It seems reasonable that the conception should be suited for other problems outside the device simulation domain.

References

[1] Van Roosbroeck, W.V. "Theory of Flow of Electrons and Holes in Germanium and Other Semiconductors" Bell Syst. Techn. J. 29, 560-607, 1950.

[2] Gummel, H.K. "A Self-Consistent Iterative Scheme for One-Dimensional Steady State Transistor Calculations" IEEE Trans. ED-11, 455-465, 1964.

[3] Unger, H.G., Schultz,W., Weinhausen, G. "Elektronische Bauelemente und Netzwerke I" Vieweg 1979.

[4] S.M. Sze "Semiconductor Devices - Physics and Technology" John Wiley and Sons 1985.

[5] Markowich, P.A. "The Stationary Semiconductor Device Equations" Springer 1985.

[6] Mock, M.S. "Analysis of Mathematical Models of Semiconductor Devices" Boole Press Dublin 1983.

[7] Mock, M.S. "Basic Theory of Stationary Numerical Models" in Engl, W.L. (Ed.) "Process and Device Modeling", 159-194, North-Holland 1986.

[8] Paul, Reinhold "Halbleitersonderbauelemente", VEB-Verlag 1981.

[9] Franz, A. et.al. "Finite Boxes-A Generalization of the Finite-Difference Method Suitable for Semiconductor Device Simulation" IEEE Trans. ED-30, No. 9, 1070-1082, 1983.

[10] Selberherr, Siegfried "Analysis and Simulation of Semiconductor Devices" Springer 1984.

[11] Bank, R.E. et al. "Computational Aspects of Semiconductor Device Simulation" in Engl, W.L. (Ed.) "Process and Device Modeling", 229-264, North-Holland 1986.

[12] Barton, T.M. et al. "Modelling of Recessed Gate MESFET Structures" in Board, K. and Owen, D.R.J., (Eds.) "Simulation of Semiconductor Devices and Processes" Vol. 2, 528-543, Pineridge Press 1986.

[13] Trottenberg, U. "Multigrid Methods" in Hackbusch, W. and Trottenberg, U. (Eds.), `Proceedings of the Conference held at Köln-Porz, Nov. 23-27, 198, Lecture Notes in Mathematics 960, Springer 1982.

Algebraic Multigrid Methods and the Schur Complement

Wolfgang Dahmen
Institut für Mathematik III
Freie Universität Berlin
Arnimallee 2-6
1000 Berlin (33)

Ludwig Elsner
Fakultät für Mathematik
Universität Bielefeld
Universitätsstraße
4800 Bielefeld 1

Abstract

In this paper we propose and discuss a general purely algebraic framework for multilevel iterative schemes for solving linear systems where the role of the 'coarse grid' operators is played by Schur complements.

1. Introduction

The idea of algebraic multigrid (AMG) is to still provide efficient but robust solvers for large scale linear systems which may not be successfully treated anymore by classical multigrid techniques. The range of application should cover for instance systems arising from finite element discretizations for irregular triangulations or even problems which do not stem from elliptic boundary value problem at all. In that sense AMG aims at serving more as a 'black box' scheme exploiting mainly information which can be extracted directly from the system matrix (cf. [St]). A series of systematic contributions towards this goal was given for instance in [BMcR], [St], [RSt]. This approach is, however, in some sense based on still mimicing (in a partly heuristic fashion) the ingredients of classical multigrid schemes for elliptic second order problems.

Instead we propose here a general purely algebraic concept which in principle applies to any linear system $Ax = b$. The idea is to approximate A by a matrix $\tilde{A}$ which can easily be factored into a product of a lower and upper 2×2 block triangular matrix so that each step of the iteration $\tilde{A}x^{j+1} = (\tilde{A}-A)x^j+b$, $j = 0,1,\ldots$ reduces to the solution of a smaller linear system involving a Schur complement. This covers special cases in [St], [RSt]. Operations which may be viewed as smoothing, restriction and prolongation arise naturally. The performance of such a scheme depends in a transparent way on the approximation to the inverse of an appropriate leading block of A.

After completion of the paper we became aware of some recent closely related work in [A_2], [AG], [AP],[Va]. In fact, the Schur complement plays a crucial role in connection with incomplete block factorizations (cf. [A_2], [AP]) and some of our observations for M-matrices could have been concluded also from these results. Also using 2×2 block triangular factorizations as preconditioners has been investigated in the special case of hierarchical finite element discretizations [AG], [Va].

Nevertheless, the present concept in its generality does not seem to be covered and we try to focus on those aspects which are independent of any specific underlying geometric model.

The paper is organized as follows. Section 2 summarizes some basic facts about Schur complements and relates them to the 'basic two level scheme' considered in [St], [RSt]. This suggests to propose in Section 3 a general two-level scheme based on Schur complements. Section 4 is devoted to discussing some variants as well as concrete realizations when A is positive definite or an M-matrix. A general multilevel scheme is finally formulated in Section 5. Corresponding convergence estimates are obtained along the lines in [HT].

2. Partitioned systems and a basic two-level scheme

Let A be a nonsingular $n \times n$ matrix with a block decomposition

$$A = \begin{pmatrix} A_{11} & A_{12} \\ A_{21} & A_{22} \end{pmatrix} \tag{2.1}$$

where A_{22} is $k \times k$. Assuming that the $(n-k) \times (n-k)$ matrix A_{11} is nonsingular the matrix

$$S = A_{22} - A_{21} A_{11}^{-1} A_{12} = (A/A_{11}) \tag{2.2}$$

is called the Schur complement of A_{11} with respect to A.

It is well known that if A_{11}^{-1} exists A is nonsingular if and only if S is nonsingular. This is an immediate consequence of the factorization

$$A = \begin{bmatrix} I_{n-k} & 0 \\ A_{21}A_{11}^{-1} & I_k \end{bmatrix} \begin{bmatrix} A_{11} & A_{12} \\ 0 & S \end{bmatrix} = \begin{bmatrix} A_{11} & 0 \\ A_{21} & I_k \end{bmatrix} \begin{bmatrix} I_{n-k} & A_{11}^{-1} A_{12} \\ 0 & S \end{bmatrix} \tag{2.3}$$

from which the following Schur-Banachiewicz formula for A^{-1} is easily derived (see Theorem 2.7 in [Ou]).

$$\begin{aligned} A^{-1} &= \begin{pmatrix} A_{11}^{-1} + A_{11}^{-1} A_{12} S^{-1} A_{21} A_{11}^{-1} & -A_{11}^{-1} A_{12} S^{-1} \\ -S^{-1} A_{21} A_{11}^{-1} & S^{-1} \end{pmatrix} \\ &= \left[\begin{pmatrix} A_{11}^{-1} & 0 \\ 0 & 0 \end{pmatrix} + \begin{pmatrix} -A_{11}^{-1} A_{12} \\ I_k \end{pmatrix} S^{-1} \left(-A_{21}A_{11}^{-1} \quad I_k \right) \right] . \end{aligned} \tag{2.4}$$

One readily concludes from (2.4) that if A is positive definite then A_{11} and S both are positive definite and likewise if A is an M-matrix then A_{11} and S are both M-matrices.

Now the factorization (2.3) or the expression (2.4) shows that the linear system

$$Ax = b \tag{2.5}$$

where $b = \begin{pmatrix} b_1 \\ b_2 \end{pmatrix}$, $b_2 \in \mathbb{C}^k$, $b_1 \in \mathbb{C}^{n-k}$, $x = \begin{pmatrix} x_1 \\ x_2 \end{pmatrix}$ may be solved by the following steps

$$\text{i) } A_{11} z = b_1 \text{ , ii) } S x_2 = b_2 - A_{21} z \text{ , iii) } A_{11} x_1 = b_1 - A_{12} x_2 . \qquad (2.6)$$

Clearly when A_{11}^{-1} is known explicitly (2.6)i) and iii) require just matrix vector multiplications so that after forming S according to (2.2) only the smaller system (2.6) ii) of order k remains to be solved.

This procedure is closely related to the 'basic two-level' scheme described in [St], [RSt] which served there as the starting point for the development of an algebraic multigrid scheme for symmetric matrices (see also [BMcR]). Following [St] let $F \subseteq \{1,\dots,n\}$ such that $i \in F$ implies $\{j \neq i : a_{ij} \neq 0\} \subseteq \{1,\dots,n\} \setminus F$ which plays the role of the 'coarse grid'. Clearly, reordering equations and unknowns if necessary so that $F = \{1,\dots,n-k\}$ A_{11} is in this case just a diagonal matrix. In the present terms the prolongation operator defined in [St] takes the form

$$I_H^h = \begin{bmatrix} -A_{11}^{-1} A_{12} \\ I_k \end{bmatrix} . \qquad (2.7)$$

As usual, the restriction is taken as the adjoint

$$I_h^H = (I_H^h)^T = \left[-A_{12}^T A_{11}^{-1} \mid I_k \right] = \left[-A_{21} A_{11}^{-1} \mid I_k \right] \qquad (2.8)$$

since here we have assumed $A_{12} = A_{21}^T$. Then the Galerkin type coarse grid operator simply turns out to be the Schur complement (A/A_{11})

$$A^H = I_h^H A I_H^h = A_{22} - A_{12}^T A_{11}^{-1} A_{12} . \qquad (2.9)$$

To describe now a multigrid cycle based on these ingredients consider the residual equation $Av = r$, $r = b - A\tilde{x}$ where $\tilde{x}$ is some approximation to the solution of (2.5). Note that one Gauss-Seidel relaxation sweep over F with starting vector $v^0 = 0$ amounts to forming $\begin{bmatrix} A_{11}^{-1} & 0 \\ 0 & 0 \end{bmatrix} r$.

The new residual is then

$$\tilde{r} = b - A\left(\tilde{x} + \begin{bmatrix} A_{11}^{-1} & 0 \\ 0 & 0 \end{bmatrix} r\right) = \begin{bmatrix} 0 \\ r_2 - A_{21} A_{11}^{-1} r_1 \end{bmatrix}$$

so that by (2.8) $I_h^H \tilde{r} = I_h^H r$. Hence a cycle consisting of one Gauss-Seidel sweep over F followed by restriction, solution by the coarse grid operator for the restricted residual and prolongation back to the fine grid amounts to forming

$$\left\{ \begin{bmatrix} A_{11}^{-1} & 0 \\ 0 & 0 \end{bmatrix} + I_H^h (A^H)^{-1} I_h^H \right\} r \qquad (2.10)$$

which in view of (2.4) is equivalent to applying (2.6)-(2.8) to the residual equation, and hence solves the system exactly.

3. A general two level scheme

Of course, in general F will be too small to allow for an efficient reduction of the size of the systems. Rather than modifying the definition of F, I_H^h, I_h^H heuristically as in [St], [RSt] to cope with this drawback the above observations suggest to consider iteration schemes of the form

$$\tilde{A}x^{j+1} = (\tilde{A} - A)x^j + b, \quad j = 0, 1, 2, \ldots \tag{3.1}$$

where

$$\tilde{A} = \begin{bmatrix} G_1^{-1} & 0 \\ A_{21} & I_k \end{bmatrix} \begin{bmatrix} I_{n-k} & G_1 A_{12} \\ 0 & \tilde{S} \end{bmatrix} = \begin{bmatrix} G_1^{-1} & A_{12} \\ A_{21} & A_{22} + A_{21}(G_1 - G_2)A_{12} \end{bmatrix} \tag{3.2}$$

and
$$\tilde{S} = A_{22} - A_{21} G_2 A_{12} .$$

Thus $\tilde{A}$ will be close to A if G_1, G_2 approximate A_{11}^{-1} well. Rewriting the iterates as

$$x^{j+1} = x^j + v^j \tag{3.3}$$

where

$$\tilde{A}v^j = r^j = b - Ax^j \tag{3.4}$$

we have in view of (2.4)

$$v^j = \left[\begin{pmatrix} G_1 & 0 \\ 0 & 0 \end{pmatrix} + \begin{pmatrix} -G_1 A_{12} \\ I_k \end{pmatrix} \tilde{S}^{-1} \left(-A_{21} G_1 \mid I_k\right)\right] r^j \tag{3.5}$$

so that one iteration step (3.3) requires for $r^j = \begin{bmatrix} r_1^j \\ r_2^j \end{bmatrix}$

$$\text{i) } z = G_1 r_1^j, \quad \text{ii) } \tilde{S} v_2^j = r_2^j - A_{12} z, \quad \text{iii) } v_1^j = G_1(r_1^j - A_{12} v_2). \tag{3.6}$$

The complete analogy of (3.5) with (2.4), (2.10), respectively suggests to view the operations

$$I_H^h = \begin{pmatrix} -G_1 A_{12} \\ I_k \end{pmatrix}, \quad I_h^H = (-A_{21} G_1 \mid I_k) \tag{3.7}$$

as prolongation and restriction operators, respectively while the multiplication of the current residual by $\begin{pmatrix} G_1 & 0 \\ 0 & 0 \end{pmatrix}$ corresponds to one relaxation sweep over part of the equations. Clearly, when A (and G_1) is symmetric, one has $I_h^H = (I_H^h)^T$.

Moreover a straightforward calculation yields

$$\tilde{S} = I_h^H A I_H^h - A_{21} (G_1 A_{11} - I) G_1 A_{12} + A_{21}(G_1 - G_2)A_{12} \qquad (3.8)$$

i.e. $\tilde{S}$ agrees up to a perturbation with the Galerkin form (see [Mc]) of the "coarse grid" equations.

Note that the task of selecting appropriate smoothers, prolongation and restriction operators is reduced now to approximating the inverse of the leading block A_{11}. This is reflected by the following estimates for the matrix $M = I - \tilde{A}^{-1}A = \tilde{A}^{-1}(\tilde{A}-A) = (I + A^{-1}(\tilde{A}-A))^{-1} A^{-1}(\tilde{A}-A)$ where

$$A - \tilde{A} = \begin{pmatrix} G_1^{-1} - A_{11} & 0 \\ 0 & A_{21}(G_1 - G_2)A_{12} \end{pmatrix}. \qquad (3.9)$$

Thus, in view of the following standard estimates for $\|M\|$ where $\|\cdot\|$ is any matrix operator norm

$$\|M\| \leq \begin{cases} \|\tilde{A}^{-1}\| \, \|\tilde{A} - A\|, \\ \dfrac{\|A^{-1}(\tilde{A} - A)\|}{1 - \|A^{-1}(\tilde{A} - A)\|} \leq \dfrac{\|A^{-1}\| \, \|\tilde{A} - A\|}{1 - \|A^{-1}\| \, \|\tilde{A}-A\|} \end{cases} . \qquad (3.10)$$

(3.9), (3.10) may be exploited to guarantee that $\|M\| < 1$.

4. Possible variants and realizations

The performance of the above schemes, i.e. the quality of the approximations G_1, G_2 will, of course, depend on the blocking of A, which will always depend on the problem at hand. In some cases the strategies proposed in [RSt], [St] might work. Hierarchical finite element discretizations would induce a natural blocking (cf. [V]). In this special context similar schemes have been extensively studied in [AG], [Va] where more detailed information on convergence properties are given.

We will continue with some general remarks on the choice of G_1, G_2.

In case sufficiently good approximations to A_{11}^{-1} are explicitly available at low cost the most natural choice would be $G_1 = G_2$. For instance in case of a strongly diagonally dominant sparse block A_{11} truncated Neumann series of low order might serve that purpose. Alternatively, one might choose other polynomial approximations to A_{11}^{-1} as described in [JMP] if some information about the spectrum of A_{11} is available. The choice $G_1 = G_2$ has the additional advantage that the rank of the iteration matrix $M = I - \tilde{A}^{-1} A$ is at most $n-k$ as can be seen from (3.9). This might be an important advantage when using $\tilde{A}$ as a preconditioner for a CG method.

However, there are situations where $G_1 \neq G_2$ is reasonable. To this end, note first that G_1 and G_2 are used for different purposes. If for instance the system $A_{11}z = c$ is easily solvable we might choose $G_1 = A_{11}^{-1}$ since solving (3.6) i), iii) does not require explicit knowledge of G_1. Again this would reduce the rank of the iteration matrix M to at most k and, as could be seen from (3.1), (3.2) the residuals r^j would satisfy $r_1^j = 0$, $j \geq 1$. On the other hand G_2 has to be given explicitly in order to form $\tilde{S}$ needed in (3.6) ii). Moreover if A is sparse one would like to have $\tilde{S}$ to be sparse as well so that one might choose G_2 to have certain sparsity patterns as well.

As a first example we will indicate next some concrete choices for G_1, G_2 when A is an M-matrix. The following observations could be also deduced for instance from [AP] . In order to keep the paper selfcontained we include a short derivation.

Our objective is to replace A_{11} by a matrix $B_{11} = G_1^{-1}$ so that systems $B_{11}z = c$ can be solved efficiently. To this end, let B_{11} be constructed from A_{11} by replacing all entries of A_{11} off the first ℓ codiagonals by zero. Hence

$$\hat{A} = \begin{pmatrix} B_{11} & A_{12} \\ A_{21} & A_{22} \end{pmatrix}$$

still has non-positive off diagonal entries. As $A \leq \hat{A}$ componentwise it is well-known that $\hat{A}$ is an M-matrix too and $0 \leq \hat{A}^{-1} \leq A^{-1}$. Recall from $[A_1]$ that the first ℓ codiagonals of the inverse of an ℓ-banded $n \times n$ matrix can be computed with $O(n)$ operations. Hence we may define G_2 to be the ℓ-banded matrix whose nonzero entries coincide with those of B_{11}^{-1} (cf. [AP]). According to our above remarks one may choose $G_1 = B_{11}^{-1}$. For this choice of

$$\tilde{A} = \begin{pmatrix} B_{11} & A_{12} \\ A_{21} & A_{22}+A_{21}(B_{11}^{-1}-G_2)\,A_{12} \end{pmatrix} \tag{4.1}$$

the two-level iteration (3.1) converges (cf. [AP]). To see this, we note first that

$$(A/A_{11}) \leq (\hat{A}/B_{11}) \leq (\tilde{A}/B_{11}) \,. \tag{4.2}$$

The first inequality follows from $A_{11}^{-1} \geq B_{11}^{-1}$. Since

$$(\tilde{A}/B_{11}) = A_{22} + A_{21}(B_{11}^{-1} - G_2)A_{12} - A_{21}\, B_{11}^{-1} A_{12} \tag{4.3}$$

and $B_{11}^{-1} - G_2 \geq 0$ the second inequality in (4.2) is immediate. Again since $B_{11}^{-1} - G_2 \geq 0$ we obtain $\tilde{A} \geq \hat{A} \geq A$.

From the representation (2.4) of $\widetilde{A}^{-1}$, $\hat{A}^{-1}$ we conclude $0 \leq \widetilde{A}^{-1} \leq \hat{A}^{-1} \leq A^{-1}$. Thus the splittings $A = \widetilde{A} - (\widetilde{A}-A) = \hat{A} - (\hat{A}-A)$ are both regular. Hence by a standard result (see e.g. [V]) $\rho(\hat{A}^{-1}(\hat{A}-A)) \leq \rho(\widetilde{A}^{-1}(\widetilde{A}-A)) < 1$ which proves the above claim.

Similar considerations can be made in the case that A is positive definite. We have already stated at the end of section two, that then (A/A_{11}) is positive definite, too. In the following $B > 0$ (≥ 0) means that B is positive (semi)definite, i.e. we consider now a partial order different from the previous one. Define for some given B_{11} and G_2 the matrices

$$\hat{A} = \begin{pmatrix} B_{11} & A_{12} \\ A_{12}^T & A_{22} \end{pmatrix} \qquad \widetilde{A} = \begin{pmatrix} B_{11} & A_{12} \\ A_{12}^T & A_{22}+A_{12}^T(B_{11}^{-1}-G_2)A_{12} \end{pmatrix}$$

as above.

If $A_{11} \leq B_{11} \leq G_2^{-1}$ holds then we infer that $0 < A \leq \hat{A} \leq \widetilde{A}$, $\widetilde{A}^{-1} \leq \hat{A}^{-1} \leq A^{-1}$ and $(A/A_{11}) \leq (\hat{A}/B_{11}) \leq (\widetilde{A}/B_{11})$ hold. The splittings $A = \hat{A}-(\hat{A}-A) = \widetilde{A}-(\widetilde{A}-A)$ both lead to convergent iteration schemes, since $\rho(\hat{A}^{-1}(\hat{A}-A)) \leq \rho(\widetilde{A}^{-1}(\widetilde{A}-A)) < 1$.

As above we wish to choose B_{11} as a band matrix satisfying $B_{11} \geq A_{11}$ in this case. A possible construction for a given $0 \leq \ell \leq n-k$ is the following:

Set

$$B_{11} = (b_{ij})_{i,j=1,\dots,n-k} \qquad b_{ij} = \begin{cases} 0; & |i-j| > \ell \\ a_{ij}; & 0 < |i-j| \leq \ell, \\ a_{ii} + \sum\limits_{\substack{s=1 \\ |s-i| > \ell}}^{n-k} |a_{is}|; & i=j. \end{cases} \tag{4.4}$$

For $\ell = 0$ the second possibility in (4.4) does not occur. In any case $B_{11} - A_{11}$ is diagonally dominant and hence $B_{11} - A_{11} \geq 0$.

A possible choice for G_2^{-1} is given by (4.4) where $\ell = 0$. Then obviously $A_{11} \leq B_{11} \leq G_2^{-1}$ if B_{11} is also chosen according to (4.4) with some $\ell \geq 0$. A further possibility is to choose $G_2^{-1} = B_{11}$. The structure of inverses of band matrices is well known (e.g. [Ro]) and this might be used to calculate G_2 efficiently.

Finally we want to mention a variant of our method which avoids the explicit formation of the Schur complement S or $\widetilde{S}$ resp..

A possible way to solve (2.6), $Sx_2 = b_2 - A_{21} z_1 = r$ e.g. when $G_1 = G_2 = A_{11}^{-1}$ is to use the following iteration:

$$A_{22}\, u^{(i+1)} = A_{21}\, A_{11}^{-1}\, A_{12}\, u^{(i)} + r \tag{4.5}$$

(or similarly the equation 3.6ii with A_{11}^{-1} replaced by G_2). This iteration can be achieved by solving linear systems with coefficient matrices A_{11} (or G_2^{-1}) and A_{22} resp. . Observe that in the cases considered above, where S is either an M-matrix or positive definite, (4.5) is convergent.

5. A multilevel scheme

A general N-level scheme may be formulated with respect to any given sequence $n = k_0 > \dots > k_N > 0$ of strictly decreasing integers. Given nonsingular $(k_\ell - k_{\ell+1}) \times (k_\ell - k_{\ell+1})$ matrices $G_1^{(\ell)}$, $G_2^{(\ell)}$, $\ell = 0,\dots,N-1$, we define $(k_\ell \times k_\ell)$-matrices $A^{(\ell)}$ by (cf. (2.2)).

$$A^{(0)} = A\,, \quad A^{(\ell+1)} = (\hat{A}^{(\ell)} / (G_2^{(\ell)})^{-1})\,, \quad \ell = 0,\dots,N-1\,, \tag{5.1}$$

where $\hat{A}^{(\ell)}$ is obtained from $A^{(\ell)}$ by replacing the leading $(k_\ell - k_{\ell+1}) \times (k_\ell - k_{\ell+1})$ block $A_{11}^{(\ell)}$ by $(G_2^{(\ell)})^{-1}$. Again $G_j^{(\ell)}$, $j = 1,2$ are supposed to approximate $(A_{11}^{(\ell)})^{-1}$ and hence will have to be determined in the course of the calculations. Moreover, in analogy to (3.7) we introduce

$$I_\ell^{\ell+1} = \left[-A_{21}^{(\ell)} G_1^{(\ell)} \;\middle|\; I_{k_{\ell+1}} \right], \quad I_{\ell+1}^{\ell} = \begin{bmatrix} -G_1^{(\ell)} A_{12}^{(\ell)} \\ I_{k_{\ell+1}} \end{bmatrix}$$

where $A^{(\ell)} = \begin{bmatrix} A_{11}^{(\ell)} & A_{12}^{(\ell)} \\ A_{21}^{(\ell)} & A_{22}^{(\ell)} \end{bmatrix}$ and $G^{(\ell)} = \begin{bmatrix} G_1^{(\ell)} & 0 \\ 0 & 0 \end{bmatrix}$.

Finally we will need the matrices

$$\tilde{A}^{(\ell)} = \begin{bmatrix} I_{k_\ell - k_{\ell+1}} & 0 \\ A_{21}^{(\ell)} G_1^{(\ell)} & I_{k_{\ell+1}} \end{bmatrix} \begin{bmatrix} (G_1^{(\ell)})^{-1} & A_{12}^{(\ell)} \\ 0 & A^{(\ell+1)} \end{bmatrix} = \begin{bmatrix} (G_1^{(\ell)})^{-1} & A_{12}^{(\ell)} \\ A_{21}^{(\ell)} & A_{22}^{(\ell)} + A_{21}^{(\ell)} (G_1^{(\ell)} - G_2^{(\ell)}) A_{12}^{(\ell)} \end{bmatrix}.$$

The iteration (3.1) may now be rephrased as

$$x^{j+1} = M_1(A^{(0)})\, x^j + s\,, \quad j = 0,1,2,\dots \tag{5.2}$$

where $s = (\tilde{A}^{(0)})^{-1} b$ and

$$M_1(A^{(0)}) = I_{k_0} - (\tilde{A}^{(0)})^{-1} A^{(0)} = I_{k_0} - (G^{(0)} + I_1^0 (A^{(1)})^{-1} I_0^1) A^{(0)}\,. \tag{5.3}$$

Determining x^{j+1} in the form (3.3) requires to solve the residual equation (3.4) $\tilde{A}^{(0)} v^j = r^j$ which means to solve the smaller system

$$A^{(1)} v = I_1^0 A^{(0)} r^j . \tag{5.4}$$

The multilevel scheme consists now of solving (5.4) only approximately by applying γ_1 steps of an analogous iteration with $A^{(0)}$ and $b = b^{(0)}$ being replaced by $A^{(1)}$ and $b^{(1)} = I_1^0 A^{(0)} r^j$, respectively. Since the right hand side is supposed to be small zero is chosen as a starting vector which is quite in the spirit of multigrid algorithms. This reduction is continued with γ_ℓ iteration steps on the ℓ-th level until reaching the level N where the corresponding system is solved exactly. This whole process may be viewed as one iteration step described by a certain iteration matrix $M_N(A^{(0)})$ depending on the sequences $\{k_j\}_{j=0}^N$, $\{\gamma_j\}_{j=1}^{N-1}$, $\{G_j^{(\ell)}\}_{\ell=1}^{N-1}$, $j=1,2$. To describe $M_N(A^{(0)})$ we will closely follow [HT].

To this end, we define general two-level iteration matrices

$$M_{\ell,\ell+1} = I_{k_\ell} - (G^{(\ell)} + I_{\ell+1}^\ell (A^{(\ell+1)})^{-1} I_\ell^{\ell+1}) A^{(\ell)} \tag{5.5}$$

generalizing (5.3). Note that $M_1(A^{(0)}) = M_{0,1}$. Since in general the j-th iterate of a scheme $x^{j+1} = Mx^j + s$ for the solution of $Ax = b$ with starting vector $x^0 = 0$ is given by $x^j = (I - M^j) A^{-1} b$ we can describe $M_{N-\ell}(A^{(\ell)})$ inductively by replacing $(A^{(\ell+1)})^{-1}$ in (5.5) by $(I_{k_{\ell+1}} - M_{N-1-\ell}^{\gamma_{\ell+1}}(A^{(\ell+1)}))(A^{(\ell+1)})^{-1}$ so that

$$M_{N-\ell}(A^{(\ell)}) = I_{k_\ell} - (G^{(\ell)} + I_{\ell+1}^\ell (I_{k_{\ell+1}} - M_{N-\ell-1}^{\gamma_{\ell+1}}(A^{(\ell+1)}))(A^{(\ell+1)})^{-1} I_\ell^{\ell+1}) A^{(\ell)} , \quad \ell = 0,\dots,N-1 \tag{5.6}$$

where $M_0(A^{(N)}) = 0$.

Note that for $k_\ell = n-\ell$, $N = n$, $G_1^{(\ell)} = (A_{11}^{(\ell)})^{-1} = G_2^{(\ell)}$, $\gamma_\ell = 1$, the above multi level scheme reduces to Gaussian elimination.

In order to derive bounds for $\|M_{N-\ell}\|$ we follow [HT] and rewrite $M_{N-\ell}(A^{(\ell)})$ given by (5.6) as a perturbation of the two level scheme (5.5).

$$\begin{aligned} M_{N-\ell}(A^{(\ell)}) &= M_{\ell,\ell+1} + I_{\ell+1}^\ell M_{N-\ell-1}^{\gamma_{\ell+1}}(A^{(\ell+1)})(A^{(\ell+1)})^{-1} I_\ell^{\ell+1} A^{(\ell)} \\ &= M_{\ell,\ell+1} + I_{\ell+1}^\ell M_{N-\ell-1}^{\gamma_{\ell+1}}(A^{(\ell+1)}) W_\ell^{\ell+1} \end{aligned} \tag{5.7}$$

where $W_\ell^{\ell+1} = (A^{(\ell+1)})^{-1} I_\ell^{\ell+1} A^{(\ell)}$. Assuming that for some norm

$$\|M_{\ell,\ell+1}\| \le \sigma , \quad \|I_{\ell+1}^\ell\| \; \|W_\ell^{\ell+1}\| \le C , \quad \ell = 0,\dots, N-1 \tag{5.8}$$

we obtain

$$\|M_{N-\ell}(A^{(\ell)})\| \leq \eta_\ell , \quad \ell = 0,\ldots, N-1 \tag{5.9}$$

where η_ℓ is defined recursively by

$$\eta_N = 0 , \quad \eta_{N-1} = \sigma , \quad \eta_\ell = \sigma + C(\eta_{\ell+1})^{\gamma_{\ell+1}} \quad \ell = N-2,\ldots,0 . \tag{5.10}$$

For instance if $\gamma_\ell = \gamma = 2$, $\ell = 0,\ldots,N-1$ one gets [HT]

$$\|M_{N-\ell}\| \leq \eta := (1-\sqrt{1-4C\sigma})/2C \leq 2\sigma \tag{5.11}$$

provided that $4C\sigma \leq 1$.

References

[A_1] O. Axelsson, A survey of vectorizable preconditioning methods for large scale finite element matrix problems, Report CNA-190, Center for Numerical Analysis, Univ. of Texas at Austin, 1984.

[A_2] O. Axelsson, A general incomplete block-matrix factorization method, Linear Algebra and its Applications 74 (1986), 179-190.

[AG] O. Axelsson, I. Gustafsson, Preconditioning and two-level multigrid methods of arbitrary degree of approximation, Math. Comp., 40 (1983), 219-242.

[AP] O. Axelsson, B. Polman, On approximate factorization methods for block matrices suitable for vector and parallel processors, Linear Algebra and its Applications 77 (1986), 3-26.

[BMcR] A. Brand, S.F. McCormick, J. Ruge, Algebraic Multigrid for automatic multigrid solution with application to geodetic computation. Submitted (1983) to SIAM J. Sci. Stat. Comp.

[HT] W. Hackbusch, U. Trottenberg, Multigrid Methods, Lecture Notes in Mathematics Vol. 960, Springer Verlag

[JMP] O.G. Johnson, C.A. Micchelli, G. Paul, Polynomial preconditioners for conjugate gradient calculations, SIAM J. Numer. Anal. 20 (1983), 362-376.

[Mc] S.F. McCormick, An Algebraic Interpretation of Multigrid Methods, SIAM J. Numer. Anal. 19, (1982), 548-560.

[Ou] D.V. Ouellette, Schur Complements and Statistics, Linear Algebra Appl. 36, (1981), 187-295.

[Ro] P. Rósza, On the inverse of Band Matrices, Integral Equations and Operator Theory 10 (1987), 82-95.

[RSt] J. Ruge, K. Stüben, Algebraic Multigrid (AMG) Arbeitspapiere der GMD 210 (1986) a chapter in the book Multigrid Methods (McCormick, ed.) Frontiers in Applied Mathematics, Vol. 5, SIAM, Philadelphia

[St] K. Stüben, Algebraic Multigrid (AMG), Experiences and Comparisons: Arbeitspapiere der GMD 23 (1983)

[V] R.S. Varga, Matrix Iterative Analysis, Prentice Hall, Englewood Cliffs, N.J., 1962.

[Va] P.S. Vassilevski, Nearly optimal iterative methods for solving finite element elliptic equations based on the multilevel splitting of the matrix, Research Report, Institute of Mathematics with Computing Center, Bulgarian Academy of Sciences, 1987.

A MULTIGRID METHOD FOR STEADY EULER EQUATIONS, BASED ON FLUX-DIFFERENCE SPLITTING WITH RESPECT TO PRIMITIVE VARIABLES

E. DICK
Department of machinery, State University of Ghent
Sint Pietersnieuwstraat 41, B-9000 Gent, Belgium

SUMMARY

A flux-difference splitting method for steady Euler equations, resulting in a splitting with respect to primitive variables is introduced. This splitting is applied to finite volumes centered around the vertices of the computational grid. The discrete set of equations is both conservative and positive.

Due to the positivity, the solution can be obtained by collective variants of relaxation methods, that can be brought into multigrid form. Two full multigrid methods are presented. As restriction operator, both use full weighting within the flow field and injection at the boundaries. Bilinear interpollation is used as prolongation. The cycle is of W-type. The first method uses symmetric successive underrelaxation, while the second uses Jacobi-iteration. In terms of cycles, the successive formulation is the most efficient while in terms of computer time, due to the vectorizability the second formulation is most efficient.

Due to the conservativity, the algebraically exact flux-difference splitting and the positivity, the solution for transonic flow shows shocks represented as sharp discontinuities, without wiggles.

INTRODUCTION

In recent years, upwind techniques based on flux-vector splitting and flux-difference splitting for solving the Euler equations have gained considerable popularity.

The flux-vector splitting approach was introduced by Steger and Warming [1] for the unsteady Euler equations. This splitting is based on the homogeneity of degree one with respect to the conservative variables ρ, ρu, ρv, ρE. It was shown by Jespersen [2] that the flux-vector splitting approach can also be used directly on the steady Euler equations to generate discrete equations that can be solved by relaxation methods in multigrid-form. The technique, however, shows some shortcomings in the treatment of shocks. In the conservative formulation, so-called undifferenced terms appear. These terms represent a loss of positivity of the discrete set of equations and cause oscillations in the vicinity of shocks.

Going back to the earlier work of Godunov [3], a remedy for the shock oscillations can be found in not splitting the flux-vectors themselves, but differences of flux-vectors. Several flux-difference splitting procedures were proposed for unsteady equations, simplifying the Godunov method. The splitting of Roe [4] is based on the homogeneity of degree two of the flux-vectors with respect to the variables $\sqrt{\rho}$, $\sqrt{\rho}\, u$, $\sqrt{\rho}\, v$, $\sqrt{\rho}\, H$. The splitting of Osher [5] is a splitting with respect to the variables $\sqrt{\gamma p/\rho}$, u, v, $\ln(p/\rho^\gamma)$. A very simple splitting based on the polynomial character of the flux-vectors with respect to the primitive variables ρ, u, v, p was proposed by Lombard et al. [6].

It was shown by Hemker and Spekreijse [7] that the Osher scheme can be used directly on steady Euler equations, leading to a conservative set of discrete equations that can be solved by relaxation techniques. Hemker and Spekreijse chose the Osher scheme, although it is the most complex of the mentioned flux-difference splitting schemes, because of its rigour in the construction of the discrete equations and the treatment of the boundary conditions.

In this paper, the flux-difference splitting of Lombard et al., with respect to the primitive variables is used on the steady Euler equations, to form a set of equations that can be solved by relaxation methods in multigrid form. In contrast to the original approach of Lombard et al., which used an approximate splitting, the splitting is done here in an algebraically exact way. This is necessary to treat the steady equations directly, avoiding the time marching necessary in the original approach. Also, due to the algebraically exact manipulation, boundary conditions can be introduced in a rigourous way.

A detailed description of the splitting technique was given by the author in [8]. In this paper, the principles of the method are summarized and the multigrid formulation is treated.

FLUX-DIFFERENCE SPLITTING WITH RESPECT TO PRIMITIVE VARIABLES

Steady Euler equations, in two dimensions, take the form :

$$\frac{\partial f}{\partial x} + \frac{\partial g}{\partial y} = 0 \ , \qquad (1)$$

where the flux vectors are :

$$f = \begin{pmatrix} \rho u \\ \rho uu+p \\ \rho uv \\ \rho Hu \end{pmatrix} , \qquad g = \begin{pmatrix} \rho v \\ \rho uv \\ \rho vv+p \\ \rho Hv \end{pmatrix} . \qquad (2)$$

ρ is density, u and v are Cartesian velocity components, $H = E+p/\rho$ is total enthalpy, p is pressure, $E = p/(\gamma-1)\rho + 1/2\, u^2 + 1/2\, v^2$ is total energy and γ is adiabatic constant.

Since the components of the flux vectors form polynomials with respect to the primitive variables ρ, u, v and p, components of flux-differences can be written as follows :

$$\Delta\rho u = \bar{u}\,\Delta\rho + \bar{\rho}\,\Delta u \;,$$

$$\begin{aligned}\Delta\rho uu{+}p &= \overline{\rho u}\,\Delta u + \bar{u}\,\Delta\rho u + \Delta p \\ &= \bar{u}^2\Delta\rho + (\overline{\rho u} + \bar{\rho}\,\bar{u})\Delta u + \Delta p \;,\end{aligned}$$

$$\begin{aligned}\Delta\rho uv &= \overline{\rho u}\,\Delta v + \bar{v}\,\Delta\rho u \\ &= \bar{u}\,\bar{v}\,\Delta\rho + \bar{\rho}\,\bar{v}\,\Delta u + \overline{\rho u}\,\Delta v \;,\end{aligned}$$

$$\begin{aligned}\Delta\rho Hu &= \overline{\rho u}\left(\tfrac{1}{2}\,\Delta u^2 + \tfrac{1}{2}\,\Delta v^2\right) + \tfrac{1}{2}\,(\overline{u^2} + \overline{v^2})\Delta\rho u + \frac{\gamma}{\gamma-1}\,\Delta pu \\ &= \tfrac{1}{2}\,(\overline{u^2} + \overline{v^2})\bar{u}\,\Delta\rho + \tfrac{1}{2}\,(\overline{u^2} + \overline{v^2})\bar{\rho}\,\Delta u + \overline{\rho u}\,\bar{u}\,\Delta u + \frac{\gamma}{\gamma-1}\,\bar{p}\,\Delta u \\ &\quad + \overline{\rho u}\,\bar{v}\,\Delta v + \frac{\gamma}{\gamma-1}\,\bar{u}\,\Delta p \;.\end{aligned}$$

where the bar denotes mean value.

With the definition of

$$\bar{q}^2 = \tfrac{1}{2}\,(\overline{u^2} + \overline{v^2}) \;,$$

the flux-difference Δf can be written as :

$$\Delta f = \begin{pmatrix} \bar{u} & \bar{\rho} & 0 & 0 \\ \bar{u}^2 & \bar{\rho}\,\bar{u} + \overline{\rho u} & 0 & 1 \\ \bar{u}\,\bar{v} & \bar{\rho}\,\bar{v} & \overline{\rho u} & 0 \\ \bar{q}^2\bar{u} & \bar{q}^2\bar{\rho} + \overline{\rho u}\,\bar{u} + \frac{\gamma}{\gamma-1}\,\bar{p} & \overline{\rho u}\,\bar{v} & \frac{\gamma}{\gamma-1}\,\bar{u} \end{pmatrix} \Delta\xi \;, \qquad (4)$$

where $\xi^T = \{\rho, u, v, p\}$.

With the definition of $\bar{\bar{u}}$ by :

$$\bar{\rho}\,\bar{\bar{u}} = \overline{\rho u} \;, \qquad (5)$$

the flux-difference Δf can also be written as :

$$\Delta f = \begin{pmatrix} 1 & 0 & 0 & 0 \\ \bar{u} & \bar{\rho} & 0 & 0 \\ \bar{v} & 0 & \bar{\rho} & 0 \\ \bar{q}^2 & \bar{\rho}\,\bar{u} & \bar{\rho}\,\bar{v} & 1/\gamma-1 \end{pmatrix} \begin{pmatrix} \bar{u} & \bar{\rho} & 0 & 0 \\ 0 & \bar{\bar{u}} & 0 & 1/\bar{\rho} \\ 0 & 0 & \bar{\bar{u}} & 0 \\ 0 & \gamma\bar{p} & 0 & \bar{u} \end{pmatrix} \Delta\xi \;. \qquad (6)$$

In the sequel, the first matrix in (6) is denoted by T.

In a similar way the flux-difference Δg can be written as :

$$\Delta g = T \begin{pmatrix} \bar{v} & 0 & \bar{\rho} & 0 \\ 0 & \bar{\bar{v}} & 0 & 0 \\ 0 & 0 & \bar{\bar{v}} & 1/\bar{\rho} \\ 0 & 0 & \gamma\bar{p} & \bar{v} \end{pmatrix} \Delta\xi \ , \tag{7}$$

where $\bar{\rho}\ \bar{\bar{v}} = \overline{\rho v}$. (8)

Any linear combination of Δf and Δg can be written as :

$$\Delta\phi = \alpha_1\ \Delta f + \alpha_2\ \Delta g = A\ \Delta\xi = T\ \tilde{A}\ \Delta\xi \ , \tag{9}$$

where :

$$\tilde{A} = \begin{pmatrix} \bar{w} & \alpha_1\ \bar{\rho} & \alpha_2\ \bar{\rho} & 0 \\ 0 & \bar{\bar{w}} & 0 & \alpha_1/\bar{\rho} \\ 0 & 0 & \bar{\bar{w}} & \alpha_2/\bar{\rho} \\ 0 & \alpha_1\ \gamma\bar{p} & \alpha_2\ \gamma\bar{p} & \bar{w} \end{pmatrix} , \tag{10}$$

with $\bar{w} = \alpha_1\ \bar{u} + \alpha_2\ \bar{v}\ ,\qquad \bar{\bar{w}} = \alpha_1\ \bar{\bar{u}} + \alpha_2\ \bar{\bar{v}}$. (11)

For the case $\alpha_1^2 + \alpha_2^2 = 1$, the eigenvalues of the matrix $\tilde{A}$ are :

$$\lambda_1 = \bar{w}\ , \qquad \lambda_2 = \bar{\bar{w}}\ ,$$

while λ_3 and λ_4 are given by :

$$(\lambda - \bar{w})(\lambda - \bar{\bar{w}}) - \gamma\bar{p}/\bar{\rho} = 0\ . \tag{12}$$

With the definition of :

$$\bar{c}^2 = \left(\frac{\bar{w} - \bar{\bar{w}}}{2}\right)^2 + \frac{\gamma\bar{p}}{\bar{\rho}}\ , \tag{13}$$

(12) can be written as :

$$\left(\lambda - \frac{\bar{w} + \bar{\bar{w}}}{2}\right)^2 - \bar{c}^2 = 0\ ,$$

so that :

$$\lambda_{3,4} = \tilde{w} \pm \bar{c}\ , \tag{14}$$

where

$$\tilde{w} = \frac{1}{2}\ (\bar{w} + \bar{\bar{w}})\ . \tag{15}$$

The matrix $\tilde{A}$, given by (10), is the discrete analogue of a linear combination of the Jacobians in the quasi-linear formulation of the Euler equations (1) :

$$\begin{pmatrix} u & \rho & 0 & 0 \\ 0 & u & 0 & 1/\rho \\ 0 & 0 & u & 0 \\ 0 & \gamma p & 0 & u \end{pmatrix} \frac{\partial \xi}{\partial x} + \begin{pmatrix} v & 0 & \rho & 0 \\ 0 & v & 0 & 0 \\ 0 & 0 & v & 1/\rho \\ 0 & 0 & \gamma p & v \end{pmatrix} \frac{\partial \xi}{\partial y} = 0 \ . \qquad (16)$$

As a consequence, the matrix T is to be seen as the discrete analogue of the transformation matrix in the transformation from conservative to primitive variables.

As can be seen from the expressions for the eigenvalues of matrix A, this matrix has the same properties as the Jacobians in the differential formulation.

The left eigenvector matrix of $\tilde{A}$ is :

$$X = \begin{pmatrix} 1/\rho & 0 & 0 & -1/\gamma p \\ 0 & \alpha_1/\bar{c} & \alpha_2/\bar{c} & (1+\delta)/\gamma\bar{p} \\ 0 & -\alpha_2/\bar{c} & \alpha_1/\bar{c} & 0 \\ 0 & \alpha_1/\bar{c} & \alpha_2/\bar{c} & -(1-\delta)/\gamma\bar{p} \end{pmatrix} , \qquad (17)$$

$$\text{where} \quad \delta = \frac{\bar{w} - \bar{\bar{w}}}{\bar{c}} \ . \qquad (18)$$

The order of the eigenvalues, used to obtain (17) is :

$$\bar{w} \ , \quad \tilde{w} + \bar{c} \ , \quad \bar{\bar{w}} \ , \quad \tilde{w} - \bar{c} \ .$$

Following the procedure of Steger and Warming [1], the matrix $\tilde{A}$ can be split into positive and negative parts by :

$$\tilde{A}^+ = X^{-1}\Lambda^+ X \ , \qquad \tilde{A}^- = X^{-1}\Lambda^- X \ , \qquad (19)$$

where $\Lambda^+ = \mathrm{diag}(\lambda_1^+, \lambda_2^+, \lambda_3^+, \lambda_4^+)$, $\Lambda^- = \mathrm{diag}(\lambda_1^-, \lambda_2^-, \lambda_3^-, \lambda_4^-)$,

with $\lambda_i^+ = \max(\lambda_i, 0)$, $\lambda_i^- = \min(\lambda_i, 0)$.

This allows a splitting of the flux-difference (9) by :

$$\Delta\phi = A^+\Delta\xi + A^-\Delta\xi \ , \qquad (20)$$

and the definition of the absolute value of this flux-difference by :

$$|\Delta\phi| = (A^+ - A^-)\Delta\xi \ . \qquad (21)$$

CONSTRUCTION OF A POSITIVE DISCRETISATION

Figure 1 shows a control volume centered around the node (i,j). Also the nodes located inside the adjacent volumes are indicated.

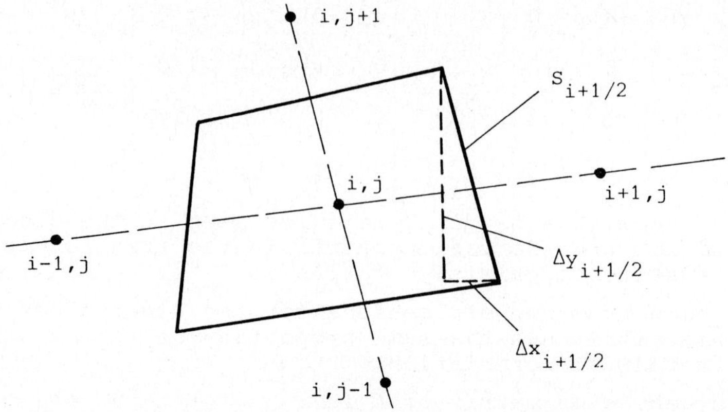

Fig. 1. Control volume centered around a vertex of the grid

When a piecewise constant interpolation of variables is chosen, the flux through the surface $S_{i+1/2}$ of the control volume can be defined in an upwind way by :

$$F_{i+1/2} = \frac{1}{2}\left(F_i + F_{i+1} - |\Delta F_{i,i+1}|\right) . \tag{22}$$

where F_i and F_{i+1} denote the fluxes computed with the values of the variables at the nodes (i,j) and (i+1,j) respectively. For simplicity, in the above, the non-varying index is omitted.

The flux-difference across the surface $S_{i+1/2}$ can be written as :

$$\begin{aligned}\Delta F_{i,i+1} &= \Delta y_{i+1/2}\, \Delta f_{i,i+1} + \Delta x_{i+1/2}\, \Delta g_{i,i+1} \\ &= \Delta s_{i+1/2}\, (\alpha_1\, \Delta f_{i,i+1} + \alpha_2\, \Delta g_{i,i+1}) ,\end{aligned} \tag{23}$$

where $$\Delta s^2_{i+1/2} = \Delta x^2_{i+1/2} + \Delta y^2_{i+1/2} ,$$

$$\alpha_1 = \Delta y_{i+1/2}/\Delta s_{i+1/2} , \qquad \alpha_2 = \Delta x_{i+1/2}/\Delta s_{i+1/2} .$$

With the notation of the previous section, this is :

$$\Delta F_{i,i+1} = F_{i+1} - F_i = \Delta s_{i+1/2}\, A_{i,i+1}\, \Delta \xi_{i,i+1} .$$

Furthermore, the matrix $A_{i,i+1}$ can be split into a positive and a negative part :

$$A_{i,i+1} = A^+_{i,i+1} + A^-_{i,i-1} . \tag{24}$$

This allows the definition of the absolute value of the flux-difference by :

$$|\Delta F_{i,i+1}| = \Delta s_{i+1/2}(A^+_{i,i+1} - A^-_{i,i+1})\Delta\xi_{i,i+1} \, . \tag{25}$$

The flux $F_{i+1/2}$ given by (22) can be written in either of the two following ways, which are completely equivalent :

$$\begin{aligned} F_{i+1/2} &= F_i + \frac{1}{2}\Delta F_{i,i+1} - \frac{1}{2}|\Delta F_{i,i+1}| \\ &= F_i + \Delta s_{i+1/2}\, A^-_{i,i+1}\, \Delta\xi_{i,i+1} \, , \end{aligned} \tag{26}$$

$$\begin{aligned} F_{i+1/2} &= F_{i+1} - \frac{1}{2}\Delta F_{i,i+1} - \frac{1}{2}|\Delta F_{i,i+1}| \\ &= F_{i+1} - \Delta s_{i+1/2}\, A^+_{i,i+1}\, \Delta\xi_{i,i+1} \, . \end{aligned} \tag{27}$$

The fluxes on the other surfaces of the control volume $S_{i-1/2}$, $S_{j+1/2}$, $S_{j-1/2}$, can be treated in a similar way as the flux on the surface $S_{i+1/2}$.

From the expressions (26) and (27) it is seen that the flux defined by (22) corresponds to an upwind flux. Indeed, when $A_{i,i+1}$ only has positive eigenvalues, the flux $F_{i+1/2}$ is taken to be F_i and when $A_{i,i+1}$ only has negative eigenvalues, the flux $F_{i+1/2}$ is taken to be F_{i+1}.

With (26) and (27), the flux balance on the control volume of figure 1 can be brought into the following form :

$$\begin{aligned} &\Delta s_{i+1/2}\, A^-_{i,i+1}(\xi_{i+1} - \xi_i) + \Delta s_{i-1/2}\, A^+_{i,i-1}(\xi_i - \xi_{i-1}) \\ &+ \Delta s_{j+1/2}\, A^-_{j,j+1}(\xi_{j+1} - \xi_j) + \Delta s_{j-1/2}\, A^+_{j,j-1}(\xi_j - \xi_{j-1}) = 0 \, . \end{aligned} \tag{28}$$

The set of equations (28) is both conservative and positive. It is conservative since it exactly expresses the sum of fluxes on the control volume to be zero.
It is positive since it can be put into the form :

$$\begin{aligned} C\xi_{i,j} &= \Delta s_{i-1/2}\, A^+_{i,i-1}\, \xi_{i-1,j} + \Delta s_{i+1/2}\, (-A^-_{i,i+1})\, \xi_{i+1,j} \\ &+ \Delta s_{j-1/2}\, A^+_{j,j-1}\, \xi_{i,j-1} + \Delta s_{j+1/2}\, (-A^-_{j,j+1})\, \xi_{i,j+1} \end{aligned} \tag{29}$$

where C is the sum of the matrix-coefficients in the right hand side and where all matrix-coefficients in the right hand side involved have non-negative eigenvalues.

As a consequence of the positivity, the set of equations of form (29) on all grid nodes can be solved by a collective variant of any scalar relaxation method. By a collective variant it is meant that in each node, all components of the vector of dependent variables ξ are relaxed simultaneously.

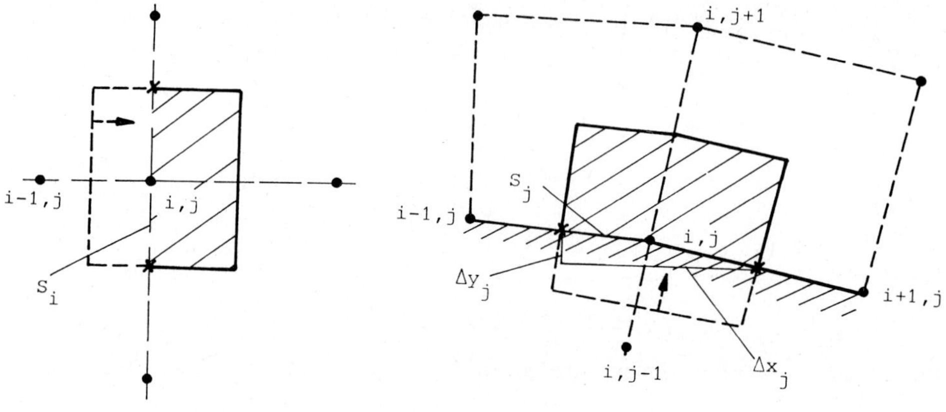

Fig. 2. Control volumes at inlet and at solid boundary

Figure 2 shows the half-volumes centered around a node at inlet and around a node on a solid boundary.

These half-volumes can be seen as the limit of complete volumes in which one of the sides tends to the boundary.

As a consequence, the flux on the side S_i of the inlet control volume can be expressed according to (27) by :

$$F_i - \Delta s_i \, A^+_{i,j} (\xi_i - \xi_{i-1}) \, , \tag{30}$$

where the matrix $A_{i,j}$ is calculated in the node (i,j).

Similarly, the flux on the side S_j of the control volume at the solid boundary can be expressed by :

$$F_j - \Delta s_j \, A^+_{i,j} (\xi_j - \xi_{j-1}) \, , \tag{31}$$

where again the matrix $A_{i,j}$ is calculated in the node (i,j).

With the definitions (30) and (31), the flux balance on the control volumes at boundaries takes the form (28) in which a node outside the domain comes in. These nodes, however, can be eliminated.

At in- and outflow boundaries, due to an assumption of nearly uniform flow, the T-matrix in considered to be constant and the set of discrete equations simplifies. For the inlet, (28) becomes :

$$\Delta s_i \, \tilde{A}^+_{i,j} (\xi_i - \xi_{i-1}) + \Delta s_{i+1/2} \, \tilde{A}^-_{i,i+1} (\xi_{i+1} - \xi_i)$$
$$+ \Delta s_{j-1/2} \, \tilde{A}^+_{j,j-1} (\xi_j - \xi_{j-1}) + \Delta s_{j+1/2} \, \tilde{A}^-_{j,j+1} (\xi_{j+1} - \xi_j) = 0 \, . \tag{32}$$

According to the eigenvector matrix given by (17), in subsonic flow, at the inflow boundary one combination of the equations in (32) exists, eliminating the node i-1.

It is easily seen that :

$$d_1^T \tilde{A}_{i,j}^{+} = 0 ,$$

where $d_1^T = (0, \alpha_1/c, \alpha_2/c, -1/\gamma p)$.

with c and p values taken at the node (i,j).

The resulting equation is to be supplemented with three boundary conditions : stagnation temperature, stagnation pressure and flow direction.

At the outflow boundary, the set of equations is similar to (32), now involving $\tilde{A}_{i,j}^{-}$. For subsonic flow, according to the eigenvector matrix (17), three combinations can now be made, eliminating the node i+1.

Clearly $$d^T \tilde{A}_{i,j}^{-} = 0 ,$$

for $$d^T = d_2^T, d_3^T \text{ or } d_4^T ,$$

where $$d_2^T = (1/\rho, 0, 0, -1/\gamma p) ,$$

$$d_3^T = (0, \alpha_1/c, \alpha_2/c, 1/\gamma p) ,$$

$$d_4^T = (0, -\alpha_2, \alpha_1, 0) .$$

The resulting equations are to be supplemented by one boundary condition. This can be the specification of the Mach number or the pressure.

At a solid boundary, according to figure 2, the condition of impermiability is :

$$\alpha_1 u_{i,j} + \alpha_2 v_{i,j} = 0 . \qquad (33)$$

As a consequence, at this boundary :

$$\bar{w} = \bar{\bar{w}} = \tilde{w} = 0 .$$

Therefore, in (17) it is seen that three combinations exist eliminating an outside node at a solid boundary.

Clearly : $$d^T \tilde{A}^{+} = 0 ,$$

for $$d^T = d_1^T, d_2^T \text{ or } d_3^T ,$$

where $$d_1^T = (\gamma p/\rho, 0, 0, -1) ,$$

$$d_2^T = (0,\ \alpha_2,\ -\alpha_1,\ 0)\ ,$$

$$d_3^T = (-c/\rho,\ \alpha_1,\ \alpha_2,\ 0)\ .$$

Similarly : $\qquad d^T\ \tilde{A}^- = 0\ ,$

for $\qquad d^T = d_1^T,\ d_2^T \text{ or } d_4^T\ ,$

with $\qquad d_4^T = (+c/\rho,\ \alpha_1,\ \alpha_2,\ 0)\ .$

Using (9) and (33) it is seen that

$$e^T\ A^+ = 0\ ,$$

for $\qquad e^T = e_1^T,\ e_2^T,\ e_3^T\ ,$

where $\qquad e_1^T = (H,\ 0,\ 0,\ -1)\ ,$

$$e_2^T = (-(u^2+v^2),\ u,\ v,\ 0)\ ,$$

$$e_3^T = (-c,\ \alpha_1,\ \alpha_2,\ 0)\ .$$

and $\qquad e^T\ A^- = 0\ ,$

for $\qquad e^T = e_1^T,\ e_2^T,\ e_4^T\ ,$

where $\qquad e_4^T = (+c,\ \alpha_1,\ \alpha_2,\ 0)\ .$

The nodal equations at a solid boundary are premultiplied by e_i^T, with $i = 1, 2, 3$ on a southern boundary and $i = 1, 2, 4$ on a northern boundary, leading to three significant equations. These are supplemented with the kinematic boundary condition (33).

NUMERICAL EXAMPLE

Figure 3 shows the well known GAMM-test case [9] for transonic flows, discretized by a grid with 24 x 8 elements. In the actual computation a twice more refined grid was used with 96 x 32 elements. Vertex based finite volumes, as indicated in figure 2, were used.

At inflow, the specification of a horizontal flow direction was used as boundary condition. At outflow the Mach number was fixed at 0.85.

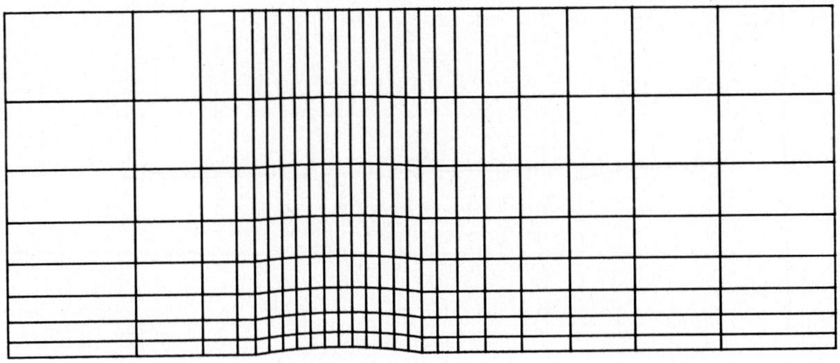

Fig. 3. Coarse computational grid

Starting from a uniform flow at Mach number 0.85, the discrete equations were solved by Jacobi relaxation with a relaxation factor 0.95.

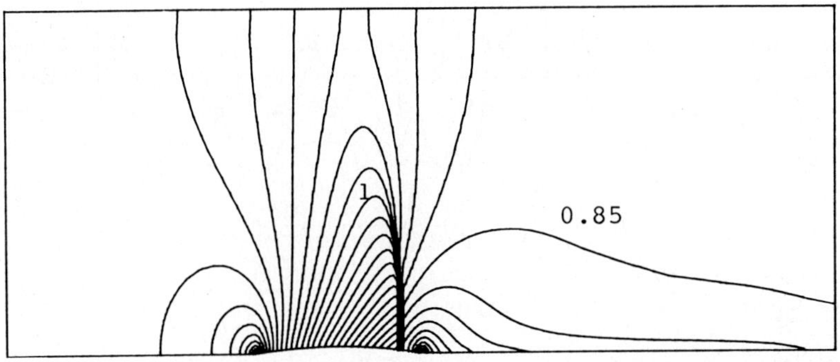

Fig. 4. Iso-Mach lines obtained at a 96 x 32 grid

Figure 4 shows the iso-Mach lines for the fully converged solution plotted by piecewise linear interpolation within the elements of the grid. Figure 5 shows the surface pressure distribution on the southern boundary.

The obtained solution coincides almost with the solution obtained from the most reliable time-marching methods reported in [9]. However, unlike most time-marching solutions, due to the guaranteed positivity everywhere, the solution has no wiggles in the shock region.

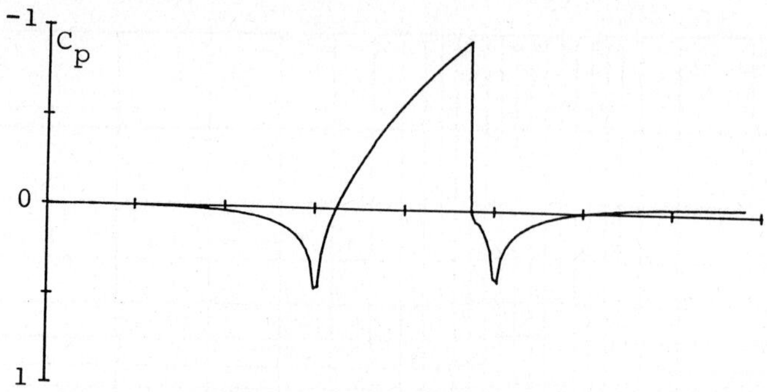

Fig. 5. Pressure distribution on the bottom

MULTIGRID FORMULATION

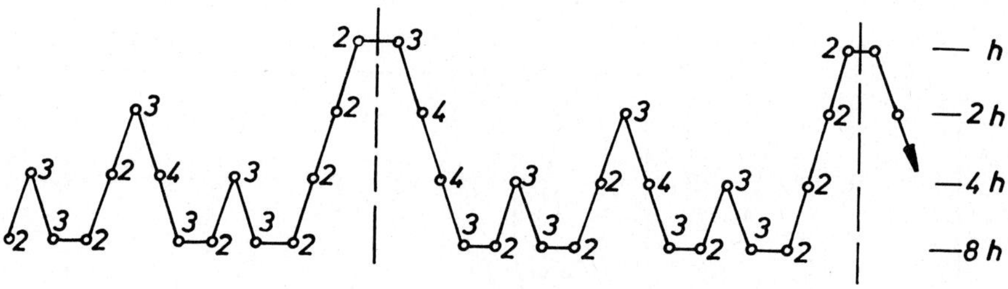

Fig. 6. Nested iteration and W-cycle in the multigrid formulation

Figure 6 shows the geometry of the multigrid method used. A full multigrid formulation with a full approximation scheme on the non-linear equations (29) is employed. The cycle is of W-form. Also the nested iteration, used as starting cycle, has W-form. Two types of relaxation methods were investigated. In the first formulation, the relaxation algorithm is symmetric successive underrelaxation with relaxation factor 0.9. The order of relaxation is the lexicographic order, i.e. going from the lower left point to the upper right point first varying the row index and then going from the upper right point to the lower left point in the reverse order. In relaxing the set of equations (29), the coefficients are formed with the latest available information, i.e. $A^{+}_{i,i-1}$ is evaluated with the function values in node (i,j) on the old level, but with the function values in node (i-1,j) on the new level. After determination of the new values in node (i,j), no updates of coefficients and no extra iterations are done. This means that the set of equations (28) is treated as a quasi-linear set. As restriction operator, full weighting is used within the flow field while injection is used at the boundaries. The prolongation operator is bilinear interpolation. Restriction is not employed for function values. As an approximation for a restricted function value, the last available value on the same grid is taken. The calculation starts from a uniform flow on the coarsest grid (12 x 4).

In figure 6, the operation count is indicated. A relaxation on the current grid is taken as one local work unit. So, the symmetric relaxation is seen as two work units. A residual evaluation plus the associated grid transfer is also taken as one work unit. Hence, the 4 in figure 6, in going down, stands for the construction of the right hand side in the FAS formulation, two relaxations and one residual evaluation. With this way of evaluating the work, the cost of the cycle is 8.6875 work units on the finest level. The cost of the nested iteration is about 4.39 work units.

Figure 7 shows the convergence behaviour of the single grid and the multigrid formulation. The residual shown is the maximum residual of all equations, after normalizing these equations, i.e. bringing the coefficient of ρ, u, v and p on 1 in the mass-, momentum-x-, momentum-y- and energy equation respectively.

A maximum residual of 10^{-4} is reached after approximately 87 work units. The convergence factor of the multigrid method, i.e. the residual reduction per work unit is about 0.915. This probably can be considered as being optimal. This is seen by the pressure distribution on the bottom obtained after the nested iteration and one cycle. Up to plotting accuracy this pressure distribution coincides with the distribution obtained after full convergence, as shown in figure 5.

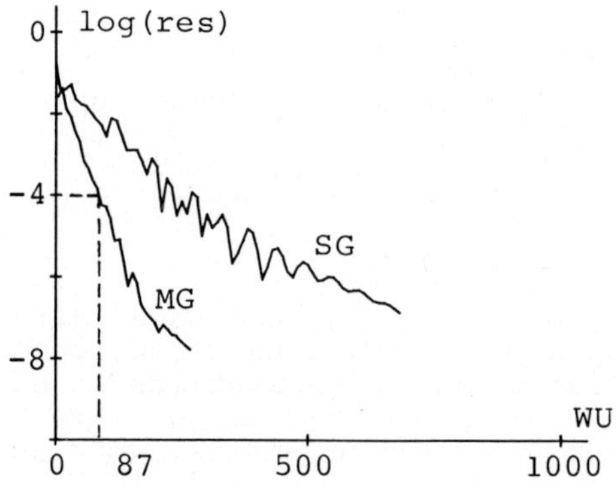

Fig. 7. Convergence behaviour for single grid and multigrid relaxation with successive symmetric underrelaxation (ω = 0.9)

The successive algorithm described above is of course highly sequential. By vectorization on the cyber 205, only some 20 % can be gained in computing time. On the cyber 205, with 2 pipe configuration, the computing time for 87 work units with the symmetric successive relaxation is about 78.89 s.

In order to maximize the vectorizability of the code, damped Jacobi iteration was tried. It was found that with the boundary conditions at inlet and outlet, as described above, multigrid convergence could not be obtained. It is to be remarked that these boundary conditions are highly non-linear. At outflow, the field equations give linear combinations of the flow variables ρ, u, v and p. A prescribed Mach number is a non-linear combination of these variables. Therefore an iterative procedure is necessary to impose this condition. At inflow, one field equation expressing a linear combination of u and p is to be combined with given stagnation pressure and stagnation temperature. These stagnation values are highly non-linear combinations of the flow variables and again an iterative procedure is necessary to impose these conditions.

The boundary condition at outlet becomes linear by imposing pressure. This is equivalent to imposing an isentropic Mach number, i.e. the Mach number which corresponds to the given pressure if the flow were isentropic. This Mach number is given by :

$$\frac{p_o}{p} = \left(1 + \frac{\gamma-1}{2} M_s^2\right)^{\frac{\gamma}{\gamma-1}} \tag{34}$$

where M_s is the isentropic Mach number and p_o is the stagnation pressure imposed at inlet.

In order to remove the iteration at inlet, the field equation was replaced by an extrapolation of Mach number. As is well known, this form of numerical boundary condition often is employed with time marching schemes. Using then isentropic expressions for pressure, density and velocity, the flow quantities can be obtained at inflow without iteration.

With these boundary conditions, multigrid convergence could be obtained using damped Jacobi iteration, if a sufficient number of relaxations was done before grid transfer. The precise number of relaxations was found not to be very critical provided it was at least equal to two.

Figure 8 shows the convergence history for Jacobi iteration using a full multigrid formulation with the same characteristics as before. As outlet boundary condition the isentropic Mach number was fixed at 0.85. The number of relaxations on each level is 4. It was found that a relaxation factor equal to 0.9 was optimal. The cost of the cycle is here 15.1875 work units on the finest level. The cost of the nested iteration is about 8.235 work units.

As is seen from figure 8, Jacobi iteration is, in terms of work units, less efficient than symmetric successive relaxation. It takes about 111.5 work units to obtain a residual equal to 10^{-4}. Also, in order to obtain a pressure distribution at the bump within plotting accuracy of the fully converged distribution, here a nested iteration and three cycles are required. Due to the complete vectorizability, the computing time necessary to obtain the same level of convergence is however much less. The computing time for 111.5 work units with the Jacobi relaxation is, for a not really completely optimized code, about 5.39 s on the cyber 205 with a 2 pipe configuration. This represents a gain of a factor of about 15 with respect to the symmetric successive relaxation.

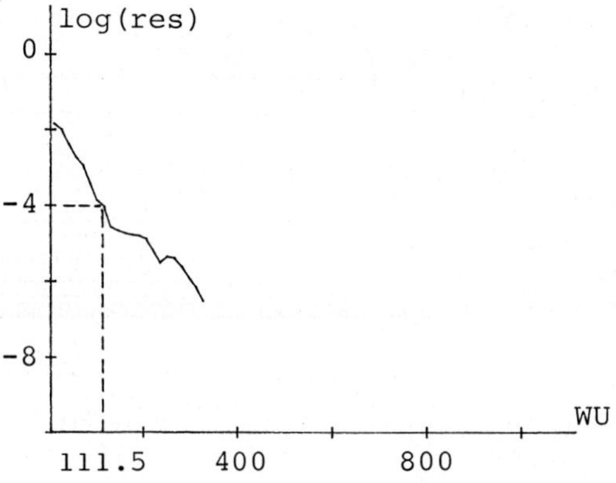

Fig. 8. Convergence behaviour for multigrid relaxation with Jacobi iteration with underrelaxation factor $\omega = 0.9$ and 4 relaxations per level

Figure 9 shows the iso-Mach lines obtained by prescription of the isentropic Mach number equal to 0.85 at outlet.

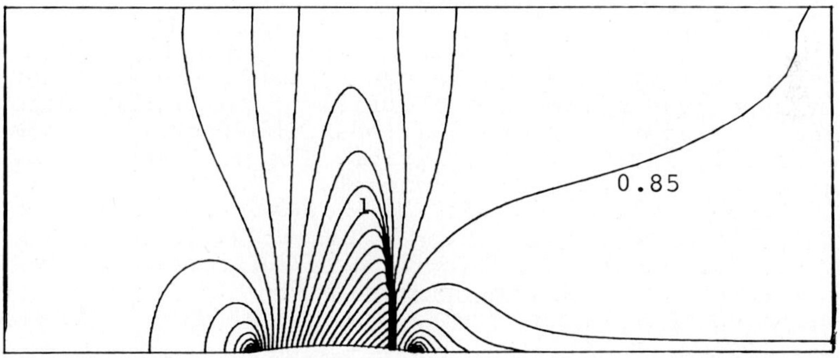

Fig. 9. Iso-Mach lines for isentropic outlet Mach number 0.85

There is of course some difference with the result obtained by prescription of the Mach number itself, as shown on figure 4. The result shown on figure 9 has as unphysical feature that the iso-Mach line 0.85 near the outlet raises above the shock. This shows that the computed flow is not really isentropic above the shock. This is clarified on figure 10, showing the iso-entropy lines.

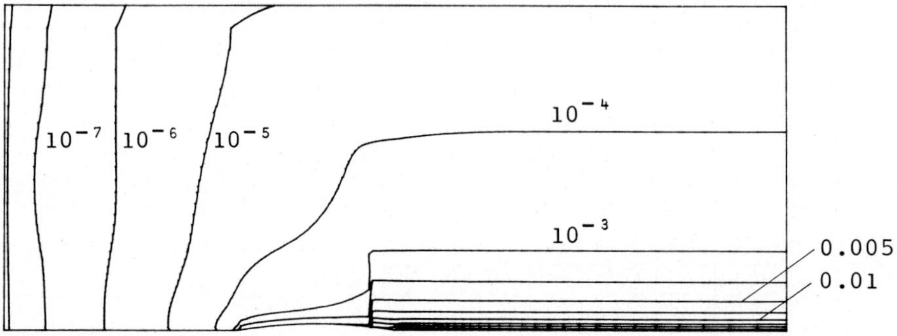

Fig. 10. Iso-entropy lines corresponding to the solution of fig. 9. The parameter plotted is $(p/p_\infty)/(\rho/\rho_\infty)^\gamma$ where p_∞ and ρ_∞ correspond to isentropic flow with $M = 0.85$

From this figure it is seen that there is not only entropy production by the shock, but also by numerical error. This illustrates the first order character of the method since this effect is not seen in the second order solutions reported in [9].

CONCLUSION

It was shown that by an adequate use of the flux-difference splitting technique a simple and efficient multigrid method can be obtained for steady Euler equations. The formulation described here is only first order accurate. The next step is of course to bring it into second order accuracy.

ACKNOWLEDGEMENT

The research reported in this paper was granted by the Belgian National Science Foundation (N.F.W.O.).

REFERENCES

[1] STEGER J.L., WARMING R.F. : "Flux vector splitting for the inviscid gasdynamic equations with application to finite-difference methods", J. Comp. Phys., 40 (1981), pp. 263-293.

[2] JESPERSEN D.C. : "A multigrid method for the Euler equations", AIAA paper 83-0124 (1983).

[3] GODUNOV S.K. : "A finite difference method for the numerical computation of discontinuous solutions of the equations of fluid dynamics", Mat. Sb., 47 (1959), pp. 271-290.

[4] ROE P.L. : "Approximate Riemann solvers, parameter vectors and difference schemes", J. Comp. Phys., 43 (1981), pp. 357-372.

[5] OSHER S. : "Numerical solution of singular perturbation problems and hyperbolic systems of conservation laws", in : "Mathematical and numerical approaches to asymptotic problems in analysis", Axelsson O., Frank L.S. and van der Sluis A., eds., North Holland, Amsterdam 1981, pp. 179-205.

[6] LOMBARD C.K., OLIGER J. and YANG J.Y. : "A natural conservative flux difference splitting for the hyperbolic systems of gas dynamics", AIAA paper 82-0976 (1982).

[7] HEMKER P.W., SPEKREIJSE S.P. : "Multigrid solution of the steady Euler equations", in : "Notes on numerical fluid dynamics, vol. 11", Braess D., Hackbusch W. and Trottenberg U. (eds.), Vieweg, Braunschweig 1985, pp. 33-44.

[8] DICK E. : "A flux-difference splitting method for steady Euler equations", J. Comp. Phys. (1988), to appear.

[9] RIZZI A., VIVIAND H. (eds.) : "Numerical methods for the computation of inviscid transonic flows with shock waves", Notes on numerical fluid dynamics, vol. 3, Vieweg, Braunschweig, 1981.

Treatment of Singular Perturbation Problems with Multigrid Methods

Joachim H. Dörfer

Mathematatisches Institut der Universität Düsseldorf

Abstract

The combination of several components of multigrid algorithms has been tested for the solution of singular perturbation problems. 1- and 2-dimensional linear problems were considered. An algorithm has been found heuristically for which the usual lower bound for the spectral radius of the iteration matrix of 1/2 does not hold. Artificial viscosity is used to obtain a stable discretisation. The algorithm is discussed and numerical results are presented.

Introduction

For the treatment of singular perturbation problems with standard multigrid methods 1/2 is a lower bound for the spectral radius of the iteration matrix [BOE81]. With some more work even 1/3 is obtainable [DEB85]. Compared to the typical spectral radii of standard multigrid methods of about 1/10 this seems to be rather bad. The problem is to develop stable methods which converge fast on the "smooth" part of the solution and at the same time show existing boundary layers as sharply as possible. The layers themselves can then be calculated by specially fitted methods (e.g. adaptive). There are basically two ways to reach this goal:

— The relaxation can be improved (ILU, linerelax., blockrelax.)

— The coarse grid correction can be improved

We restrict ourselves to pointwise Gauss-Seidel relaxation with direction independent numbering of the points. All other components of the multigrid method are varied:

Discretisation, coarse grid operator calculation, restriction and interpolation.

The resulting versions are at first tested on a 1-dimensional model problem. Only the versions that show up to be stable are transferred to the 2-dimensional case. The results of the tests are that there is only one version which can solve the model problem with even uncontinuous coefficients at an effectiveness comparable to standard multigrid methods in the unperturbed case. In particular: More relaxations yield a much reduced spectral radius of the iteration matrix.

Basic facts

We assume the reader to be familiar with the ideas and basic facts of multigrid methods. (see e.g. [HTR81],[HTR85], [HAC85]) This section therefore deals with the specifics of singular perturbation.

Used Notation

Ω_h discrete Grid in Ω of meshsize h
L_h discrete difference-operator on Ω_h
$I_{2h}^h : \Omega_{2h} \longrightarrow \Omega_h$ Interpolation operator
$I_h^{2h} : \Omega_h \longrightarrow \Omega_{2h}$ Restriction operator
u_h, d_h, f_h grid functions on Ω_h
$M_{l,h}$ Iteration Matrix of l-level algorithm
with h meshsize of the finest grid

Consider a linear elliptic boundary value problem $L_\varepsilon u = f$ on a domain Ω with the differential operator L_ε depending on some parameter $\varepsilon > 0$. We call this problem singularly perturbed if the limiting operator

$$L_0 := \lim_{\varepsilon \to 0} L_\varepsilon$$

is not elliptic or elliptic of a lower order than L_ε.

Central differences

For small values of ε standard discretisations can lead to a solution of the discrete problem which has nothing to do with the solution of the original problem: Discretisation of

$$\varepsilon u_{xx} + 2u_x = 0 \ on \ [0, \infty[, \ u(0) = 1, \ u(\infty) = 0$$

with central differences yields as discrete solution

$$u_{\varepsilon,h}(ih) = \left(\frac{\varepsilon - h}{\varepsilon + h}\right)^i.$$

This is an O(h^2)-discretisation. For ih fixed and $\frac{h}{\varepsilon} \to 0$:

$$|u_{\varepsilon,h}(ih) - u_\varepsilon| = \left| \left(\frac{\varepsilon - h}{\varepsilon + h} \right)^i - \left(e^{\frac{-2h}{\varepsilon}} \right)^i \right| \leq C \left(\frac{h}{\varepsilon} \right)^2$$

with C independent of i, h and ε.
But the solution of the reduced difference equation is

$$u_{0,h}(ih) = \lim_{\varepsilon \to 0} u_{\varepsilon,h}(ih) = (-1)^i.$$

Upstream

Discretisation of the first derivative with the upstream scheme

$$u_x \doteq \frac{1}{h}[-1\ 1\ 0]$$

and Taylor expansion of L_h yields

$$L_h = \varepsilon u_{xx} + 2u_x - hu_{xx} + O(h^2)$$

$$L_{2h} = \varepsilon u_{xx} + 2u_x - 2hu_{xx} + O(h^2)\,.$$

If $u_{xx} = 0$: $L_h u \sim L_{2h} u$ as it should be, but if $u_x = 0$ and $\varepsilon \ll h$:

$$L_h u = -hu_{xx} \sim 1/2 \cdot L_{2h} u\,.$$

In this case the computed correction is only half the correct value:

$$L_{2h} v_{2h} = d_{2h} \Rightarrow L_h v_h = 1/2 \cdot d_h \text{ instead of } d_h\,.$$

From this follows $\rho(M_{2,h}) \geq 1/2$.

Model problem

We chose the convection-diffusion equation for our tests.

$$-\varepsilon \Delta u + c \cdot \nabla u = f \textit{ on the unit square}$$

$$u \equiv g \textit{ on the boundary}\ ,\ \varepsilon > 0\ ,\ c = (c_1, c_2)\,.$$

With $c = (cos\varphi, sin\varphi)$ the equation describes a flow with it's direction given by the angle φ.
Simplification (1D):

$$-\varepsilon u'' + c \cdot u' = f \textit{ on }]0,1[$$

$$u(0) = a\ ,\ u(1) = b\,.$$

Special components

In [DOE86] the following components of the multigrid algorithm were tested on the model problems stated above for constant coefficients:
Discretisation: Central, upstream, artificial viscosity, Il'in's scheme
Coarse Grid operator: Standard, Galerkin type
Relaxation: pointwise Gauss-Seidel
Numbering: lexicographic, red-black, four colours
Interpolation: linear, using the grid equation, local transformation
Restriction: injection, full weighting, adjoint of the interpolation
One combination showed up to yield good spectral radii for the MG iteration matrix:

- **Artificial viscosity discretisation**
- **Galerkin type coarse-grid operator**
- **4 colour relaxation**
- **Interpolation using the grid equation**
- **Full weighting restriction**

These components are briefly described:

Artificial viscosity

Having fixed the changing ratio of ε and h as the reason for the behavior of the standard discretisation it is natural to try to keep it constant. e.g. $\varepsilon' := max(\varepsilon, \beta h)$, $\beta \geq 1$. With this new value central differences are used. Here the order of approximation is one since ε' is an O(h)-approximation of ε. Enlarging ε physically means to add viscosity to the flowing medium. For all ε, $h > 0$ the resulting operators are weakly diagonally dominant. The value of β is not critical; we chose it to be 1.5 This is according to Brandt's results from smoothing analysis.

Galerkin type coarse grid operator

The coarse-grid operator is defined using the grid transfer operators:

$$L_{2h} := I_h^{2h} \cdot L_h \cdot I_{2h}^h \,.$$

When using arbitrary operators one can easily run into trouble obtaining even coarse grid operators which are the 0 operator (e.g. bilinear interpolation and full weighting restriction yield unstable operators). To test for possible instabilities the calculation was iterated for the model problem with $c = (cos\varphi, sin\varphi)$ for $\varphi \in [0, 2\pi]$. The limiting stencils showed no instabilities. Two special stencils are presented here with the center element scaled to 1:

$$\varphi = 0: \qquad\qquad \varphi = \tfrac{\pi}{4}:$$

$$\begin{bmatrix} -\frac{1}{4} & \frac{1}{4} & 0 \\ -1 & 1 & 0 \\ -\frac{1}{4} & \frac{1}{4} & 0 \end{bmatrix} \begin{bmatrix} 0 & 0 & 0 \\ -\frac{1}{2} & 1 & 0 \\ 0 & -\frac{1}{2} & 0 \end{bmatrix}.$$

Four colour relaxation

Ω_h is split into four subsets Ω_{00}, Ω_{10}, Ω_{01} and Ω_{11} in such a way that the relaxation becomes independent of the marching direction. Two colours as with red-black relaxation are not sufficient since we have 9 point stencils. The indices correspond to the directions of the grid lines (see [HTR81]).

Interpolation using the grid equation

Special about this interpolation is that the fine grid points which aren't coarse grid points at the same time are computed from $L_h u_h = f_h$. In the onedimensional case this means that they have defect 0 and if the coarse grid correction solves exactly, the interpolated solution stays exact. Generalisation to the 2d case is straightforward but doesn't have the old properties exactly.
We distinguish three categories of fine grid points:

$$\begin{bmatrix} \circ & & \circ \\ \bullet & & \bullet \\ \circ & & \circ \end{bmatrix} \begin{bmatrix} \circ & \bullet & \circ \\ & & \\ \circ & \bullet & \circ \end{bmatrix} \begin{bmatrix} \circ & & \circ \\ & \bullet & \\ \circ & & \circ \end{bmatrix}.$$

In the first two cases we sum up the rows respectively columns of the 9 point difference stencil to proceed with the resulting 3 point stencil as in the 1d case. The points belonging to the third category can then be interpolated using the 9 point stencil itself, since the values of all surrounding points are known. f_h is set to zero.

Full weighting restriction

see [HTR81]

$$I_h^{2h} \doteq \frac{1}{16}\begin{bmatrix} 1 & 2 & 1 \\ 2 & 4 & 2 \\ 1 & 2 & 1 \end{bmatrix}.$$

Observations

This version was then generalized to treat variable, even uncontinuous cofficients. We always used the F-cyle since in all tests we obtained nearly the same results as with the W-cycle at much less work. Looking at the numerical results we see the following:

- The spectral radius stays small for all calculated directions of flow, 1/2 does not hold as a lower bound. (see Figure 3)
- The spectral radius is nearly independent of the meshsize of the finest grid. (For $h \geq \frac{1}{16}$)
- The spectral radius is independent of the number of levels (The 2-level algorithm performs as good as the 3-level, 4-level, ...-algorithm)
- Efficiency is nearly constant for 2 — 6 relaxation per level. In particular the time needed for relaxation multiplied by the spectral radius is a constant if we do 3 or more relaxations per level. (Even 2 relaxations are appropriate for good performance)
- For continuous coefficients the spectral radii match the ones for constant coefficients.
- If we deal with discontinuous coefficients, things get worse if the step does not occur along gridpoints.

The scope of further research is focused on this problem as well as on the use of defect correction to obtain an $O(h^2)$-approximation to the solution of the original problem.

Numerical Results

Spectral radius of MG iteration matrix $M_{l,h}$:

$$\rho(M_{l,h}) = \lim_{n\to\infty} \frac{\|U_h - u_h^n\|}{\|U_h - u_h^{n-1}\|}.$$

Computed approximation:

$$\rho_n := \frac{\|U_h - u_h^n\|}{\|U_h - u_h^{n-1}\|}.$$

Computations were done using the following global parameters:

$$max.\ 7\ levels,\ 2\ prerelaxations,\ 1\ postrelaxation, \varepsilon = 10^{-6},\ h = \frac{1}{128}.$$

After 100 Iterations the value of ρ_n has nearly become a constant.

The following problems were considered:

Table 1: Different choices of **c**

c_1	c_2	f	u	ρ_{100}	Remarks
0	0	0	0	.029	$\Delta u = 0$
$\cos\phi$	$\sin\phi$	0	0	$\leq .033$	$0 \leq \phi \leq 2\pi$
1	0	1	x	.029	
1	0	-1	$-x$	.028	
0	1	1	y	.029	
0	1	-1	$-y$	.028	
$\frac{1}{2}\sqrt{2}$	$\frac{1}{2}\sqrt{2}$	$\frac{x+y}{\sqrt{2}}$	xy	.026	
$\frac{1}{2}\sqrt{2}$	$\frac{1}{2}\sqrt{2}$	$\sqrt{2}$	$x+y$	.027	
x	y	0	0	.028	
$1\ if\ x \leq \frac{1}{2}$ $0\ else$	0	0	0	.033	step = gridpoint
$2\ if\ x \leq \frac{1}{10}$ $0\ else$	0	0	0	.061	step $\neq$ gridpoint

Table 2: Different numbers of grids / Different finest grids

# relax	ρ_{100}	h_{fine}	h_{coarse}	# relax	ρ_{100}	h_{fine}	h_{coarse}
3	.033	1/128	1/2	3	.033	1/64	1/2
3	.033	1/128	1/4	3	.033	1/32	1/2
3	.033	1/128	1/8	3	.033	1/16	1/2
3	.033	1/128	1/16	3	.010	1/8	1/2
				3	.002	1/4	1/2

Table 3: Different numbers of relaxations

#Rel	ρ_{100}	time	time - time(# Rel = 0)	Remarks
0	1.00	0.2	0.0	These are the
1	.240	0.3	0.1	Results the
2	.058	0.4	0.2	following
3	.033	0.5	0.3	figures are
4	.025	0.6	0.4	derived from
5	.020	0.7	0.5	
6	.016	0.8	0.6	

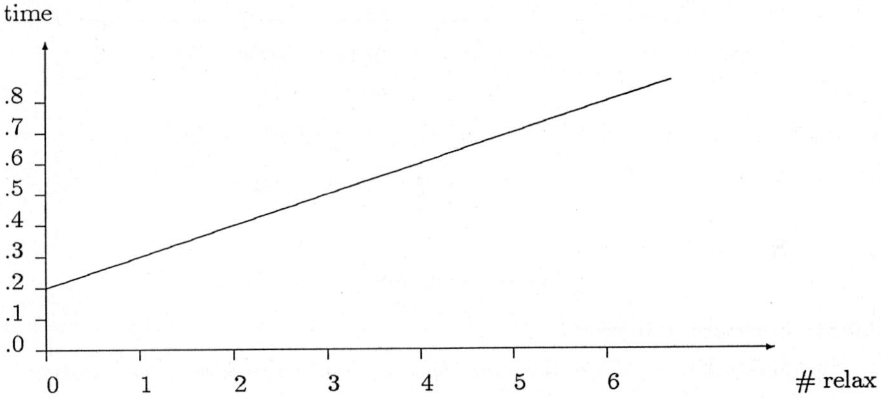

Figure 1

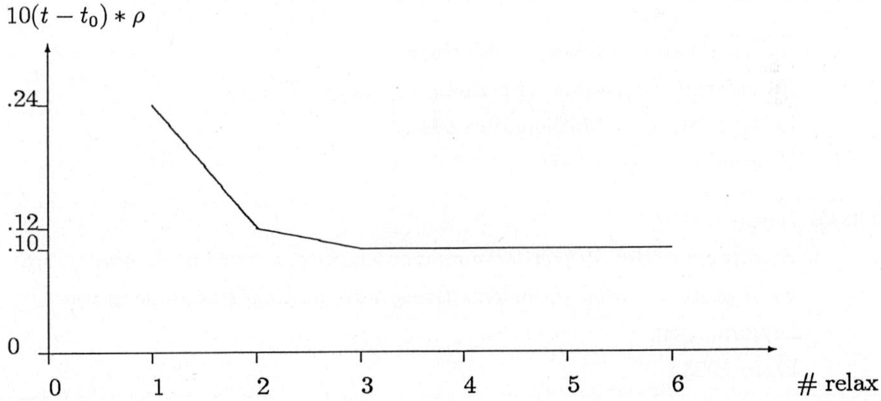

Figure 2

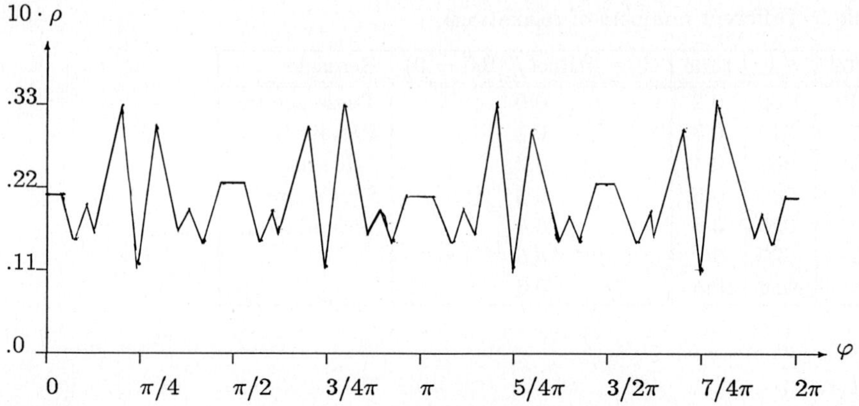

Figure 3

References

[BOE81] Christoph Börgers
Mehrgitterverfahren für eine Mehrstellendiskretisierung der Poisson-Gleichung und für eine zweidimensionale singulär gestörte Aufgabe
Diplomarbeit
Bonn 1981.

[BGM76] C. M. Brauner, B. Gay, J. Mathieu
Singular Perturbations and Boundary Layer Theory
Lecture Notes in Mathematics 594
Proceedings, Lyon 1976 .

[DEB85] Bernd Debus
Ansatz spezieller Mehrgitterkomponenten für ein zweidimensionales, singulär gestörtes Modellproblem: Grobgitter- und Glättungsoperatoren
Diplomarbeit
Bonn 1985 .

[DOE86] Joachim Dörfer
Mehrgitterverfahren bei singulären Störungen
Diplomarbeit
Düsseldorf 1986 .

[HTR85] W. Hackbusch, U. Trottenberg (Eds.)
Multigrid Methods II
Lecture Notes in Mathematics 1228
Proceedings, Köln 1985.

[MIL80] J. J. H. Miller (Ed.)
Boundary and Interior Layers — Computational and Asymptotic Methods
Boole Press
Proceedings, Dublin 1980.

[HAC85] Wolfgang Hackbusch
Multi-Grid Methods and Applications
Springer Verlag
Berlin, Heidelberg 1985.

[HTR81] W. Hackbusch, U. Trottenberg (Eds.)
Multigrid Methods
Lecture Notes in Mathematics 960
Proceedings, Köln-Porz 1981.

The Frequency Decomposition Multi-Grid Algorithm

Wolfgang Hackbusch
Praktische Mathematik
Christian-Albrechts-Universität zu Kiel
Olshausenstr. 40
D-2300 Kiel, FRG

Summary

Multi-grid methods are known as very fast solvers of a large class of discretised partial differential equations. However, the multi-grid method cannot be understood as a fixed algorithm. Usually, the components of the multi-grid iteration have to be adapted to the given problem and sometimes the problems are modified into those acceptable for multi-grid methods. In particular, the smoothing iteration is the most delicate part of the multi-grid process.

An iteration is called a *robust* one, if it works for a sufficiently large class of problems. Attempts have been made to construct robust multi-grid iterations by means of sophisticated smoothing processes (cf. Wesseling [6], [3, p.222]). In particular, robust methods should be able to solve singular perturbation problems. Examples of such problems are the anisotropic equations of the next subsection, the convection-diffusion equation and others (cf. Hackbusch [3,§10]).

To overcome the problem of robustness we propose a new multi-grid variant. It is called the *frequency decomposition multi-grid algorithm* since different parts of the frequency spectrum are treated by different respective coarse-grid corrections. This explains that we need more than one coarse grid and further prolongations and restrictions. It is to be emphasized that the smoothing procedure may be very simple (e.g. the Gauß-Seidel iteration). Nevertheless, we claim that the resulting multi-grid algorithm is suited not only to the anisotropic equations described below but also for many other singular perturbation problems. We describe the application to the anisotropic problem since this simplifies the choice of the prolongations and restrictions. Other applications possibly require the matrix-dependent prolongation (cf. Hackbusch [3,§10.3])

1. The Anisotropic Equation and Standard Multi-Grid Methods

1.1 The Anisotropic Equation

We consider the boundary value problem

(1.1.1a) $$-\alpha \frac{\partial^2}{\partial x^2} u(x,y) - \beta \frac{\partial^2}{\partial y^2} u(x,y) + \tfrac{1}{4} u(x,y) = f(x,y) \quad \text{in } \Omega=(-1,1)\times(-1,1),$$

(1.1.1b) u periodic, i.e. $u(-1,y)=u(1,y),\ u(x,-1)=u(x,1)$ for $x,y\in(-1,1)$.

The only restictions on α and β are

(1.1.2) $\alpha, \beta \geqslant 0.$

Under condition (1.1.2) the boundary value problem (1.1.1a,b) is positive definite. The absolute term $u(x,y)$ is added to avoid a singularity for $\alpha=\beta=0$. Eq. (1.1.1a) represents a singular perturbation problem in the cases (1.1.3a) or (1.1.3b):

(1.1.3a) $\alpha \to 0,\ \beta \geqslant 1,$

(1.1.3b) $\beta \to 0,\ \alpha \geqslant 1.$

Note that in the limiting cases $\alpha=0$ and $\beta=0$ Eq (1.1.1a) is no more elliptic.

For the discretisation we choose a step size

$$h = 1 / N \qquad (N \in \mathbb{N})$$

The standard discretisation with respect to this grid size yields the difference star

$$(1.1.4) \qquad L_h = h^{-2} \begin{bmatrix} & -\beta & \\ -\alpha & h^2 + 2\alpha + 2\beta & -\alpha \\ & -\beta & \end{bmatrix}.$$

There is a practical reason why a solver should be tested with the anisotropic equation

$$(1.1.5) \qquad L_h u = f.$$

Even when the *isotropic* Poisson equation is to be solved, an *anisotropic* discrete equation may result from a rectangular grid with different widths in the x- and y-directions or from a (non-conform) body-fitted grid.

1.2 Standard Multi-Grid Method

A standard multi-grid method would use the following components: the nine-point prolongation and the nine-point restriction (1.2.1), the Galerkin coarse-grid matrix (1.2.2) and $\nu = 2$ smoothing steps by the red-black Gauß-Seidel iteration:

$$(1.2.1) \qquad p = \frac{1}{4} \begin{bmatrix} 1 & 2 & 1 \\ 2 & 4 & 2 \\ 1 & 2 & 1 \end{bmatrix}, \qquad r = p^* = \frac{1}{16} \begin{bmatrix} 1 & 2 & 1 \\ 2 & 4 & 2 \\ 1 & 2 & 1 \end{bmatrix},$$

$$(1.2.2) \qquad L_{\ell-1} = r\, L_\ell\, p,$$

where $L_\ell = L_h$ and $L_{\ell-1} = L_{2h}$ if we set $h_\ell = h$ and $h_{\ell-1} = 2h$. For the notation (1.2.1) we refer to [3]. From the same reference [p.31] we take the description of the standard multi-grid algorithm:

```
        procedure MGM(ℓ,u,f); integer ℓ;   array u,f;
        if ℓ=0 then u:= L₀⁻¹f else
        begin integer j; array v,d;
(1.2.3)   u := 𝒮ℓ^ν(u,f); d:=r(Lℓu-f); v:=0;
          for j:=1 step 1 until γ do MGM(ℓ-1,v,d);
          u:=u-pv
        end;
```

$\mathcal{S}_\ell^\nu(u,f)$ is the result of ν steps of the smoothing procedure $\mathcal{S}_\ell$ (e.g. the Gauß-Seidel iteration) applied to Eq. (1.1.5) with starting iterate u. The V-cycle corresponds to $\gamma = 1$, the W-cycle to $\gamma = 2$.

Typical results of the multi-grid algorithm (1.2.3) with $\nu = 2$, $\gamma = 2$ and red-black Gauß-Seidel applied to the anisotropic scheme (1.1.4) ($h_0 = 2/1, \ldots, h_4 = 2/16$) are shown in Table 1.2.1:

Table 1.2.1 Averaged convergence rate of multi-grid with *point-wise* red-black Gauß-Seidel smoothing

α	1	1	1	1	1	1	0.5	0.01
β	1	0.7	0.5	0.1	0.01	0.0001	1	1
rate	0.05	0.09	0.15	0.53	0.78	0.82	0.15	0.78

While the unperturbed problem $\alpha = \beta = 1$ leads to fast convergence, ratios α/β smaller or larger than 1 yield unsatisfactory results. The reason is the "wrong" smoother. According to Criterion 10.1.1 from [3] one has to apply a smoother which is fast or exact for the limiting cases $\alpha=0$ or $\beta=0$. For $\alpha=0$ the difference star (1.1.4) reduces to a 3-point scheme in x-direction and the x-line Gauß-Seidel iteration is an exact solver. Indeed, the multi-grid iteration with x-line Gauß-Seidel yields good results whenever $\alpha \geqslant \beta$ as shown in Table 1.2.2.

Table 1.2.2 Averaged convergence rate of multi-grid with x-line (block-wise) Gauß-Seidel smoothing

α	1	1	1	1	1	1	0.5	0.001
β	1	0.7	0.5	0.1	0.01	0.0001	1	1
rate	0.05	0.027	0.03	0.025	0.024	0.0	0.15	0.76

1.3 Attempts towards Robustness

The results of Table 1.2.2 show that a x-line Gauß-Seidel smoother leads to a robust multi-grid in the sense that all anisotropic problems with $\alpha \geqslant \beta$ can be solved with uniform (more precisely: uniformly bounded) convergence speed. To include the case $\alpha \leqslant \beta$ one has to add also the y-line Gauß-Seidel iteration. The resulting smoother is the alternating line Gauß-Seidel smoother (cf. Stüben-Trottenberg [5]). Since its robustness is limited (cf. §10.1.2 of [3]), other smoothing iterations have been constructed by means of incomplete (line-)LU decompositions (cf. Wesseling [6], §10.1.3 of [3], Wittum [7]). Although the ILU smoothers proved to be robust also with respect to other classes of problems (e.g. to convection diffusion equations) there is **no** extension of these methods to the three-dimensional multi-grid applications.

2. Fourier Analysis

The term "smoothing" means that "high frequencies" are damped by the iteration $\mathcal{S}_\ell$. A precise analysis can be given by Fourier analysis. For practical reasons, the Fourier analysis can be performed only for the two-grid version of algorithm (1.2.3). Because of the periodic boundary conditions the complex eigenfunctions $e^{\nu\mu}(x,y)=\frac{1}{2}e^{\pi i(\nu x+\mu y)}$ are considered for all frequencies $1-N\leqslant\nu,\mu\leqslant N$ with $N=N_\ell:=1/h_\ell$. For details we refer to §2.4 and §8.1.2 of [3].

In Fig 2.1 we represent one quarter of the frequencies. Region I contains the low frequencies in both directions, region II consists of frequencies low w.r.t. y and high w.r.t. x, etc. The coarse-grid correction is restricted to region I: Error components with at least one oscillatory direction cannot be corrected by the uniformly coarsened grid. These components must be reduced by the smoothing process. But e.g. the pointwise Gauß-Seidel iteration does not (efficiently) work in region II if $\alpha<<\beta$. An typical error which is neither reduced by the coarse-grid correction nor by the pointwise Gauß-Seidel process in this situation is the function depicted in Fig 2.2.

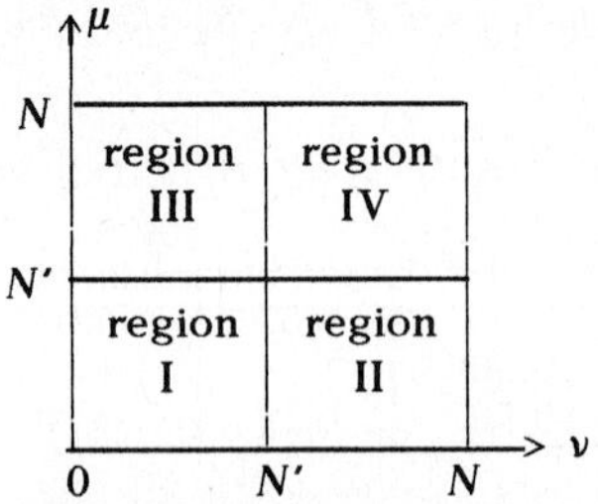

Fig 2.1 Spectrum decomposition into 4 different regions

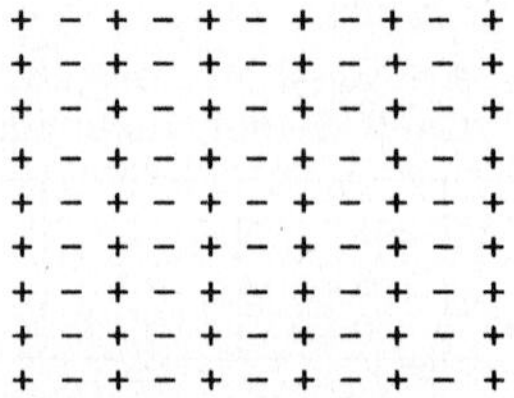

Fig 2.2 Example of an oscillatory function in x-direction

3. The Frequency Decomposition Multi-Grid Method

3.1 Construction of the Coarse Grids

Fig 3.1.1 shows a standard coarse grid of grid size $h_{\ell-1} = 2h_\ell$. Shifting this grid by h_ℓ into x-direction or y-direction or both directions, we obtain three further grids denoted in Fig.3.1.2-4. The shift vector is h_ℓ times (0,0), (1,0), (0,1), (1,1). Therefore these indices are used to denote the four different grids: $\Omega_{00}^{\ell-1}, \Omega_{10}^{\ell-1}, \Omega_{01}^{\ell-1}, \Omega_{11}^{\ell-1}$. The set of these four indices is denoted by

$$I = \{ (0,0), (0,1), (1,0), (1,1) \}.$$

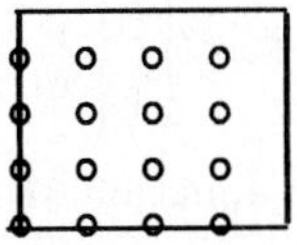

Fig. 3.1.1 $\Omega_{00}^{\ell-1}$

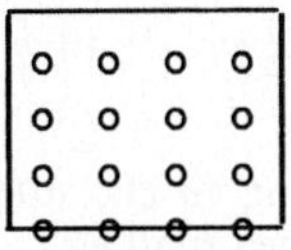

Fig. 3.1.2 $\Omega_{10}^{\ell-1}$

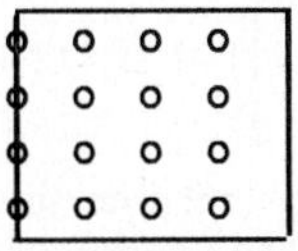

Fig. 3.1.3 $\Omega_{01}^{\ell-1}$

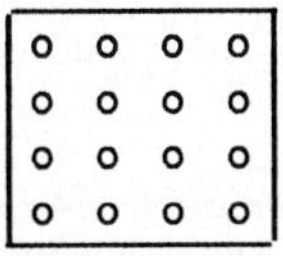

Fig. 3.1.4 $\Omega_{11}^{\ell-1}$

Remark 3.1.1 The fine grid Ω^ℓ is the union of the coarse grids $\Omega_\iota^{\ell-1}$, $\iota \in I$.

3.2 Prolongations and Restrictions

Each grid is associated with a prolongation

$$p_\iota : \Omega_\iota^{\ell-1} \rightarrow \Omega^\ell \qquad (\iota \in I).$$

For $\iota = (0,0)$ p_ι is the standard one, for the other indices p_ι represents no interpolation:

(3.2.1a,b) $$p_{00} = \frac{1}{4}\begin{bmatrix} 1 & 2 & 1 \\ 2 & 4 & 2 \\ 1 & 2 & 1 \end{bmatrix}, \qquad p_{10} = \frac{1}{4}\begin{bmatrix} -1 & 2 & -1 \\ -2 & 4 & -2 \\ -1 & 2 & -1 \end{bmatrix},$$

(3.2.1c,d) $$p_{01} = \frac{1}{4}\begin{bmatrix} -1 & -2 & -1 \\ 2 & 4 & 2 \\ -1 & -2 & -1 \end{bmatrix}, \qquad p_{11} = \frac{1}{4}\begin{bmatrix} +1 & -2 & +1 \\ -2 & 4 & -2 \\ +1 & -2 & +1 \end{bmatrix}$$

(cf. (1.2.1)). The range of the prolongations p_ι, $\iota \neq (0,0)$, contains also high frequencies. E.g. the application of p_{10} to a constant coarse-grid function yields the oscillatory function depicted in Fig. 2.2.

Remark 3.2.1 The span of range(p_ι) for all $\iota \in I$ is the space of all fine-grid functions.

Remark 3.2.2 For different indices $\iota, \varkappa \in I$ range(p_ι) and range$(p_\varkappa)$ are orthogonal.

As in the standard case (cf. (1.2.1)) the restriction is defined as the adjoint of the prolongation:

(3.2.2) $$r_\iota = p_\iota^* \qquad (\iota \in I).$$

Remark 3.2.2 implies

Remark 3.2.3 $r_\iota p_\varkappa = 0$ holds for different indices $\iota, \varkappa \in I$.

3.3 Coarse-Grid Matrices

In each coarse grid $\Omega_\iota^{\ell-1}$ we will have to solve a coarse-grid equation with a coarse-grid matrix $L_\iota^{\ell-1}$. As in (1.2.2) they are defined by the Galerkin product

(3.3.1) $$L_\iota^{\ell-1} = r_\iota L^\ell p_\iota \qquad (\iota \in I)$$

from the fine-grid matrix L^ℓ. Because of (3.2.2), positive definiteness of L^ℓ implies positive definiteness of all coarse-grid matrices (3.3.1).

The matrix L^ℓ of the anisotropic difference scheme (1.1.4) yields the standard coarse-grid matrix

(3.3.2a) $$L_{00}^{\ell-1} = \alpha h_{\ell-1}^{-2} \frac{1}{8} \begin{bmatrix} -1 & 2 & -1 \\ -6 & 12 & -6 \\ -1 & 2 & -1 \end{bmatrix} + \beta h_{\ell-1}^{-2} \frac{1}{8} \begin{bmatrix} -1 & -6 & -1 \\ 2 & 12 & 2 \\ -1 & -6 & -1 \end{bmatrix} + E, \quad E = \frac{1}{64} \begin{bmatrix} 1 & 6 & 1 \\ 6 & 36 & 6 \\ 1 & 6 & 1 \end{bmatrix}.$$

The terms become more transparent when they are written as (tensor) products of one-dimensional difference schemes:

(3.3.2a') $$L_{00}^{\ell-1} = \alpha h_{\ell-1}^{-2} \frac{1}{8} \begin{bmatrix} 1 \\ 6 \\ 1 \end{bmatrix} [-1\ 2\ -1] + \beta h_{\ell-1}^{-2} \frac{1}{8} [1\ 6\ 1] \begin{bmatrix} -1 \\ 2 \\ -1 \end{bmatrix} + E, \quad E = \frac{1}{8} \begin{bmatrix} 1 \\ 6 \\ 1 \end{bmatrix} \frac{1}{8} [1\ 6\ 1]$$

Obviously, this difference scheme is consistent to the differential equation (1.1.1a). For the other indices $\iota \in I$ the coarse-grid matrices read as

(3.3.2b) $$L_{10}^{\ell-1} = \alpha h_{\ell-1}^{-2} \frac{1}{8} \begin{bmatrix} 1 \\ 6 \\ 1 \end{bmatrix} [3\ 10\ 3] + \beta h_{\ell-1}^{-2} \frac{1}{8} [1\ 6\ 1] \begin{bmatrix} -1 \\ 2 \\ -1 \end{bmatrix} + E,$$

(3.3.2c) $$L_{01}^{\ell-1} = \alpha h_{\ell-1}^{-2} \frac{1}{8} \begin{bmatrix} 1 \\ 6 \\ 1 \end{bmatrix} [-1\ 2\ -1] + \beta h_{\ell-1}^{-2} \frac{1}{8} [1\ 6\ 1] \begin{bmatrix} 3 \\ 10 \\ 3 \end{bmatrix} + E,$$

(3.3.2d) $$L_{11}^{\ell-1} = \alpha h_{\ell-1}^{-2} \frac{1}{8} \begin{bmatrix} 1 \\ 6 \\ 1 \end{bmatrix} [3\ 10\ 3] + \beta h_{\ell-1}^{-2} \frac{1}{8} [1\ 6\ 1] \begin{bmatrix} 3 \\ 10 \\ 3 \end{bmatrix} + E,$$

The first term in (3.3.2b), the second in (3.3.2c), and both in (3.3.2d) are no difference operators at all. This proves

Remark 3.3.1 The schemes $L_\iota^{\ell-1}$ for $\iota \neq (0,0)$ are not consistent to the differential problem.

The Gerschgorin theorem shows that e.g. the eigenvalues of $[3\ 10\ 3]$ are between 4 and 16. Estimating in this manner the eigenvalues of L_ι one obtains

Remark 3.3.2 The spectral condition numbers of the schemes (3.3.2b-d) can be estimated by

(3.3.3a) $$\mathrm{cond}(L_{10}^{\ell-1}) \leq [16\alpha + 4\beta + h_{\ell-1}^2]/[8\alpha + h_{\ell-1}^2],$$

(3.3.3b) $$\mathrm{cond}(L_{01}^{\ell-1}) \leq [16\beta + 4\alpha + h_{\ell-1}^2]/[8\beta + h_{\ell-1}^2],$$

(3.3.3c) $$\mathrm{cond}(L_{11}^{\ell-1}) \leq [16(\alpha+\beta) + h_{\ell-1}^2]/[2(\alpha+\beta) + h_{\ell-1}^2].$$

Hence, $\mathrm{cond}(L_{11}^{\ell-1}) = O(1)$ for all values of α and β, while $\mathrm{cond}(L_{10}^{\ell-1}) = O(h_{\ell-1}^2)$ may happen for $\alpha = O(h_{\ell-1}^2)$ and $\beta = O(1)$. However, if α and β are of order $O(1)$, all condition numbers are $O(1)$.

3.4 The Multiple Coarse-Grid Correction

Using all coarse grids we are lead to the multiple coarse-grid correction

(3.4.1) $$u_\ell \mapsto u_\ell - \sum_{\iota \in I_0} p_\iota (L_\iota^{\ell-1})^{-1} r_\iota (L^\ell u_\ell - f_\ell),$$

where I_0 is a subset of I. At least I_0 has to contain the index $(0,0)$. If $(0,0)$ is the only index in I_0, (3.4.1) represents the standard coarse-grid correction as used in (1.2.3). The purpose of the additional terms in (3.4.1) is to correct also oscillatory errors from the regions II to IV of Fig.2.1. More precisely, the index $\iota=(1,0)$ corresponds to region II, $\iota=(0,1)$ to III, and $\iota=(1,1)$ to IV.

3.5 Smoothing Process

In order to demonstrate that the robustness is a consequence of the multiple coarse-grid correction (3.4.1) and not of a suitably chosen smoothing iteration we choose the pointwise damped Jacobi iteration or the red-black (pointwise) Gauß-Seidel iteration as smoother. Of course, the choice of more sophisticated smoothing processes can only improve the convergence.

3.6 The Frequency Decomposition Two-Grid Iteration

The FD two-grid method consists of pre-smoothing (ν steps) according to §3.5 followed by the new coarse-grid correction (3.4.1). The two-grid method depends on the choice of I_0. For $I_0=\{(0,0)\}$ one obtains the standard two-grid method.

The FD two-grid method can be analysed by the Fourier method. The resulting rates of convergence (spectral radii) are given in Table 3.6.1. The first column of numbers belonging to $I_0=\{(0,0)\}$ contains the two-grid rates of the *standard* two-grid method. Adding the second correction $(1,0)$ obviously does not improve the convergence rate, while the convergence is much better when the third correction $(0,1)$ is contained in the index set I_0 (last three columns). We have shown the results only for $\alpha=1$. By symmetry we get the same results for interchanged values of α and β, when we also interchange the indices $(1,0)$ and $(0,1)$.

Table 3.6.1 Exact two-grid rates in the case of two red-black Gauß–Seidel iterations as smoother ($h=1/16$)

α	β	I_0				
		$\{(0,0)\}$	$\{(0,0),(1,0)\}$	$\{(0,0),(0,1)\}$	$\{(0,0),(1,0),(0,1)\}$	I
1	1	0.062	0.062	0.062	0.039	0.083
1	0.7	0.119	0.119	0.060	0.060	0.098
1	0.5	0.197	0.197	0.088	0.089	0.121
1	0.2	0.479	0.479	0.175	0.178	0.198
1	0.1	0.678	0.678	0.232	0.228	0.241
1	0.05	0.817	0.817	0.260	0.253	0.261
1	0.01	0.954	0.954	0.239	0.232	0.236
1	0.001	0.988	0.988	0.097	0.096	0.096
1	$1_{10}-4$	0.992	0.993	0.015	0.049	0.049
1	$1_{10}-5$	0.992	0.993	0.002	0.049	0.049

3.7 Necessary Coarse-Grid Corrections

The second column in Table 3.6.1 corresponding to $I_0=\{\ (0,0),\ (1,0)\ \}$ indicates that the use of the additional coarse-grid correction w.r.t. $\iota=(1,0)$ yields no improvement. Therefore $\iota=(1,0)$ is called an *unnecessary coarse-grid correction.* Uniform convergence is achieved when the correction $\iota=(0,1)\in I_0$ is added. Hence, we call $\iota=(0,1)$ a *necessary coarse-grid correction*. In all of the last three columns the index $\iota=(0,1)$ is contained in I_0 and ensures uniform convergence for all $\beta\leqslant\alpha$.

In the reverse case of $\alpha\leqslant\beta$ the roles are interchanged: $(1,0)$ is necessary, while $(0,1)$ is without any effect. However, it is important to note that the unnecessary corrections do not deteriorate the process. In the case of $I_0=\{(0,0),(1,0),(0,1)\}$ or $I_0=I$ the necessary correction is present, whether $\alpha\geqslant\beta$ or $\beta\geqslant\alpha$. Accordingly, the convergence is uniform for *all* values of α and β. $\iota=(1,1)$ is always unnecessary for the model problem (1.1.1), but it can be become necessary for other equations.

In Remark 3.3.1 L_ι for $\iota\neq(0,0)$ was called inconsistent. However, for $\alpha<<\beta$ the α-term in (1.1.4b) can be neglected and $L_{10}^{\ell-1}$ almost coincides with the consistent discretisation $L_{00}^{\ell-1}$.

Remark 3.7.1 In the same measure as the scheme L_ι becomes a consistent one, its condition number increases to $O(h_{\ell-1}^{-2})$ and the coarse-grid correction in (3.4.1) w.r.t. the index ι becomes necessary.

3.8 The Frequency Decomposition Multi-Grid Iteration

The FD multi-grid algorithm is obtained from the two-grid algorithm by replacing the exact solution in the coarse-grid correction (3.4.1) by the recursive application of the same method. In the following program the parameter ℓ from algorithm (1.2.3) is eliminated by the matrix $L=L_\ell$. The termination of the recursive process is usually expressed by $\ell=0$. Here we use $\dim(L)\le N_0$, where $\dim(A)$ is the number of rows and columns in the matrix A. The following program performs one step of the FD multi-grid iteration for solving $Lu=f$.

```
          procedure FDMGM(L,u,f); array L,u,f;
          if dim(L)≤N0 then u:= L^-1 f else
          begin integer j; array v,d;
(3.8.1)       u:= S^ν(L,u,f);   d:=Lu-f;
              for ι∈I0 do
              begin v:=0;
                    for j:=1 step 1 until γ do FDMGM(rι L pι, v, rι d);
                    u:=u-pι v
          end end;
```

$\mathcal{S}^\nu(L,u,f)$ is the result of ν steps of the smoothing procedure $\mathcal{S}$ applied to $Lv=f$ with starting value u. The loop "for $\iota\in I_0$ do ..." can be performed in parallel. The product $r_\iota L p_\iota$ in the recursive call of $FDMGM$ is assumed to be computed in a prephase once for all.

The algorithm (3.8.1) performs more coarse-grid corrections than necessary. One may try to restrict the loop "for $\iota\in I_0$ do ..." to those indices which are necessary. This would require special criteria selecting necessary and unnecessary coarse-grid corrections (cf. Section 3.7). In particular there is the following *a priori* criterion. Let $L=L^\ell$. At level $\ell-2$ matrices $L^{\ell-2}_{\iota\varkappa}:=r_\varkappa r_\iota L^\ell p_\iota p_\varkappa$ arise. They have a condition number $O(1)$ if $\varkappa$ and ι are different and also different from $(0,0)$. Therefore, matrices $L^{\ell-1}_\iota$ with $\iota\ne(0,0)$ involve only coarse-grid corrections w.r.t. $(0,0)$ and ι at the lower levels. The corresponding algorithm is given by the following program $FD2$, which has an additional parameter $\iota\in I$.

```
(3.8.2a)  procedure FD2(L,ι,u,f); array L,u,f;
(3.8.2b)  if dim(L)≤N0 then u:= L^-1 f else
(3.8.2c)  begin integer j; array v,d; set J;
(3.8.2d)      u:= S^ν(L,u,f);   d:=Lu-f;
(3.8.2e)      J:=if ι=(0,0) then I0 else {(0,0),ι};
(3.8.2f)      for ϰ∈J do
(3.8.2g)      begin v:=0;
(3.8.2h)            for j:=1 step 1 until γ do
(3.8.2i)                FD2(rϰ L pϰ, if ϰ=(0,0) then ι else ϰ, v, rϰ d);
(3.8.2j)            u:=u-pϰ v
(3.8.2k)  end end;
```

The original problem (1.1.5) has to be solved by $FD2(L^\ell,(0,0),u,f_\ell)$. The other parameters ι appear only in (3.8.2i). The meaning of ι is as follows. The products (3.3.1) generate a tree of matrices. $L=L^\ell$ produces L_ι $(\iota\in I)$ at level $\ell-1$, $L_{\iota\varkappa}$ $(\iota,\varkappa\in I)$ at level $\ell-2$ and so on. The general matrix at level $\ell-k$ is $L=L_{\iota_1\iota_2\ldots\iota_k}$. If all indices ι_j coincide with $(0,0)$, the second parameter in $FD2(L,\iota,u,d)$ is $\iota=(0,0)$. A parameter $\varkappa\ne(0,0)$ in $FD2(L,\varkappa,u,d)$ indicates that the set $\{\iota_1,\iota_2,\ldots,\iota_k\}$ equals $\{(0,0),\varkappa\}$. Any other case is avoided by the choice of J in (3.8.2e).

The number "2" in the name $FD2$ is due to the fact that algorithm (3.8.2) is adapted to the two-dimensional case. In three dimensions the matrix $L=L_{\iota_1\iota_2\ldots\iota_k}$ is unnecessary if the indices $\iota_1,\iota_2,\ldots,\iota_k$ contain three elements $\iota,\varkappa,\lambda$ such that $(0,0)$ and $\iota,\varkappa,\lambda$ are pairwise different. Therefore, the procedure needs two parameters $\iota,\varkappa$

instead of only one in (3.8.2a).

```
            three-dimensional variant of FD2:
(3.8.3a)    procedure FD3 (L,ι,λ,u,f); array L,u,f;
(3.8.3b)    if dim(L)≤N_0 then u:= L^{-1}f else
(3.8.3c)    begin integer j; array v,d; set J;
(3.8.3d)        u:= 𝒮^ν(L,u,f); d:=Lu-f;
(3.8.3e)        J:=if λ=(0,0) then I_0 else {(0,0),ι,λ};
(3.8.3f)        for κ∈J do
(3.8.3g)        begin v:=0;
(3.8.3h)            for j:=1 step 1 until γ do
(3.8.3i)              FD3(r_κ L p_κ, if ι=(0,0) then κ else ι,
                          if κ∈{(0,0),ι} then λ else κ, v, r_κ d);
(3.8.3j)              u:=u-p_κ v
(3.8.3k)    end end;
```

4. Computational Complexity

First we assume that the FD multi-grid algorithm is performed with coarse-grid corrections from I_0 with *fixed* I_0. We use the following notations:

- N_ℓ: numbers of unknowns at level ℓ.
- C_S: the operations per smoothing step at level ℓ is $\leq C_S N_\ell$.
- C_D: the computation $(u_\ell, f_\ell) \mapsto d_\ell := L_\ell u_\ell - f_\ell$ requires $\leq C_D N_\ell$ operations.
- C_r: the computation $d_\ell \mapsto r_\iota d_\ell$ for one $\iota \in I$ requires $\leq C_r N_\ell$ operations.
- C_p: the computation $v_\ell \mapsto p_\iota v_\ell$ for one $\iota \in I$ requires $\leq C_p N_\ell$ operations.
- $C_\ell = C_\ell(\nu, I_0)$: one performance of the FD multi-grid iteration at level ℓ requires $\leq C_\ell N_\ell$ operations.

The work at level $\ell = 0$ is neglected ($C_0 := 0$), since its contribution is of minor importance. The standard two-dimensional case for algorithm (3.8.1) is covered by

Remark 4.1 Let $N_{\ell-1} \leq N_\ell / 4$. If $\gamma = 1$ (V-cycle) and $\# I_0 < 4$ in algorithm (3.8.1), the number C_ℓ is uniformly bounded:

(4.1a) $$C_\ell \leq 4\,[\nu C_S + C_D + \# I_0 (C_r + C_p)] \,/\, [4 - \# I_0].$$

If $\gamma = 1$ (V-cycle) and $\# I_0 = 4$ or $\gamma = 2$ (W-cycle) and $I_0 = 2$, C_ℓ behaves logarithmically:

(4.1b) $$C_\ell \leq [\nu C_S + C_D + \# I_0 (C_r + C_p)] \log(N_\ell / N_0) \,/\, \log 4.$$

Proof. The estimate $C_\ell N_\ell \leq \nu C_S N + C_D N_\ell + \# I_0 (C_r N_\ell + C_p N_\ell + \gamma C_{\ell-1} N_{\ell-1})$ yields $C_\ell \leq C_1 + \gamma \# I_0 C_{\ell-1}$ with $C_1 := \nu C_S + C_D + \# I_0 (C_r + C_p)$. Hence $C_\ell \leq C_1 \sum_{k=0}^{\ell-1} [(\gamma \# I_0)/4]^k$. In the first case the right-hand side in (4.1a) is the infinite geometric sum. In the second case the sum equals $\ell C_1 \leq C_1 \log(N_\ell / N_0) / \log 4$. QED

The statement in three dimensions is similar: One has to replace $\# I_0 < 4, = 4, = 2$ by $\# I_0 < 8, = 8, = 4$. In the case (4.1a) the FD multi-grid method is still of optimal order $O(N_\ell)$. The case (4.1b) is slightly worse, but still fully acceptable. The situation improves, when algorithm (3.8.2) is used.

Remark 4.2 Let $N_{\ell-1} \leq N_\ell / 4$. The operation count for algorithm (3.8.2) with $\iota = (0,0)$ is

(4.2a) $$C_\ell \leq \tfrac{2}{3}(1 + \# I_0)(\nu C_S + C_D) + \tfrac{4}{3}(2 \# I_0 - 1)(C_r + C_p) \qquad \text{for } \gamma = 1 \text{ (V-cycle)},$$

(4.2b) $$C_\ell \leq 2\,[\nu C_S + C_D + \# I_0 (C_r + C_p) + \tfrac{1}{2}(\# I_0 - 1)(\nu C_S + C_D + 2(C_r + C_p)) \log(N_\ell / N_0) / \log 4] \qquad \text{for } \gamma = 2 \text{ (W-cycle)}.$$

Hence, the V-cycle is an $O(N_\ell)$-algorithm, while the W-cycle is still $O(N_\ell \log N_\ell)$, independent of the choice of I_0.

Proof. Let C_ℓ be the operation count for procedure (3.8.2) with $\iota = (0,0)$ and C_ℓ' the corresponding number for $\iota \neq (0,0)$. From the first of the two recursive inequalities

(4.3a) $C'_\ell \le \nu C_S + C_D + 2(C_r + C_p + C'_{\ell-1}/4)$,

(4.3b) $C_\ell \le \nu C_S + C_D + \# I_0 (C_r + C_p) + \frac{\gamma}{4}[C_{\ell-1} + (\# I_0 - 1)\, C'_{\ell-1}]$.

one concludes

$$C'_\ell \le 2[\nu C_S + C_D + 2C_r + 2C_p] \qquad \text{for } \gamma = 1,$$

$$C'_\ell \le [\nu C_S + C_D + 2C_r + 2C_p] \log(N_\ell / N_0) / \log 4 \qquad \text{for } \gamma = 2.$$

Inserting these inequalities into (4.3b) one obtains a recursive inequality for C_ℓ yielding (4.2a,b). QED

The result for three dimensions is still better: The V- *and* W-cycle are an $O(N_\ell)$-algorithm.

Remark 4.3 Let $N_{\ell-1} \le N_\ell/8$ for the 3D-algorithm (3.8.3). The number C_ℓ is bounded by a constant.

Proof. For the dominant case of $\iota_1 \ne \iota_2$ different from $(0,0)$, the estimate (4.3a) becomes $C'_\ell \le \text{const} + 3\gamma C'_{\ell-1}/8$. Since $3\gamma/8 < 1$ for $\gamma = 1, 2$, the result follows. QED

The coarse-grid corrections in (3.4.1) for different $\iota \in I_0$ are completely independent. Therefore they can be computed in parallel. The resulting operation count for a parallel computer is identical with the complexity of a standard multi-grid algorithm ($\# I_0 = 1$) on a usual computer. Frederickson and McBryan [1] avoided idle processors by using the multiple coarse-grid correction (3.4.1). However, in their approach all restrictions and prolongations are close to the standard ones.

5. Numerical Results of the FD Multi-Grid Method

The results in the following Table 5.1 should indicate that the observed multi-grid rates are close to the convergence radii determined for the two-grid method. Here the V-cycle is applied with a step size $h = 1/16$. The numbers are the mean values obtained from 10 iterations. For the W-cycle the standard analysis could be applied to prove multi-grid convergence from two-grid convergence (cf. Hackbusch [3, p 161]). However the analysis of the V-cycle is much more difficult.

Table 5.1 Multi-grid convergence speed for the V-cycle algorithm *FDMGM* with red-black Gauß-Seidel smoothing ($\nu = 2$) for $h = 1/16$

α	β	I_0			
		{(0,0)}	{(0,0),(1,0)}	{(0,0),(0,1)}	{(0,0),(1,0),(0,1)}
1	1	0.073	0.086	0.088	0.073
1	0.1	0.643	0.643	0.271	0.257
1	0.001	0.752	0.757	0.138	0.122
1	1_{10^-5}	0.767	0.767	0.078	0.100

Acknowledgement. The author thanks Mr. H. Gerull (Kiel) for preparing the numerical data for Tables 3.6.1 and 5.1.

References

[1] Frederickson, P.O. and O.A. McBryan: Parallel superconvergent multigrid. Techn. Report 7/87, Cornell University, 1987

[2] Hackbusch, W. A new approach to robust multi-grid methods. Proceedings, ICIAM Paris, June 1987 (to be published by SIAM). Also in Bericht 8708, Christian-Albrechts-Universität Kiel, July 1987

[3] Hackbusch, W. *Multi-Grid Methods and Applications*. Springer Heidelberg, 1985

[4] Hackbusch, W. and U. Trottenberg (eds.): Multi-Grid Methods, Proceedings. Lecture Notes in Mathematics 960. Springer Berlin-Heidelberg, 1982

[5] Stüben, K. and U. Trottenberg: Multi-grid methods: fundamental algorithms, model problem analysis and applications. In [4] 1-176

[6] Wesseling, P.: Theoretical and practical aspects of a multigrid method. SIAM J Sci Statist Comput 3 (1982) 387-407

[7] Wittum, G.: On the robustness of ILU-smoothing. In these proceedings.

ON GLOBAL MULTIGRID CONVERGENCE FOR NONLINEAR PROBLEMS

W. Hackbusch[1] and A. Reusken[2]

[1]Institut für Informatik und Praktische Mathematik, Christian-Albrechts-Universität, Olshausenstr. 40, D-2300 Kiel, Germany

[2]Mathematical Institute, University of Utrecht, Budapestlaan 6, Utrecht, The Netherlands

SUMMARY

We present a modification of the Nonlinear Multigrid Method. This new algorithm converges globally for nonlinear problems that result from finite element discretization of some class of nonlinear boundary value problems.

1. INTRODUCTION

There are different strategies for solving a (large) system of nonlinear equations using multigrid techniques. Well-known are the following methods (for a suitable class of problems) :

(1) Newton iteration combined with a linear MG method for solving the Jacobian system occurring in the Newton iteration. In general this method converges only locally, and if most of the work is needed for solving the Jacobian system (so computing the Jacobian is not too much work) then this method is characterized by "linear-MG efficiency" locally.

(2) Damped Newton iteration combined with a linear MG method. The difference with (1) is that this method converges (more) globally.

(3) Nonlinear Multigrid Method (or FAS). In general this method converges locally and then has "linear-MG efficiency". In this algorithm computation of the Jacobian is not needed.

With respect to (1) we note that there is an extensive literature both on (approximate) Newton methods and on linear MG solvers. Concerning (2) we mention the work of Bank and Rose in [1]. For an introduction into the Nonlinear Multigrid Method and for references concerning this algorithm we refer to [2]. In [5,6] one can find quantitative convergence statements for this method.

In this paper we introduce the Damped Nonlinear Multigrid Method (DNMGM). We will present the method and give some numerical results. The method converges globally (for the class of problems that we describe in §2) and has "linear-MG efficiency" locally. Proofs of these statements and a profound analysis of the method will be published elsewhere (see [3]).

In the last section of this paper we present a simple modification of the classical linear MG algorithm. This modified iteration is just the DNMGM applied to a linear problem. It turns out that this new algorithm has nice (unexpected) convergence properties compared to the classical MG method.

2. THE PROBLEM CONSIDERED AND ITS DISCRETIZATION

In this section we define a class of nonlinear problems and their Ritz-Galerkin discretizations. We start with a general setting, but for ease we then consider a more concrete situation.

Let H be a real Hilbert space with dual space H'. The inner product on H is denoted by $(\cdot,\cdot)$ and the corresponding norm by $\|\cdot\|$.
Let $n:H\to H'$ be a (non)linear operator that satisfies the following conditions
2.1 - n is Fréchet-differentiable on H ($Dn(u)$ bounded linear operator $H\to H'$)
- $\forall\, u\in H : (v,w)\to Dn(u)(v)(w)$ defines a symmetric bilinear form on $H\times H$
- $\exists\, \gamma>0 : \forall u,v\in H : Dn(u)(v)(v) \geq \gamma\|v\|^2$
- for every bounded subset A of H there is a constant Γ_A such that for all $v,w\in A : \| Dn(v) - Dn(w)\| \leq \Gamma_A \| v - w \|$.

Remark 2.2. Using 2.1 and Theorem 4.1(5) in [4] one easily verifies that n is a gradient operator: there is a $\varphi:H\to\mathbb{R}$ such that $D\varphi = n$ (F-derivative !). Also from 4.3(12) in [4] it follows that $n:H\to H'$ is a bijection. Moreover, n is a homeomorphism (use that $(n(u) - n(v))(u - v) \geq \gamma\| u - v\|^2$).

Continuous problem. In the remainder of this paper we assume some fixed $f\in H'$ and we are interested in (computable) approximations of u^*, with u^* such that $n(u^*)(v) - f(v) = 0$ for all $v\in H$.

In [5] it is proved that the conditions 2.1 are satisfied in the situation:
$H = H_0^1(\Omega)$ (Sobolev space; Ω bounded domain in $\mathbb{R}^2$ with smooth boundary)
$n(u)(v) = \int_\Omega a\nabla u\cdot\nabla v\,dx + \int_\Omega b(g\circ u)v\,dx$, with a,b smooth functions on Ω, $a>0$, $b\geq 0$, $g\in C^1(\mathbb{R})$ such that n maps $H_0^1(\Omega)$ into $H_0^1(\Omega)'$, $g'\geq 0$ and g' satisfies a "suitable" (cf. [5]) Lipschitz condition.
Now, for ease, in the remainder we take $H = H_0^1(\Omega)$ (with $(u,v) = \int_\Omega \nabla u\cdot\nabla v\,dx$) and we assume that n corresponds to the variational formulation of some boundary value problem.

Discretization. (cf. 3.1 and 3.6 in [2]). Let $(S_k)_{k\geq 0}$ be a sequence of standard finite element spaces. Let U_k be the space of coefficient vectors corresponding to S_k. On U_k we use a scaled euclidean inner product denoted by $\langle\cdot,\cdot\rangle$ with corresponding norm $\|\cdot\|_2$. The natural isomorphism $U_k \to S_k$ is denoted by P_k. Standard Ritz-Galerkin discretization now yields discrete problems $N_k(u_k^*) = f_k$ (k=0,1,...) where $N_k : U_k \to U_k$ is given by $N_k = P_k^* n P_k$ and $f_k = P_k^* f$ (adjoint w.r.t. appropriate inner products, see [2]).
We use natural prolongations $p=p_k : U_{k-1} \to U_k$ and restrictions $r=r_k=p_k^*$ such that $N_{k-1}=rN_k p$.

Remark 2.3. Due to 2.1, $N_k : U_k \to U_k$ is a homeomorphism. Also for $k \in \mathbb{N}$ and $m \in U_k$ there is a functional $\varphi_k^m : U_k \to \mathbb{R}$ such that $D\varphi_k^m(u)(w) = \langle N_k(u) - m , w\rangle$.

3. DAMPED NONLINEAR MULTIGRID METHOD

In this section we give a modification of the classical Nonlinear Multigrid Method. This modification converges globally for the class of problems that we considered in §2 (see §4).

Relaxation. We use nonlinear Jacobi-kind of relaxations. Let $A_k : U_k \to U_k$ be the symmetric positive definite operator satisfying $\langle A_k u , v\rangle = (P_k u , P_k v)$ for all $u,v \in U_k$ (A_k is just the standard Poisson discretization).
Let W_k be some symmetric positive definite operator that satisfies $W_k \geq A_k$ (e.g. $W_k = \alpha_k I$ with suitable α_k). We consider relaxations of the kind:

$$R_k(u,m) = u - \beta_{u,m} W_k^{-1}(N_k(u) - m).$$

Here $\beta_{u,m} \in \mathbb{R}$ is some positive "steplenght" parameter which should be such that $\varphi_k^m(R_k(u,m)) < \varphi_k^m(u)$.
Using Taylor expansion it is clear that this is equivalent with the condition $\beta_{u,m} \int_0^1 \langle DN_k(u - t\beta_{u,m} W_k^{-1} d) W_k^{-1} d , W_k^{-1} d\rangle \, dt < 2\langle d , W_k^{-1} d\rangle$ where $d := N_k(u) - m$.
One possibility for $\beta_{u,m}$ is the following:
if $\alpha := \max\{ \| W_k^{-\frac{1}{2}} DN_k(v) W_k^{-\frac{1}{2}} \|_2 \mid v \in U_k , \| W_k(u-v)\|_2 \leq 2\|d\|_2 \}$ then

$\beta_{u,m} = \varepsilon \min\{1 , 1/\alpha\}$ for any $\varepsilon \in]0,2[$ would suffice.
We note that in the classical NMGM we also need a quantity like α to get a "reasonable" relaxation operator.

Damping parameter. (cf. also remark 3.1). In the algorithm below we need a damping parameter $\psi(k,m,u,v)$ which is defined as follows ($k \in \mathbb{N}$, $m,u \in U_k$, $v \in U_{k-1}$) :

$$\psi(k,m,u,v) = \min\{ 2 , \frac{< r(N_k(u) - m) , v>}{< DN_k(u)\, pv , pv> + E(k,u,v)} \}$$

where E(k,u,v) should be such that

$$| < (\int_0^1 DN_k(u-2t\, pv)\, dt - DN_k(u))\, pv , pv > | \leq E(k,u,v) .$$

As an example :

$E(k,u,v) = \Gamma_A \| P_{k-1} v \|^3$ with $A = \{ w \in S_k \mid \| P_k u - w \| \leq 2 \| P_{k-1} v \| \}$.

<u>The DNMGM</u>. Let there be given:

- $k \in \mathbb{N}$, $u,m \in U_k$
- on the 0-level a solver Φ (e.g. exact or some iterations of R_0)
- $\sigma_l \in]0,\infty[$ for $l=1,..,k$, $\tau \in \mathbb{N}$, and $\nu_1,\nu_2 \in \mathbb{N}$
- $\hat{u}_l \in U_l$ and $\hat{f}_l := N_l(\hat{u}_l)$ for $l=0,..,k-1$.

Now we define the following

```
PROCEDURE DNMGM(k,u,m); INTEGER k; ARRAY u,m;
IF k=0 THEN u:=Φ(u,m) ELSE
BEGIN INTEGER i; ARRAY dd,d,v; REAL ψ;
   u:=R_k^ν1(u,m);
   d:=r*(N_k(u) - m);
   dd:=f̂_k-1 + σ_k*d ;
   v:=û_k-1; FOR i:=1 (1) τ DO DNMGM(k-1,v,dd);
   v:=(v - û_k-1)/σ_k ;
   ψ:=ψ(k,m,u,v) ;
   u:=u - ψ*p*v ;
   u:=R_k^ν2(u,m)
END ;
```

<u>Remark 3.1</u>. With respect to the computation of $\psi(k,m,u,v)$ we note the following :

- $\|v\|_2$ will often be small (v is a correction vector)
- $d=r(N_k(u) - m)$ and v are already computed, so $< r(N_k(u) - m) , v>$ reduces to computing an inner product in U_{k-1}
- the vector pv should be computed anyway
- if it is known a priori how to evaluate (cheaply) the linear operator

$v \to r\,DN_k(u)\,pv$, then direct computation of $\langle DN_k(u)\,pv\,,\,pv\rangle =$ $= \langle r\,DN_k(u)\,pv\,,\,v\rangle$ is possible. Otherwise one might use the difference approximation $DN_k(u)\,pv \approx (N_k(u+t\,pv) - N_k(u))/t$ for suitable t; then only $N_k(u+t\,pv)$ need to be computed (because $N_k(u)$ has already been computed).

- E(k,u,v) can be neglected if

(1) we have a weak nonlinearity (Γ_A small), or

(2) $\|P_{k-1}v\|$ small enough (note that $\langle DN_k(u)\,pv\,,\,pv\rangle = O(\|P_{k-1}v\|^2)$, $E(k,u,v) = O(\|P_{k-1}v\|^3)$.

4. GLOBAL CONVERGENCE AND LOCAL EFFICIENCY

In this section we give a theorem about global convergence of the DNMGM and we give the main idea of the proof. We also shortly discuss the local efficiency of our algorithm. Proofs and further details can be found in [3].

Theorem 4.1. (global convergence)

Take $k \in \mathbb{N}$ and assume that there are given $\tau \geq 0$, ν_1 , ν_2 with $\nu_1 + \nu_2 \geq 1$, $\sigma_1 \in]0,\infty[$ and $\hat{u}_{1-1} \in U_{1-1}$ for 1=1,..,k. For the DNMGM with this choice of the parameters the following holds:

for all $m \in U_k$ and all $u \in U_k$: $\lim_{i\to\infty} \|DNMGM^i(k,u,m) - N_k^{-1}(m)\|_2 = 0$.

Idea of the proof. Solving $N_k(u_k^*) = m$ corresponds to minimizing an (energy) functional $\varphi_k^m : U_k \to \mathbb{R}$ (cf. 2.3).

For the relaxation we have:

$$\varphi_k^m(R_k(u,m)) < \varphi_k^m(u)$$

(such a descent also holds for nonlinear Gauss relaxation !).

For the coarse grid correction $u \to u - \psi\,pv$ we can prove that pv is a descent direction and besides our choice of ψ gives a suitable steplength.

This results in:

$$\varphi_k^m(u - \psi\,pv) \leq \varphi_k^m(u) \quad .$$

So we conclude that the DNMGM is a descent iteration with respect to the natural functional φ_k^m. Using this the global convergence can be proved.

Remark. It is evident that Theorem 4.1 expresses a strong robustness of the DNMGM: (due to the damping parameter ψ) for every $m \in U_k$ the algorithm corresponding to $N_k(u_k)=m$ will converge from any starting vector and with an arbitrary choice of the parameters τ, $(\hat{u}_l)_{0\leq l\leq k-1}$, $(\sigma_l)_{1\leq l\leq k}$, ν_1 and ν_2 (provided $\nu_1+\nu_2\geq 1$).

Local efficiency. To be able to prove the usual MG efficiency we have to make additional assumptions that are correlated to the well-known "Approximation property" and "Smoothing property". Then we can prove that "locally" (in some ball with centre the continuous solution u^*) the DNMGM, with "suitably" chosen parameters (e.g. σ_l small enough, $\hat{u}_l$ close to u_k^*, $\tau=2$, $\nu_1=\nu_2=1$) , has a contraction number smaller than one which is independent of the level. The following two aspects are important for the proof. First, in [5,6] we proved local quantitative convergence statements for the classical NMGM. Secondly, in the two grid situation $\psi\to 1$ if $\Gamma_B\,\mathrm{diam}(B)\downarrow 0$, where $B\subset H$ is some ball which contains the starting vector $P_k u$, the discrete solution $P_k u_k^*$ and the continuous solution u^*; so "locally" the DNMGM is just a perturbation of the NMGM.

5. NUMERICAL RESULTS

We consider the following problem with solution $u^*=0$:

$$\begin{cases} -0.1\,\Delta u + \arctan(10u) = 0 & \text{in } \Omega =]0,1[\times]0,1[\\ u = 0 & \text{on } \partial\Omega . \end{cases}$$

We use finite element dicretization with piecewise linear functions on triangles with sides h_k, h_k, $\sqrt{2}\,h_k$; $h_0=\frac{1}{2}$, $h_{k+1}=\frac{1}{2}h_k$.

We use Richardson relaxation :

$$R_k(u,f) = u - \beta_k(N_k(u) - f) \quad \text{with} \quad \beta_k = h_k^2(0.1(4+\cos(\pi h_k)) + 10\,h_k^2)^{-1} .$$

Furthermore: standard seven point prolongation and restriction

finest grid corresponding to $h_3 = 1/16$

coarsest grid corresponding to $h_0 = \frac{1}{2}$ (one point)

$\nu_1=\nu_2=1$, $\tau=2$ (W-cycle) , $\sigma_l=1$

starting vector u_α: trivial injection into U_3 of the function $(x,y)\to\alpha\, e^x \sin(\pi x)\, y\,(1-y)^2$

$\hat{u}_l$: trivial injection of the starting vector into U_l.

We give results for two situations : starting vector "close" to the solution ($\alpha=0.1$) and starting vector "far" from the solution ($\alpha=1$). In both cases we take the smooth starting vector u_α because the behaviour of the coarse grid correction is illustrated clearly by how it reduces a smooth error.

Below we give the ratios of the iteration errors for the first 8 iterations of the classical NMGM ($\psi=1$) and the DNMGM ($\psi \in]0,2[$). For these computations we took into account the factor E(k,u,v) (cf. §3). However the results do not really change if E(k,u,v) is neglected (due to the fact that $\| P_{k-1}v \|$ is small, cf. 3.1).

(1) : $\alpha=0.1$; then $\| u_\alpha \|_2 = 1.18\,e{-2}$

ratios of the iteration errors (w.r.t. $\|\cdot\|_2$)									
	NMGM	0.03	0.15	0.20	0.22	0.24	0.25	0.26	0.27
	DNMGM	0.04	0.09	0.19	0.22	0.24	0.25	0.26	0.27

Remarks:- behaviour for NMGM and DNMGM is about the same
- ratios, after starting effects, of about 0.25 agree with what is expected from linear MG (using damped Jacobi for Poisson equation)

(2) : $\alpha=1$ ($\| u_\alpha \|_2 = 1.18\,e{-1}$)

ratios of the iteration errors (w.r.t. $\|\cdot\|_2$)									
	NMGM	0.63	0.12	1.10	1.03	1.21	1.00	1.22	0.96
	DNMGM	0.32	0.32	0.18	0.20	0.31	0.30	0.32	0.31

Remarks:- the NMGM fails to converge !
- for DNMGM the ratios of about 0.30 are satisfactory ; in this case the coarse grid approximations $\hat{u}_1$ are worse than in (1), this is why we get 0.30 instead of 0.25 .

6. WHAT HAPPENS IN THE LINEAR CASE ?

We give some easy modification of the classical <u>linear</u> MG algorithm. This modified algorithm is just the DNMGM applied to some symmetric positive definite linear problem. For this new algorithm we observed, in the V-cycle case, much faster convergence than we expected. To explain this is subject of current research.

Consider $L_k u = f_k$ (k=0,1,...) where $L_k : U_k \to U_k$ is some positive definite symmetric linear operator (e.g. standard Poisson discretization).

We now give the following modification of the classical MGM which we call: Multigrid Method with Natural Correction (MGNC ; cf. 6.1).

```
PROCEDURE MGNC(k,u,f); INTEGER k; ARRAY u,f;
IF k=0 THEN u:=L_0^{-1}f  ELSE
BEGIN INTEGER i; ARRAY v,d; REAL ψ;
   u:=R_k^ν(u,f);
   d:=r*(L_k*u - f);
   v:=0 ;  FOR i:=1 (1) τ DO MGNC(k-1,v,d);
   ψ:=<d , v>/< L_{k-1}v , v>;
   u:=u - ψ*p*v ;
   u:=R_k^ν(u,f)
END;
```

Interpretation. After relaxation one wants to find the solution $v^* \in U_{k-1}$ of $L_{k-1}v = d$. Another equivalent formulation is: find $v^* \in U_{k-1}$ such that $\varphi_{k-1}(v^*) = \min\{ \varphi_{k-1}(v) \mid v \in U_{k-1} \}$ with $\varphi_{k-1}(v) := \frac{1}{2}\langle L_{k-1}v , v\rangle - \langle d , v\rangle$. After τ correction cycles we end up with an approximation v of v^*. It seems natural to minimize φ_{k-1} on the line $\alpha \to \alpha v$ ($\alpha \in \mathbb{R}$). This leads to : $\min\{ \varphi_{k-1}(\alpha v) \mid \alpha \in \mathbb{R} \} = \varphi_{k-1}(\psi v)$ with $\psi = \langle d , v\rangle / \langle L_{k-1}v , v\rangle$.

Remarks:- ψ is very cheaply computable
- in TG situation : $\psi=1$
- in MG situation one can prove : $\psi \geq 1$.

Numerical results. As an example we take the standard five-point discretization of the Poisson equation :

$$\begin{cases} \Delta u(x,y) = 4 & \text{in } \Omega =]0,1[\times]0,1[\\ u(x,y) = x^2+y^2 & \text{on } \partial\Omega \end{cases} .$$

With:- finest grid corresponding to $h_4 = 1/32$
- coarsest grid corresponding to $h_0 = \frac{1}{2}$
- one Gauss-Seidel checker-board pre- and post-smoothing
- standard nine-point prolongation and restriction
- starting vector $u = 0$.

Below we give the ratios of the iteration errors for the first 7 iterations of the classical MGM ($\psi=1$) and for MGNC ($\psi \geq 1$) (the results for the MGM are comparable to the results in §4.4.1 of [2]) .

ratios of the iteration errors (w.r.t. $\|\cdot\|_2$)	MGM V-cycle	0.110	0.109	0.112	0.114	0.116	0.117	0.118
	MGM W-cycle	0.031	0.049	0.055	0.057	0.059	0.061	0.062
	MGNC V-cycle	0.035	0.047	0.050	0.053	0.054	0.055	0.056

These numerical results and other results with Jacobi smoothing, seven-point prolongation and restriction and other starting vectors (all for the Poisson equation) suggest the following :
the error reduction for MGNC using V-cycles is about the same as for the classical MGM with W-cycles.

REFERENCES

1 BANK, R.E., ROSE, D.J.: Analysis of a multilevel iterative method for nonlinear finite element equations. Math. of Comp. 39 (1982).

2 HACKBUSCH, W.: Multi-Grid Methods and Applications. Springer-Verlag, Berlin 1985.

3 HACKBUSCH, W., REUSKEN, A.: Global convergence and local efficiency of a multigrid method for nonlinear problems. Preprint University of Utrecht, in preparation.

4 JEGGLE, H.: Nichtlineare Funktionalanalysis. Teudner, Stuttgart 1979.

5 REUSKEN, A.: Convergence of the Multigrid Full Approximation Scheme for a class of elliptic mildly nonlinear boundary value problems. To appear in Numerische Mathematik (1988).

6 REUSKEN, A.: Convergence of the multilevel Full Approximation Scheme including the V-cycle. Preprint University of Utrecht (1987). Submitted to Numerische Mathematik.

Multigrid Methods for the Solution of the Compressible Navier-Stokes Equations

D. Hänel, W. Schröder, G. Seider
Aerodynamisches Institut, RWTH Aachen, West-Germany

1. Introduction

The numerical solution of the Navier-Stokes equations requires a large amount of computational work, much more than a corresponding inviscid solution does. The reason for this is the more complex structure of the Navier-Stokes equations, and mainly the existence of different characteristic length scales which have to be resolved. In the past years a large number of so-called high resolution schemes for the solution of the Euler equations have been developed and later adjusted to the Navier-Stokes equations e. g. [1, 2, 3]. In general they result in an improved accuracy and efficiency. Nevertheless, for complex viscous flow problems further concepts have to be investigated to improve the convergence. One promising concept is the multigrid method of which the fundamentals were formulated by Brandt [4]. This idea was taken up by many authors for the solution of the Euler and Navier-Stokes equations as well. Examples for multigrid applications in explicit and implicit Euler solutions are given by Cima, Johnson [5], Ni [6], Jameson [7] and by Hemker[8], Mulder[9] and others. Navier-Stokes applications can be found e. g. in the paper of Shaw, Wesseling [10], Thomas et al [11] and Schröder, Hänel [12].

The present paper continues the study of multigrid solutions of the time-dependent Navier-Stokes equations as presented in [12, 13]. The method of solution is based on a flux-splitting upwind scheme for the Euler part and central discretization for the viscous part. The discrete problem is solved by two methods, an implicit relaxation method and an explicit time-stepping Runge-Kutta method. The Full Approximation Storage (FAS) multigrid concept [14] is applied to both methods but in different ways. The direct multigrid method was used for the explicit scheme to advance the solution in space and time, whereas the indirect method was used to accelerate the matrix inversion of the implicit method. In the following sections the basic methods of solution will be explained and the implementation of the multigrid methods discussed. The influence of the multigrid methods on the convergence is presented for different viscous flow problems.

2. Governing Equations and Spatial Discretization

The governing equations are the time-dependent Navier-Stokes equations for a compressible fluid. For the sake of brevity, the equations and the methods of solution will be described in the following for 2-D Cartesian coordinates only. Applications in general, curvilinear meshes are given e. g. in [12, 15]. Then, the equations in conservative, Cartesian form read

$$U_t + \left(F - \frac{1}{Re} S\right)_x + \left(G - \frac{1}{Re} T\right)_y = 0 . \qquad (1)$$

Herin U is the vector of the conservative variables, F, G are the Euler fluxes and S, T represent the viscous terms where Re is the Reynolds number.

In the present consideration the gas is assumed to be perfect, and the flow to be laminar.

To preserve the conservative properties in the discretized space, the equations (1) are applied to a finite control volume $\Delta x \cdot \Delta y$. In general a node-centered arrangement of the control volume (Fig. 1a) was used, but for comparison the cell-centered grid (Fig. 1b) considered as well. Both arrangements are widely used.

The conservative discretization requires the evaluation of the fluxes at the cell interfaces $(i \pm \frac{1}{2}, j)$ and $(i, j \pm \frac{1}{2})$ of the defined control volume around a grid point (i,j). This results in a set of difference equations approximating the Navier-Stokes equations (1)

$$\frac{\Delta U}{\Delta t} + \mathrm{Res}\,(U) = 0 \tag{2}$$

where $\Delta U/\Delta t$ is the discrete time derivative, defined later, and Res (U) coresponds to the discretized steady-state operator:

$$\mathrm{Res}\,(U)_{i,j} = \delta_x (F - \frac{1}{Re} S) + \delta_y (G - \frac{1}{Re} T) \tag{2a}$$

with $\delta_x f = \dfrac{f_{i+1/2,j} - f_{i-1/2,j}}{\Delta x}$; $\delta_y f = \dfrac{f_{i,j+1/2} - f_{i,j-1/2}}{\Delta y}$.

The essential properties of the method are determined by the definition of the fluxes at the cell-interfaces. In this paper the Euler fluxes are approximated by upwind differences and the viscous terms by central differences. The discretization will be explained briefly in the following for one component.

To apply upwind differencing the Euler fluxes are split according to the sign of the eigenvalues by van Leer's splitting concept [16]. This results in

$$F = F^+ + F^- \tag{3}$$

with $F^\pm$ defined in [16]. Then, the split fluxes at the cell interfaces are given by, e. g.

$$F^\pm_{i+1/2} = F^\pm(U^\pm_{i+1/2}) \, . \tag{4}$$

The backward or forward extrapolated variables $U^+_{i+\frac{1}{2}}$ and $U^-_{i+\frac{1}{2}}$ are calculated from

$$U^+_{i+1/2} = U_i + \frac{1}{4}\varphi_i \left[(1+\kappa)\Delta U^+_i + (1-\kappa)\Delta U^-_i\right] \tag{5a}$$

$$U^-_{i+1/2} = U_{i+1} - \frac{1}{4}\varphi_{i+1} \left[(1+\kappa)\Delta U^-_{i+1} + (1-\kappa)\Delta U^+_{i+1}\right] \tag{5b}$$

with $\Delta U^+_i = U_{i+1} - U_i \qquad \Delta U^-_i = U_i - U_{i-1}$.

For $\varphi = 0$ the scheme is first order accurate and for $\varphi = 1$ higher order accurate. The parameter $\kappa = 1$ yields a central scheme $0(\Delta x^2)$, $\kappa = \frac{1}{3}$ upwind-biased schemes of $0(\Delta x^2)$ and $0(\Delta x^3)$ and $\kappa = -1$ a fully upwind scheme of $0(\Delta x^2)$. In general, the value $\kappa = 0$ is used here. To achieve an almost monotonic solution behaviour, φ is substituted by a flux limiter $\varphi_i = \varphi_i(\Delta U^+_i, \Delta U^-_i)$. The flux limiter by van Albada et al. was applied. [17]

The viscous terms $S_{i+\frac{1}{2}}$ which have the form $S = \hat{\mu} U_x$, are formulated centrally by

$$S_{i+1/2} = \hat{\mu}_{i+1/2} \frac{U_{i+1} - U_i}{\Delta x} \, . \tag{6}$$

The other terms of equation (2) are treated accordingly.

3. Methods of Solution

Two different solution concepts are considered for multigrid applications.The first method is an unfactored implicit scheme with an iterative matrix inversion (relaxation) each time step. The second method applied to the Navier-Stokes equations is an explicit time stepping Runge-Kutta scheme similiar to that in [19]. Both methods make use of the space descretization described before.

3. 1 Implicit Relaxation Method

Starting from the discretized Navier-Stokes equations (2) the implicit backward Euler formulation is used for the time derivative. After the time linearization of the fluxes the difference equation reads

$$\left[\frac{1}{\Delta t} + \delta_x A + \delta_y B - \frac{1}{Re}(\delta_x C + \delta_y D)\right] \Delta U^n = -Res(U^n) \qquad (7)$$

with $\Delta U^n = U^{n+1} - U^n$

where the superscript n denotes the time level, and A, B, C, D are the Jacobian of the corresponding fluxes F, G, S and T ,e. g.

$$A = \frac{\partial F}{\partial U} .$$

According to the flux splitting concept the Jacobian of the Euler terms split up in two terms, like

$$\delta_x A = \delta_x A^+ + \delta_x A^-$$

and are treated in the same way as the fluxes, using the MUSCL approach. For steady-state calculations, where time consistency is not required, the solution matrix of equation (7) is simplified using first order upwinding only. Thus the matrix to be inverted consists of tridiagonal block systems in each direction.

Even for this case the implicit scheme (7) requires the inversion of a large system of difference equations. Since the solution matrix is diagonally dominant resulting from the upwind differences, an iterative relaxation procedure is employed to the unfactored scheme allowing large time steps. Then the iterative procedure from the time level n to n+1 reads

$$\left[\frac{1}{\Delta t} + \delta_x A + \delta_y B - \frac{1}{Re}(\delta_x C + \delta_y D)\right] \Delta U^\nu = -Res\ U^n \qquad (8)$$

where the superscript v is the iteration index and $\Delta U^\nu = U^{n+1,\nu} - U^n$. The iteration of equation (8) is performed by either a collective point or line Gauß-Seidel relaxation in alternating directions. The iterative procerdure is stopped if $\max |\Delta U^{\nu+1} - \Delta U^\nu| \leq \varepsilon$ where ε is a small number. For more details see [12].

3.2 Runge-Kutta Time-Stepping Scheme

The second method which has been applied to the Navier-Stokes equations is the explicit Runge-Kutta method. Considering the equation (2) as a semi-discrete approximation of the time-dependent Navier-Stokes equations, the time discretization can be carried out as a sequence of intermediate steps in the sense of the classical Runge-Kutta method. At present a version of the Runge-Kutta method is used which has been successfully applied to Euler and Navier-Stokes equations by many investigators e. g. [7, 18, 19]. In contrast to the present work, central differencing with artificial damping was used in these papers.

For a N-step Runge-Kutta method the scheme for equation (2) reads

$$\begin{aligned} U^{(0)} &= U^n \\ &\vdots \\ \Delta U^{(l)} &= -\alpha_l \Delta t \,\mathrm{Res}(U^{(l-1)}) \\ U^l &= U^{(0)} + \Delta U^{(l)} \\ &\vdots \\ U^{n+1} &= U^N . \end{aligned} \quad \left.\right\} l=1, \dots, N \tag{9}$$

Here, a five-step scheme (N=5) was used and adapted especially to the upwind scheme with the coefficients α_l= .059, .14, .275, .5, 1.0 and a maximum CFL number of 3.5 .The scheme is second order accurate in time.

To accelerate the convergence to the steady-state solution local time steps were used which are dictated by the local stability limit and constant Courant number.The local time stepping allows a faster signal propagation, and thus faster convergence. A second acceleration technique considered here is the implicit residual smoothing [18], for which a Runge-Kutta step of Eq (9) reads:

$$(1-\varepsilon\delta_{xx})(1-\varepsilon\delta_{yy})\,\Delta U^{(l)} = -\alpha_l \Delta t\, \mathrm{Res}(U^{(l+1)}) \tag{10}$$

with smoothing coefficients ε of 0(1),the CFL number can be increased by a factor of two or three.

4. Multigrid Formulations for the Implicit and Explicit Scheme

In the present paper the multigrid methods will be discussed for two basic solution concepts, an explicit and an implicit scheme. The explicit scheme requires less computational work per time step, that, however, becomes small due to the stability restriction, whereas the implicit scheme allows large time steps but with a larger amount of computational work per time step. Both concepts are widely used for the solution of the Navier-Stokes equations and show no significant difference in the large computational effort to achieve a sufficiently resolved Navier-Stokes solution. Therefore, for both concepts the multigrid principle was considered as a tool to reduce the amount of computational work.

Because of the different structures of the implicit and the explicit scheme the multigrid method is employed in different ways. In the present implicit relaxation method the multigrid method is used to accelerate the iterative matrix inversion each time step. Thus it is part of the relaxation procedure for each time step, and therefore called indirect multigrid method. For the explicit scheme the multigrid concept is applied in space and time, as well. It directly influences the solution in time and therefore is called direct method. A common requirement for the application of both multigrid formulations is the property of smoothing out the high frequency error components by the scheme. This requirement is satisfied in principle by the use of an upwind scheme. The common basic multigrid concept for both methods is the Full Approximation Storage (FAS) [14]which is employed in the present study and will be described in the next section.

4. 1 Full Approximation Storage Multigrid Concept

For accelerating schemes Brandt [14] has proposed the Full Approximation Storage (FAS) multigrid concept. This concept is suited to nonlinear equations and therefore adopted in the present solution of the Navier-Stokes equations. Consider a grid sequence G_k, $k=1,..,m$ with the step sizes $h_k=2h_{k+1}$ etc. A finite difference approximation on the finest grid G_m may be

$$L_m U_m = 0 \tag{11}$$

where $L_m U_m$ corresponds to the discretized conservation equation (2). If the solution on the fine grid is sufficiently smooth with respect to the high frequency solution components, equation (12) may be approximated on a coarser grid G_{k-1} by a modified difference approximation

$$L_{k-1} U_{k-1} = \tau^m_{k-1} \tag{12}$$

where τ is the "fine to coarse defect correction" and refered to the "discretization error" in the following. It maintains the truncation error of the fine grid G_m on the coarser grids G_{k-1} and is defined by

$$\tau^m_{k-1} = \tau^m_k + L_{k-1}(I^{k-1}_k U_k) - \mathrm{II}^{k-1}_k (L_k U_k) \tag{13}$$

where I and II are restriction operators from grid G_k to G_{k-1} which are applied to the variable U_k and the difference approximation $L_k U_k$, as well. These operators can be used as injection or full weighting operators and will be explained later.

On the coarse grid G_{k-1} equation (12) is solved and the transfer (13) is repeated for the next coarser grid until the coarsest grid is reached. After some solution steps on the coarsest grid the solution is interpolated back to the finer grids with some solution steps in between.

According to the FAS scheme only the correction between the "old" fine grid and the "new" coarse grid solution is transfered to the fine grid e. g.:

$$U_{k+1} = U_{k+1} + I^{k+1}_k (U_k - I^k_{k+1} U_{k+1}). \tag{14}$$

Herein I is the interpolation operator, where so far only bilinear interpolation is used, other interpolations like quadratic or characteristic interpolation have been tested, but have shown no advantage. The described multigrid procedure corresponds to the so-called V-cycle and is used in the present paper.

A further important factor of influence on the multigrid convergence is the treatment of the boundary approximation on the coarser grids. Consider a boundary approximation on the finest grid G_m:

$$C_m U_m + g_m = 0 \tag{15}$$

which may be e. g. a Neumann boundary condition. If eq. (16) is applied directly to the coarser grid, a large truncation error would occur. To avoid this, a coarse grid correction can be defined similar to and consistent with the FAS procedure (13). Then the boundary approximation on the coarse grid G_{k-1} reads:

$$C_{k-1} U_{k-1} + g_{k-1} = \tau_{B\,k-1}^{\ m} \tag{16}$$

where

$$\tau_{B\,k-1}^{\ m} = \tau_{B\,k}^{\ m} + C_{k-1}(I^{k-1}_k U_k) + g_{k-1} - I^{k-1}_k (C_k U_k + g_k). \tag{17}$$

However, this formulation increases the programming complexity and makes the multigrid procedure rely on the (often) changing flow problems. Therefore, as long as the multigrid method is considered only as an acceleration procedure for the fine grid solution, simplifications of the coarse grid boundary conditions can be used with advantage. One approach is to use frozen boundary conditions on the coarse grid while updating them only on the fine grid. This is a reasonable simplification, e. g. for a node-centered scheme, where the nodes (variables) are defined on the boundary itself. In schemes, like a cell-centered scheme [18], where the boundary is located between two nodes, the coarse grid boundary conditions can be used as for the fine grid (without correction), but then, only their change on the coarser grids is transfered to the fine grid using the FAS interpolation. Both boundary approaches are used presently.

4. 2 Indirect Multigrid Method

The indirect multigrid method is connected with an implicit scheme with the aim to reduce the large computational work of the matrix inversion each time step. Similar concepts for the Euler equations were used e. g. in [8, 9].In the present paper this concept is applied to the iterative matrix inversion of the previously described upwind relaxation method. By use of the upwind discretization the scheme becomes sufficiently dissipative and the iterative procedure corresponds to a solution of a discrete quasi-elliptic system, which guarantees an efficient use of the multigrid method. The relaxation scheme (8) formulated on the finest grid G_m reads

$$LHS_m^n \Delta U_m^\nu + Res_m(U_m^n) = 0 \tag{18}$$

where LHS^n is the left hand side operator matrix of eq. (8) as a function of U^n, and where ΔU^ν is the correction variable to be calculated by the iterative procedure. On a coarser grid G_{k-1} the relaxation procedure is approximated by

$$LHS_{k-1}^n \Delta U_{k-1}^\nu + Res_{k-1}(U_{k-1}^n) = \tau_{k-1}^m \tag{19}$$

with the discretization error

$$\tau_{k-1}^m = \tau_k^m + LHS_{k-1}^n (I_k^{k-1} \Delta U_k^\nu) + Res_{k-1}(I_k^{k-1} U_k^n) - II_k^{k-1}[LHS_k^n \Delta U_k^\nu + Res_k U_k^n]. \tag{19a}$$

After some relaxation sweeps on every grid level including the coarsest one, the correction ΔU^ν(not $U^{n+1,\nu}$) is interpolated back to the finer grids according to Eq (15)

$$\Delta U_{k+1}^\nu = \Delta U_{k+1}^\nu + I_k^{k+1}(\Delta U_k^\nu - I_{k+1}^k \Delta U_{k+1}^\nu) \tag{20}$$

,followed by one relaxation sweep on each grid to smooth the interpolation errors. The V-cycle is completed after the finest grid is reached and in general the next time step is carried out after one or two V-cycles.
The present scheme was developed for a node-centered mesh and therefore the simple point-to-point injection could be used for the restriction I of ΔU^ν. The volume-weighted restriction II over the nine neighbouring fine grid cells is employed for the difference operator and the residual. Bilinear interpolation is used in Eq. (20).

The boundary conditions were employed explicitly with respect to the iterative level on the finest grid only and remain unchanged on the coarser grid. The present method extended to 2-D, curvilinear grids is described by the authors in [12]. The method was applied and tested for different flow ranges. In the following the convergence behaviour of the multigrid method is demonstrated for three typical solutions of the Navier-Stokes equations. First the subsonic boundary layer flow was computed. The domain of integration and the corresponding boundary conditions are sketched in Fig. 2a. The history of convergence for three different fine meshes are given in Fig. 2b to 2d for the single gird (SG) and multigrid (MG) method. Always based on the same coarsest grid, two-, three-, and four-grid methods have been built up for the same integration domain. In general the rate of convergence of the single grid method decreases with increasing number of grid points, whereas for the multigrid method the rate is nearly independent of the number of grid points. In the second example the viscous supersonic flow past a wedge was computed. The results in Fig. 3 a to 3d are arranged in the same way as for the previous example. Again the rate of convergence of the multigrid method remains nearly independent of the grid point number.

As a third example the flow over a NACA 0012 airfoil was computed for subsonic (Fig. 4) and supersonic inflow (Fig. 5). The Fig. 4a and Fig. 5a show the lines of constant Mach numbers for each flow case and Fig. 4b and Fig. 5b give the history of convergence. Both cases show a reliable convergence to the steady-state, where the supersonic case is slightly decelerated until the captured shock has reached its final position. The computations were carried out on a 169x49 C-mesh and three grid levels were used for the multigrid method. Using the Gauß-Seidel line relaxation in alternating directions the residual reduction factor, averaged over ten time steps, was .386 for the subsonic case and .719 for the supersonic case, which confirms the good convergence properties of the indirect multigrid method.

4.3 Direct Multigrid Method

For this method the multigrid procedure is employed directly on the time-dependent solution. In contrast to the indirect method the time advances also on the coarser grids. Applied to an explicit scheme the advantage of multigrid is twofold: first, computational time is saved because of the smaller number of grid points on the coarser grids and secondly, the time steps restricted by the numerical stability can be chosen larger on the coarser grids. In principle the direct method can be used for an implicit method as well.

In the present paper the direct method was applied to the explicit Runge-Kutta time stepping scheme. In a similar manner this was already done by Jameson [7] but for a centrally discretized steady-state operator. Here, the upwind scheme, the same as used in the implicit relaxation method, is employed for the steady-state operator. The use of an upwind scheme changes the frequency properties of the Runge-Kutta method and in consequence the multigrid behaviour. Fig. 6a and 6b demonstrate the different frequency properties of a central and an upwind scheme by means of the amplification factor versus wave angle for a scalar convection equation. Herein a 5-step Runge-Kutta scheme was adapted for a maximum Courant number which is 4.0 for the central and 3.5 for the upwind scheme. The upwind scheme shows the lower Courant number but the better high frequency damping, which is of advantage for the multigrid treatment.

The present paper concerns with steady-state solutions and therefore the FAS multigrid procedure was employed only to the steady-state residual. For a time-consistent multigrid treatment an additional coarse grid correction in time must be added. Corresponding investigations are in progress. For steady-state calculations the coarse grid correction on a grid G_{K-1} reads for one Runge-Kutta step Eq. (9):

$$U_{k-1}^{(l)} = U_{k-1}^{(0)} - \alpha_l \Delta t_{k-1} (Res_{k-1} U_{k-1}^{(l-1)} - \tau_{k-1}^{m}) \tag{21}$$

where the discretization error is given by

$$\tau_{k-1}^{m} = \tau_{k}^{m} + Res_{k-1} I_k^{k-1} U_k - \mathrm{II}_k^{k-1} Res_k U_k \,. \tag{21a}$$

The transfer from coarse to fine grids was carried out by bilinear interpolation according to Eq. (14).

The choice of suitable formulations for the boundary conditions on the coarse grids and of the restriction operators depends on the mesh arrangement. Here, two widely used mesh arrangements, the cell-centered (Fig. 1a) and the node-centered (Fig. 1b) were investigated.

For the cell-centered schemes where the coarse grid nodes do not coincide with fine grid nodes, the coarse grid values are updated with four fine grid cells that make up one coarse grid cell. This yields

$$I_k^{k-1} U_k = \sum_{4cells} U_k \cdot Vol_k \,/ \sum_{4cells} Vol_k \tag{22}$$

and

$$II_k^{k-1} Res_k U_k = \sum_{4cells} Res_k U_k \,. \tag{23}$$

The boundary conditions on the coarser grids were employed in the same way as for the finest grid. However, since the boundary is located between two grid nodes, only their change is interpolated to the fine grid.

A different formulation was used in the node-centered scheme since the nodes coincides on coarse and fine grids and are located on the boundaries as well. Then for the restriction operator simple point-to-point injection was used for the variables and the residuals:

$$I_k^{k-1} U_k = U_k \;,\; II_k^{k-1} Res_k U_k = Res_k U_k \,. \tag{24}$$

Consequently, the boundary conditions remained unchanged on coarser grids. Other formulations may be used for both grid arrangements. The present formulation was found to be the simplest but still efficiently working formulation.

The behaviour of the direct multigrid procedure for the upwind Runge-Kutta scheme was investigated for different viscous flow problems. In the following, results are presented for the computation of a subsonic boundary layer flow over a flat plate. Since this flow represents typical Navier-Stokes problems and since it is easy to accomplish, a number of comparative studies were conducted for this problem. The integration domain and boundary conditions correspond to those sketched in Fig. 2a, the Mach number is .5 and the Reynolds number is 10^4. In the following, the convergence histories are shown, where the maximum residual is plotted as function of work units (WU), which correspond to the work of one Runge-Kutta time step (5-stages) on the finest grid used.

First, comparison was made between the cell-centered (CC) and the node-centered NC) grid arrangement (Fig. 1a and b) for the multigrid acceleration with 3 grid levels (3G). The Fig. 7 demonstrates that nearly the same speed of convergence for the two grid arrangements can be achieved, if the multigrid parameters are properly adjusted.

A further point of interest was the multigrid behaviour for different spatial discretizations of the Euler terms. In Fig. 8 a comparison for three different Euler solvers is shown. The three schemes are
- the present flux splitting formulation (VL)
- the flux-difference splitting scheme by Harten, Lax, Leer [20], and modified by Einfeldt [21] (HLL)
- the common central differencing with the 4th order damping (C), like in [18].

The result shows similar improvements of the convergence, which indicates a sufficiently robust multigrid behaviour. The results of Fig. 7 and 8 are important for the multigrid extension of existing algorithms. The latter and the following test examples were computed on a node-centered grid.

The influence of implicit residual smoothing, an acceleration technique successfully used e. g. in [18], on the multigrid convergence is shown in Fig. 8. The difference with smoothing ($\varepsilon = 2$, CFL$=9$) and without smoothing ($\varepsilon = 0$, CFL$=3$) is

small, which could also be observed for other test problems. It indicates that residual averaging, does not lead to an improvement of the convergence for all schemes, although a higher CFL-number can be used.
An efficient multigrid acceleration should give a convergence rate which is almost independent of the number of grid points. To prove this behaviour, the same flow problem was computed with different numbers of grid points using single grids and multigrids. The histories of convergence for these computations are shown in Fig. 10. The results indicate a dependence of the convergence on the grid size for the single - as for the multigrid method. For the latter this may be a fact of the simplified treatment of the multigrid elements. However, a partially large gain in computational time can be stated for the multigrid method in all cases. Therefore, as a consequence of the present investigation, the direct multigrid is a promising acceleration technique and will be implemented into existing non-multigrid algorithms for general coordinates in the next future.

5. Conclusions

The time-dependent Navier-Stokes equations were solved by two different methods, an implicit relaxation method and an explicit Runge-Kutta time-stepping method. In both methods the space discretization is based on flux-split upwind discretization for the Euler terms and central discretization for the viscous terms. To both the FAS multigrid concept was applied for accelerating the convergence to the steady-state. The formulation of the FAS method, and different ways to implement the boundary conditions on the coarser grids were discussed. For the implicit scheme the indirect multigrid method was employed to accelerate the matrix inversion each time step. This method proved to be an efficient and reliable solution method which was documented for different laminar flow cases. For the explicit method of solution the multigrid concept was implemented as a direct method to accelerate the convergence in time and space on the coarse and fine grids. Different factors of influence on the multigrid behaviour were considered.

6. References

[1] Thomas, J. L., Walters, R. W.: Upwind Relaxation Algorithms for the Navier-Stokes Equations. AIAA 7th CFD Conf., 87-1501 CP, Cincinatti, 1985.

[2] Klopfer, G. H., Yee, H. C.: Viscous Hypersonic Shock-on-Shock Interaction on Blunt Cowl Lips. AIAA paper 88-02333, Reno, 1988.

[3] Bardina, J., Lombard, C. K.: 3-D Hypersonic Flow Simulations with the CSCM Implicit Upwind Navier-Stokes Method. AIAA paper 87-1114, 1987.

[4] Brandt, A.: Multi-Level Adaptive Solutions to Boundary-Value Problems, Mathematics of Computation, Vol. 31, No. 138, Apr. 1977, pp. 333-390.

[5] Chima, R. V., Johnson, G. M.: Efficient Solution of the Euler and Navier-Stokes Equations with a Vectorized Multiple-grid Algorithm. AIAA Journal, Vol.23, No. 1, Jan 1985, pp. 23-32.

[6] Ni, R. H.: A Multiple Grid Scheme for Solving the Euler Equations. AIAA Journal, Vol. 20, No. 11, Nov. 1982, pp. 1565-1571.

[7] Jameson, A.: Solution of the Euler Equations for Two Dimensional Transonic Flow by a Multigrid Method. Presented at the International Multigrid Conference, Copper Mountain, 1983.

[8] Hemker, P. W., Spekreijse, S. P.: Multigrid Solution of the Steady Euler Equations. Notes on Numerical Fluid Mechanics, 11, Vieweg, 1985.

[9] Mulder, W. A.: Multigrid Relaxation for the Euler Equations. J. Comp. Phys., 60, 1985, pp. 235-252.

[10] Shaw, G., Wesseling, P.: Multigrid Method for the Compressible Navier-Stokes Equations. Rep. of the Dep. of Math. and Inf., Nr. 86-13, Univ. Delft, 1986.

[11] Anderson, W. K., Thomas J. L., Rumsey, C. L.: Extension and Application of Flux-Vector Splitting to Calculation on Dynamic Meshes. AIAA paper 87-1152 CP, 1987.

[12] Schröder, W., Hänel, D.: An Unfactored Implicit Scheme with Multigrid Acceleration for the Solution of the Navier-Stokes Equations. Computers and Fluids, Vol. 15, 1987, pp. 313-316.

[13] Schröder, W., Hänel, D.: A Multigrid Relaxation Scheme for the Navier-Stokes Equations. Proc. of the Third Copper Mountain Conference on Multigrid Methods, 1987.

[14] Brandt, A.: Guide to Multigrid Development. In: Lecture Notes in Mathematics Vol. 960. pp. 220-312, Springer Verlag Berlin, 1981.

[15] Schwane, R., Hänel, D.: Computation of Viscous Supersonic Flow around Blunt Bodies. 7th GAMM Conf. on Num. Math. in Fluid Mech., Louvain-La-Neuve, 1987.

[16] van Leer, B.: Flux-Vector Splitting for the Euler Equations. In: Lecture Notes in Physics, Vol. 170, pp. 507-512, Springer Verlag Berlin, 1982.

[17] van Albada, G. D., van Leer, B., Roberts, W. W.: A Comparative Study of Computational Methods in Cosmic Gas Dynamics, Astron. Astrphys. 108, 1982, pp. 76-84.

[18] Jameson, A., Schmidt, W., Turkel, E.: Numerical Solutions of the Euler Equations by Finite Volume Methods Using Runge-Kutta Time-Stepping Schemes. AIAA paper 81-1259, 1981.

[19] Jameson, A.: A Nonoscillatory Shock Capturing Scheme Using Flux Limited Dissipation. Lectures in Applied Mathematics, 22, pp. 345-370, 1985.

[20] Harten, A., Lax, D., van Leer, B.: On Upstream Differencing and Godunov Schemes for Hyperbolic Conservation Laws, SIAM Review 25, 1983.

[21] Einfeldt, B.: On Godunov-Type Methods for Gas Dynamics, to be published in SIAM J. on Numer. Anal.

Acknowledgement:
This research was supported by the DFG (Deutsche Forschungsgemeinschaft) Schwerpunktprogramm: Finite Approximationen in der Strömungsmechanik.

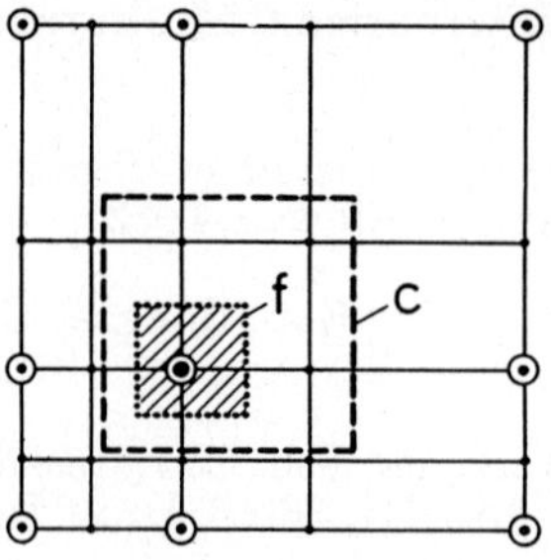

Fig. 1a: Node-Centered grid.

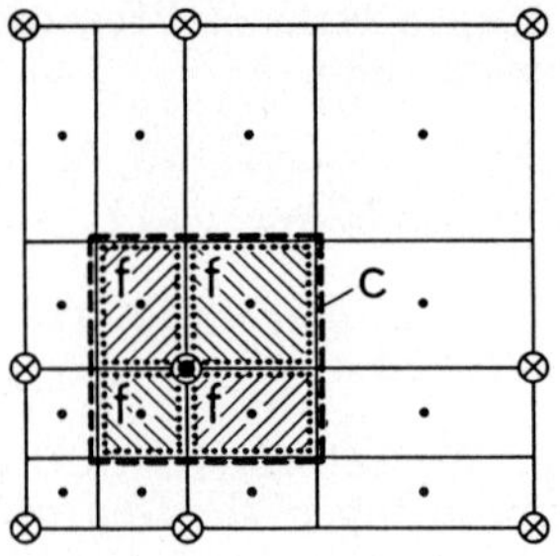

Fig. 1b: Cell-Centered grid.

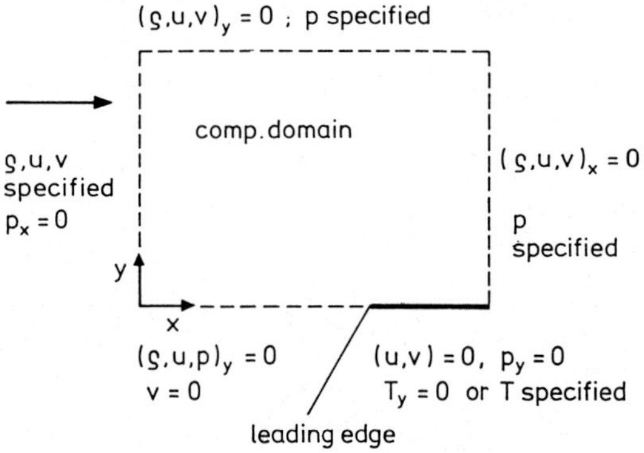

Fig. 2a : Viscous subsonic flow past a flat plate.

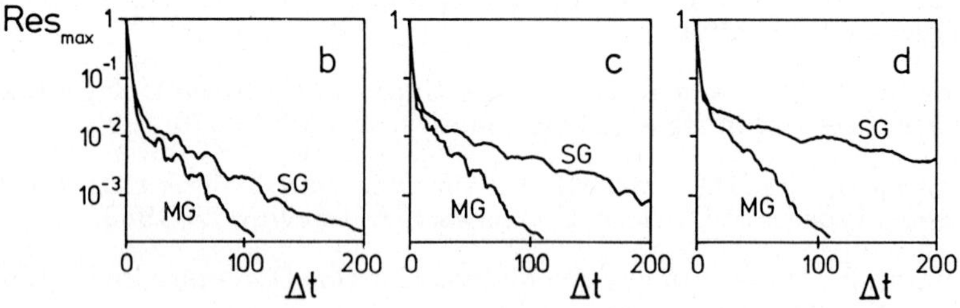

Fig. 2: Maximum residual as function of the time steps for the viscous flow past a flat plate; $Re = 10^4$, $Ma = 0.9$; Multigrid (MG) and single (SG) method.
b)fine grid 65x65 grid points
c)fine grid 17x17 grid points
d) fine grid 33x33 grid points

Fig. 3a: Viscous supersonic flow past a wedge.

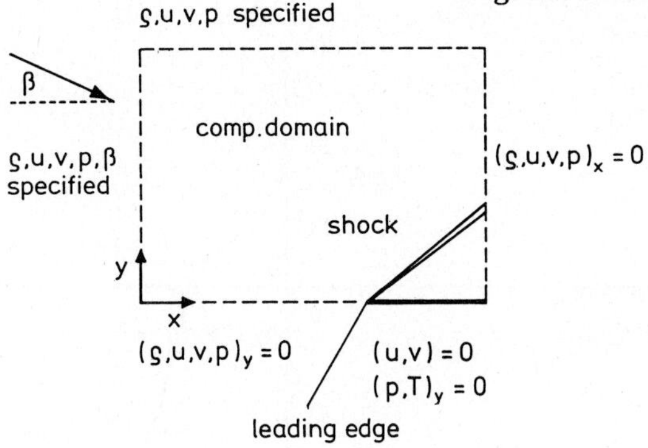

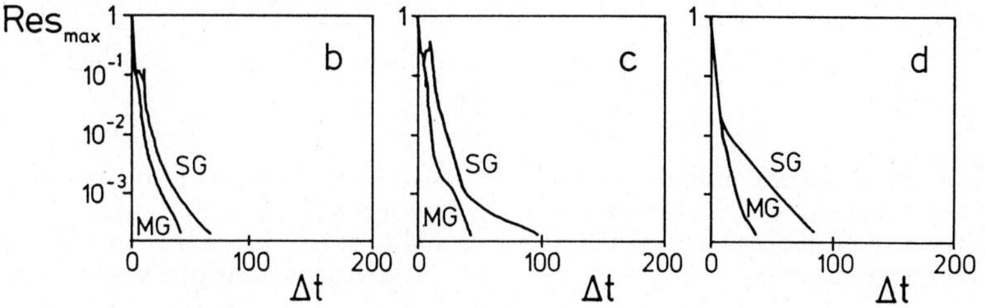

Fig.3: Maximum residual as function of the time steps for the viscous flow past a wedge; $Re=10^4$, $Ma=2.0$, $\beta=10°$; multigrid (MG) and single grid (SG) method.
b)fine grid 65x65 grid points
c)fine grid 17x17 grid points
d) fine grid 33x33 grid points

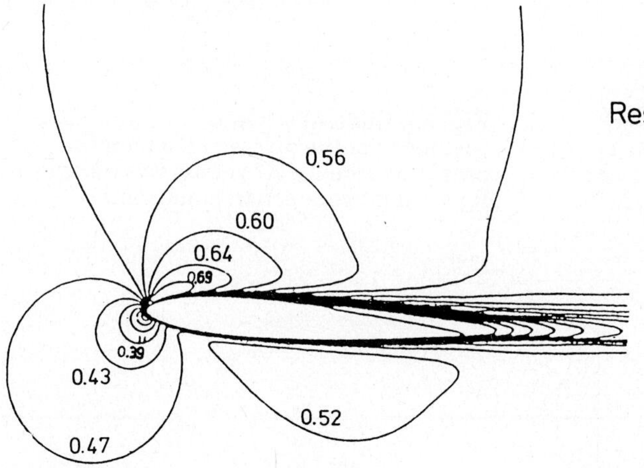

Fig. 4a: Laminar subsonic flow over an NACA 0012 airfoil ($Ma_\infty=0.5$, $Re_\infty=10^4$, $\alpha=5°$). Mach contours.

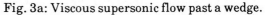

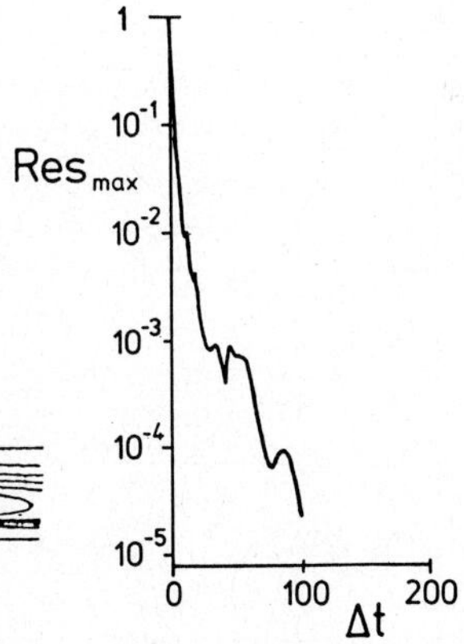

Fig. 4b: Laminar subsonic flow over an NACA 0012 airfoil ($Ma_\infty=0.5$, $Re_\infty=10^4$, $\alpha=5°$). Maximum residual versus time step.

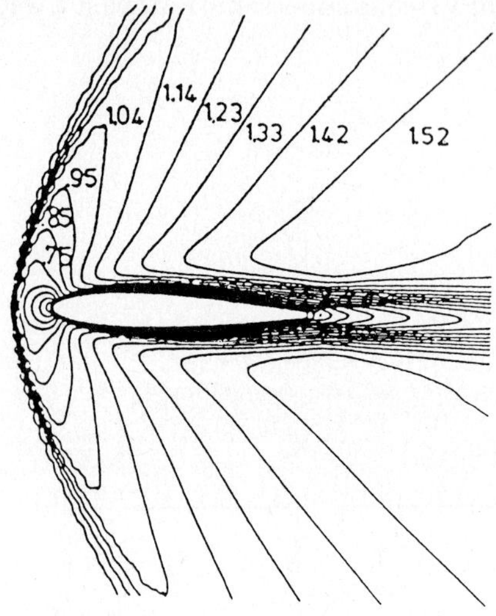

Fig. 5a: Laminar subsonic flow over an NACA 0012 airfoil ($Ma_\infty = 1.5$, $Re_\infty = 10^3$, $\alpha = 0°$). Mach contours.

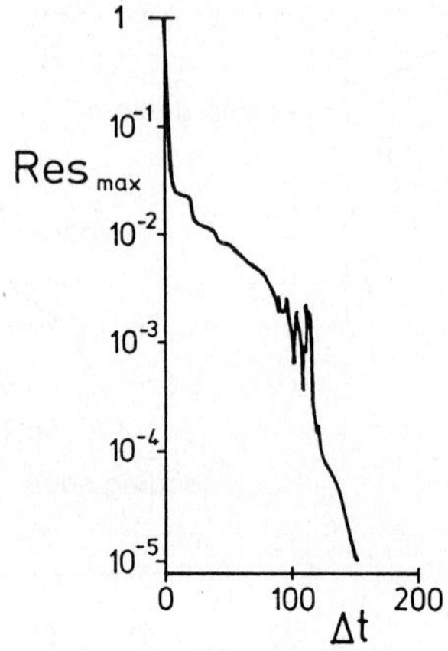

Fig. 5b: Laminar subsonic flow over an NACA 0012 airfoil ($Ma_\infty = 1.5$, $Re_\infty = 10^3$, $\alpha = 0°$). Maximum residual versus time step.

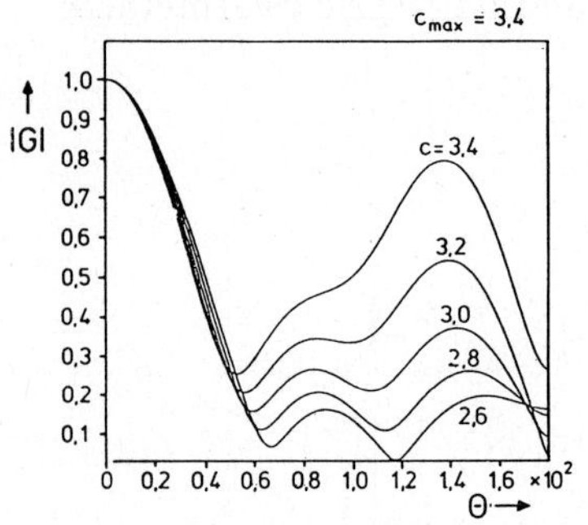

Fig. 6a: Central scheme. Influence of the discretization on the amplification factor versus wave angle for a linear convection equation.

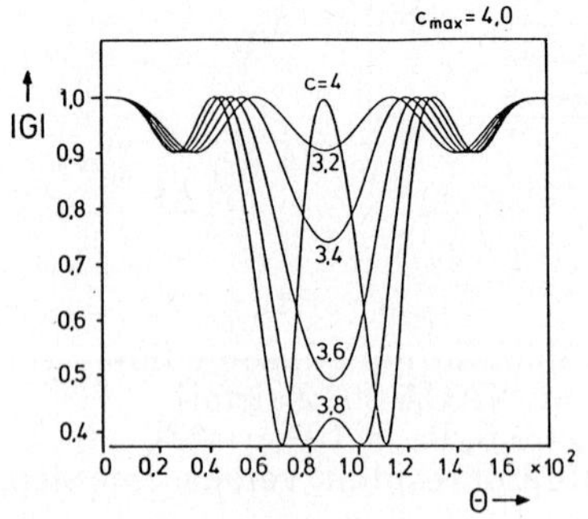

Fig. 6b: Upwind scheme. Influence of the discretization on the amplification factor versus wave angle for a linear convection equation.

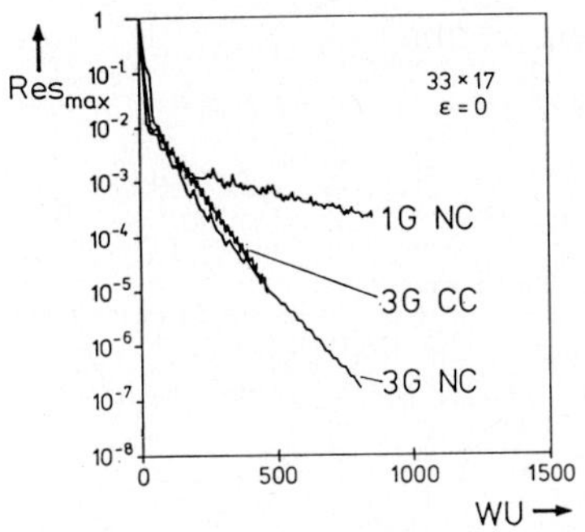

Fig. 7: Maximum residual
as function of work units.
Navier-Stokes solutions
for boudary layer flow
Ma = 5, Re = 10^4
Influence of grid arrangement
NC: node-centered grid
CC: cell-centered grid

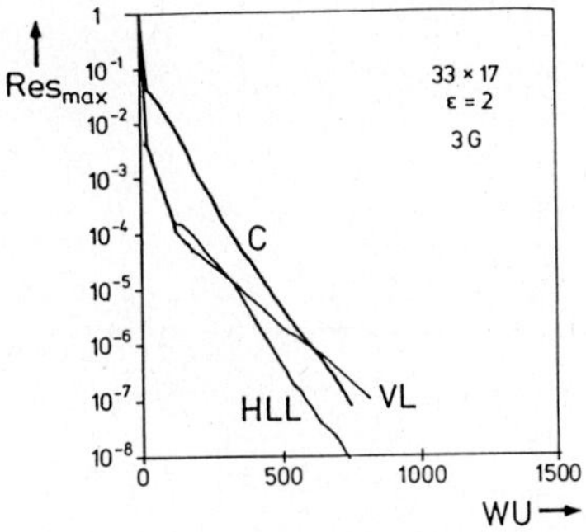

Fig. 8: Legend see Fig. 7.
Influence of different Euler solver
VL: van Leer's flux splitting $O(h^2)$
HL: flux difference splitting $O(h^2)$
by Harten, Lax, van Leer
C: central differences $O(h^2)$

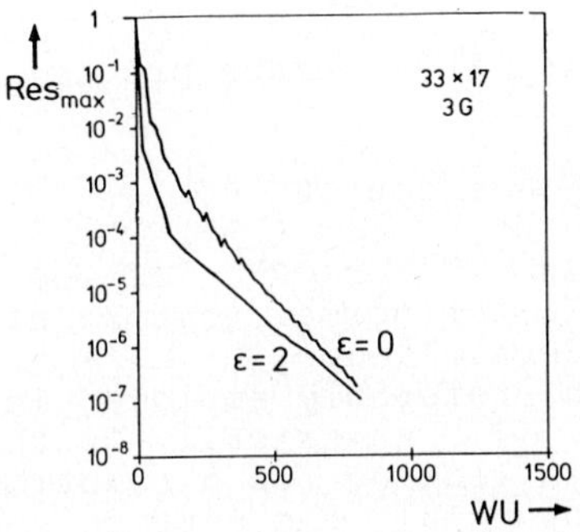

Fig. 9: Legend see Fig. 7.
Influence of implicit residual smoothing
ε = 0 without, CFL = 3
ε = 2 with, CFL = 9

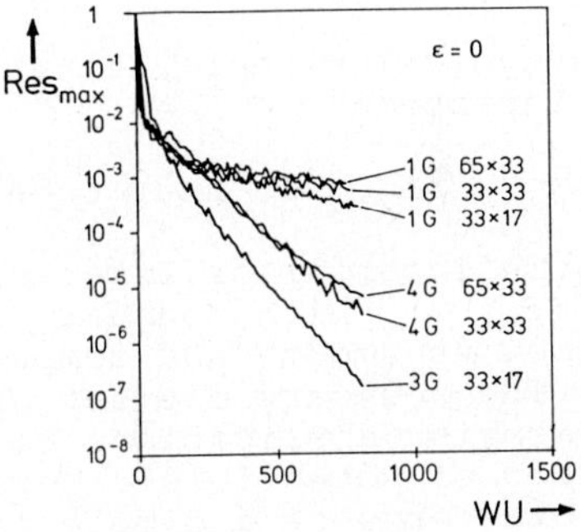

Fig. 10: Legend see Fig. 7.
Influence of different numbers
of grid points for single grid
and multigrid computation.

ON MULTIGRID METHODS OF THE FIRST KIND FOR SYMMETRIC BOUNDARY INTEGRAL EQUATIONS OF NONNEGATIVE ORDER

F.K. Hebeker
THD, Fachbereich Mathematik
Schloßgartenstraße 7, D-6100 Darmstadt

SUMMARY

Multigrid methods of the first kind are applied for solving iteratively some symmetric algebraic systems that occur in the numerical treatment of symmetric and strongly elliptic boundary integral equations of nonnegative order. Using a damped Jacobi relaxation scheme, the iterative method proves to be unconditionally convergent in case of smooth boundaries and, moreover, convergent for a sufficiently large number of smoothing steps in case of general Lipschitz boundaries. As a numerical example, the Neumann problem of the Laplacean is treated by the "first kind" boundary element approach.

1. INTRODUCTION

As a typical example, let us consider the Neumann problem of the Laplacean

$$\Delta u = 0 \quad \text{in } \Omega \text{ or } \Omega^c , \quad \frac{\partial u}{\partial n} = f \quad \text{on } \partial\Omega , \qquad (1)$$

in some interior or exterior domain (Ω or Ω^c) of the $\mathbb{R}^d$ ($d \geq 2$), with boundary $\partial\Omega$ (outer normal vector n) assumed as smooth and compact in Secs. 1.-3.

The boundary element approach preferably employed in the engineering applications to solve (1) numerically is the "method of potentials". Here one is looking for a solution of (1) in terms of a simple layer potential, the surface source of which is determined by a boundary integral equation of the second kind (numerically solved by means of the "panel method"). This now classical approach is of advantageous evidence, both from the mathematical and the numerical point of view (e.g. [10]). Specifically, the particularly fast multigrid methods of the second kind are applicable [9] - again in case of the Neumann problem of Stokes' flow, but not for that of linear elasticity theory, e.g.

Additional to the latter, this conceptual disadvantage of the second kind approach should be realized: the resulting algebraic system is nonsymmetric whereas the original boundary value problem, written in variational form, is a symmetric one. To retain symmetry, several authors investigate the "first kind" boundary element method: here the solution of (1) is represented in terms of a double layer potential

$$u(x) = \int_{\partial\Omega} \frac{\partial}{\partial n_z} \gamma(x-z)\psi(z)do_z \ , \qquad (2)$$

where γ denotes the well known fundamental solution of the Laplacean in $\mathbb{R}^d$ and ψ the wanted surface source. By adapting (2) to the given Neumann data, this first kind boundary integral equation appears:

$$\int_{\partial\Omega} \frac{\partial^2}{\partial n_x \partial n_z} \gamma(x-z)\psi(z)do_z = f(x) \quad \text{on} \quad \partial\Omega \ . \qquad (3)$$

See [20],[8],[11],[18],[6],[15],[7] for more details.

At first glance, for practical reasons, little seems to be gained with (3) : we have a hypersingular integral, defined as a part-fini integral only. Moreover, for a first kind equation, multigrid methods are developed up to date for some very special cases only [9],[22],[24]. However, these problems are not too bad, as it is shown below, and it is just the symmetry of the strongly elliptic integral operator of (3), written in variational form, which allows to construct multigrid methods with essentially the same advantageous properties as multigrid solvers for elliptic boundary value problems. Moreover, as Hsiao [15] recently pointed out, (3) has good stability properties.

This contribution is organized as follows: In Sec. 2, a Galerkin scheme is investigated to solve a slightly modified version of (3) numerically. Here we apply the general theory of asymptotic convergence and stability due to Hsiao and Wendland [18] . In Sec. 3, this theory again shows up to be basic for the construction of multigrid solvers for a rather large class of problems including (3) . In Sec. 4, we extend this to the case of general Lipschitz boundaries by employing a recent regularity result by Costabel [4] . Finally, a numerical example is carried out that essentially confirms the theoretical assertions of Secs. 1.-3.

2. GALERKIN SCHEME AND ASYMPTOTIC CONVERGENCE ANALYSIS

In this section we reconsider some results by [18] . Let us consider the equivalent boundary integral equation on $\partial\Omega$

$$(D\psi)(x) := \int_{\partial\Omega} \frac{\partial^2}{\partial n_x \partial n_z} \gamma(x-z)\psi(z)do_z + \int_{\partial\Omega} \psi(z)do_z = f(x) \ , \qquad (4)$$

where the supplementary integral operator serves to suppress some fictitious eigensolutions. Inspection of the principal symbol shows that D is a strongly elliptic pseudo-differential operator of order $+1$ that maps

$$D : H^s(\partial\Omega) \to H^{s-1}(\partial\Omega)$$

continuously (any $s \in \mathbb{R}$), where $H^s(\partial\Omega)$ denote the usual boundary Sobolev-Slobodetski spaces with norms $\|\cdot\|_s$.

Moreover, D is symmetric and positive on its energy space $H^{1/2}(\partial\Omega)$: $\langle D\phi,\psi\rangle = \langle D\psi,\phi\rangle$ and

$$\langle D\phi,\phi\rangle \geq \beta \|\phi\|^2_{1/2} \quad \text{(with } \beta > 0\text{)} \tag{5}$$

hold for all $\phi,\psi \in H^{1/2}(\partial\Omega)$. Here $\langle\cdot,\cdot\rangle$ denotes the duality pairing on $H^{-1/2}(\partial\Omega) \times H^{1/2}(\partial\Omega)$, extending the innerproduct of $H^0(\partial\Omega)$. And the positivity is implied by both the Gårding's inequality and the injectivity of D .

Let $H_h \subset H^{1/2}(\partial\Omega)$ be a regular family of finite element spaces on the boundary. In the sense of Babuska and Aziz [1] we assume (with quantities c not depending on ϕ,h) :
a) convergence property : with $1 \leq m < \ell$, let $-\ell \leq t \leq s \leq \ell$, $-m \leq s,\ t \leq m$. Then there holds

$$\inf_{\chi\in H_h} \|\chi-\phi\|_t \leq ch^{s-t}\|\phi\|_s \quad \text{for all } \phi\in H^s(\partial\Omega); \tag{6}$$

b) inverse assumption: with $t \leq s$ and $|t|,|s| \leq m$ there holds

$$\|\chi\|_s \leq ch^{-(s-t)}\|\chi\|_t \quad \text{for all } \chi \in H_h . \tag{7}$$

By employing these spaces for a Galerkin-type boundary element scheme: search for a $\psi_h \in H_h$ so that

$$\langle D\psi_h,\chi\rangle = \langle f,\chi\rangle \quad \text{for all } \chi \in H_h , \tag{8}$$

we obtain this optimal asymptotic error estimate [18] :

$$\|\psi_h - \psi\|_t \leq ch^{s-t}\|\psi\|_s , \tag{9}$$

with $1-\ell \leq t \leq s \leq \ell$, $t \leq m$, $-\ell \leq 1/2 \leq m$, and $-m \leq 1/2 \leq s$. Combining (9) with the a priori estimates for elliptic pseudodifferential operators [23] implies the specialized Aubin-Nitsche estimate

$$\|\psi_h - \psi\|_0 \leq ch\|f\|_0 \tag{10}$$

that will play a fundamental role as "approximation property" in the convergence analysis of the subsequent multigrid method. Further, by using (7) , it is easy to show that

$$\|D_h\|_{0\leftarrow 0} \leq \frac{1}{\omega} h^{-1} \tag{11}$$

holds for the spectral norm of the boundary element stiffness matrix D_h . The latter estimate serves as a basic tool to establish the "smoothing property" of our multigrid scheme.

Finally, we remark that Hsiao [15] recently proved that

the boundary element scheme (8) has good stability properties, when the mesh size h is suitably associated to the number of significant digits available in the computations.

3. MULTIGRID METHOD OF THE FIRST KIND

By generalization, we assume that D is a strongly elliptic (system of) pseudodifferential operator(s) of order $2\alpha \geq 0$, which is symmetric and positive on its energy space $H^{\alpha}(\partial\Omega)$ (resp. the product space) in the sense of (5), where $\alpha = 1/2$. Restricting ourselves to the case of a two-stage model of a multigrid scheme with doubled mesh size, let

$$H_{2h} \subset H_h \subset H^{\alpha}(\partial\Omega) \tag{12}$$

be a hierarchy of regular boundary element spaces with properties (6) and (7). Then the specialized Aubin-Nitsche estimates

$$\| \psi_{jh} - \psi \|_o \leq ch^{2\alpha} \| f \|_o \tag{13}$$

hold for the respective Galerkin approximates ψ_{jh} [18]. Here $j=1$ or $j=2$ will be constantly used in the sequel.

The functions $\phi_{jh} \in H_{jh}$ are identified with a coefficient vector $\hat{\phi}_{jh} = (\hat{\phi}_{jh}{}^i)$ (usually the values of ϕ_{jh} at some nodel points) by means of the bijective mapping

$$I_{jh} : \mathbb{R}^{N_{jh}} \ni \hat{\phi}_{jh} \rightarrow \phi_{jh} \in H_{jh} , \tag{14}$$

where $N_{jh} = \dim H_{jh}$. The discrete vector norms on $\mathbb{R}^{N_{jh}}$ are defined by

$$|\hat{\phi}_{jh}|_o = (h^{d-1} \sum_i (\hat{\phi}^i_{jh})^2)^{1/2} , \tag{15}$$

so that uniform equivalence

$$\frac{1}{c} |\hat{\phi}_{jh}|_o \leq \| \phi_{jh} \|_o \leq c|\hat{\phi}_{jh}|_o \quad \text{for all} \quad \phi_{jh} \in H_{jh} \tag{16}$$

holds (with a quantity c not depending on h, ϕ_{jh}).

By suitable scaling, Galerkin's equations are written as

$$D_{jh}\hat{\psi}_{jh} = \hat{f}_{jh} ,$$

where $D_{jh} = I^*_{jh} DI_{jh}$ and $\hat{f}_{jh} = I^*_{jh} f$ (I^*_{jh} adjoint to I_{jh}). Again, the inverse assumption (7) implies

$$\| D_{jh} \|_{o \leftarrow o} \leq \frac{1}{\omega} h^{-2\alpha} \tag{17}$$

Finally, the prolongation p and restriction r are chosen canonically,

$$p = I_h^{-1} I_{2h} , \quad r = p^* , \tag{18}$$

and the damped Jacobi relaxation scheme

$$S_h(\hat{\phi}_h, \hat{f}_h) = \hat{\phi}_h - \omega h^{2\alpha}(D_h \hat{\phi}_h - \hat{f}_h) , \tag{19}$$

with an admissible ω from (17) , is taken as a smoothing operator. This determines a two-stage multigrid method of the first kind as follows (with symbolised S_h^{ν} to denote ν-fold application of S_h):

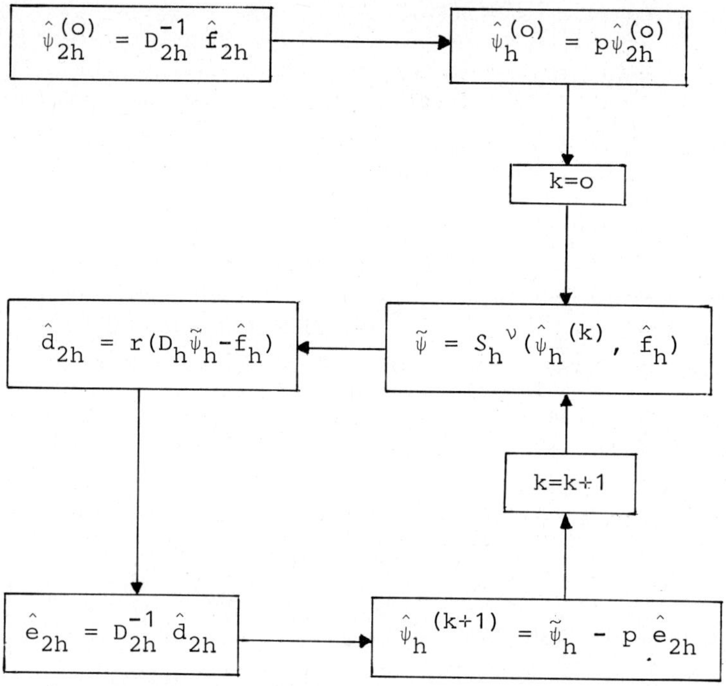

Fig.1: Two-stage multigrid method. (20)

The convergence analysis of finite element multigrid methods (cf. [9]) carries over to our more general case, and we obtain

<u>Theorem 1</u>: *Let* $\partial\Omega$ *being sufficiently smooth and the assumptions of this section hold. Then the multigrid method (20) converges whenever* $\nu \geq 1$, *with convergence rates*

$$\rho \le \begin{cases} q^{\nu} & : \text{ if } \nu \text{ small} \\ c\nu^{-1} & : \text{ if } \nu \gg 1 . \end{cases}$$

Here $q < 1$ *and* c *depend on* ω *, but not on* h.

Among the applications of this multigrid scheme of the first kind are the Neumann problems of the linear equations or systems of mathematical physics (e.g. (1)), in particular in those cases where the principal part of D_h has neither a "Toeplitz property" nor a logarithmic kernel (see [9], [22], [24]).

In this paper we do not dwell on the question of numerical work of algorithm (20), since this has to be investigated in the frame of more realistic multilevel multigrid methods. Concerning the computational expense, the major part is caused by the matrix-vector operation $D_h \hat{\phi}_h$ for the smoothing steps. Eventually, here some costs could be saved by taking advantage of the diagonaldominance of D_h.

4. CASE OF LIPSCHITZ BOUNDARIES

Due to some recent regularity result by [4], which depends on some sharp (but nevertheless elementary) regularity assertion by [19] on the Dirichlet problem, one is in a position to carry over parts of Theorem 1 to the (practically interesting) case of a general Lipschitz boundary $\partial\Omega$, which is locally the graph of a Lipschitz function. Now the spaces $H^s(\partial\Omega)$ are defined in a unique (invariant) way for $|s| \le 1$ only.

Let us consider the symmetric and positive operator D refering to Neumann's problem (1). By [4], one has some degree of regularity of D that results in an optimal asymptotic error estimate

$$\| \psi_h - \psi \|_{1/2} \le c h^{1/2} \| f \|_0 \qquad (21)$$

of Galerkin's approximate $\psi_h \in H_h \subset H^{1/2}(\partial\Omega)$. The H_h are assumed as regular boundary element spaces with properties (6)-(7), where the indices are restricted now to the interval $[-1,1]$. Also, (11) continues to hold.

For the multigrid analysis we introduce the discrete s-norms,

$$|\hat{\phi}_{jh}|_s = |D_{jh}^{s} \hat{\phi}_{jh}|_0 \qquad (s \in \mathbb{R}). \qquad (22)$$

Since symmetry and positivity of D carry over to D_{jh}, all powers D_{jh}^{s} $(s \in \mathbb{R})$ are well defined. Further, the operator norms $\| \cdot \|_{s \leftarrow t}$ are defined on the discrete spaces by

$$\| A_{jh} \|_{s\leftarrow t} = \sup |A_{jh} \hat{\phi}_{jh}|_s \, |\hat{\phi}_{jh}|_t^{-1} , \tag{23}$$

where A_{jh} denotes a (N_{jh}, N_{jh})-matrix. The iteration matrix of (20),

$$M_h = (D_h^{-1} - pD_{2h}^{-1} r) \, D_h (I_h - \omega h D_h)^{\nu} , \tag{24}$$

is estimated as follows:

$$\| M_h \|_{1/2\leftarrow 1/2} = \| D_h^{1/2} M_h D_h^{-1/2} \|_{0\leftarrow 0}$$

$$\leq \| D_h^{1/2} (D_h^{-1} - pD_{2h}^{-1} r) \|_{0\leftarrow 0} \, \| D_h^{1/2} (I_h - \omega h D_h)^{\nu} \|_{0\leftarrow 0} .$$

From (11) and Corollary 6.2.3 of [9] we infer

$$\| D_h^{1/2} (I_h - \omega h D_h)^{\nu} \|_{0\leftarrow 0} \leq c(\nu h)^{-1/2} .$$

On the other hand, this uniform equivalence property:

$$\frac{1}{c} |\hat{\phi}_{jh}|_{1/2} \leq \| \phi_{jh} \|_{1/2} \leq c |\hat{\phi}_{jh}|_{1/2}$$

(extending (16) ; cf. [9], 141f.) associated with (21) implies

$$\| D_h^{1/2} (D_h^{-1} - pD_{2h}^{-1} r) \|_{0\leftarrow 0} \leq ch^{1/2}$$

(cf. Lemma 6.3.20 of [9]). By summarizing, we have

Theorem 2: *Let* $\partial\Omega$ *being Lipschitzean and the assumptions of this section hold. Then, with a sufficiently large number* ν *of smoothing steps, the multigrid method (20) converges, with convergence rate*

$$\rho \leq c\nu^{-1/2} . \tag{25}$$

The quantity c *depends on* ω, *but not on* h .

This result even holds in case of strongly elliptic second order differential systems, e.g. for the linear system of elasticity theory, whenever Gårding's inequality holds on all $H^1(\Omega)$. For Stokes' system in 3D , it seems to be an open question (depending on some "Rellich's identity", cf. [19]).

5. AN EXAMPLE

To test this first kind multigrid scheme, let us consider the Neumann problem (1) in the plane. In variational form, integral equation (4) looks as

$$\int_{\partial\Omega}\int_{\partial\Omega}\frac{\partial^2}{\partial n_x \partial n_z}\log|x-z|\psi(z)\phi(x)ds_z ds_x + \int_{\partial\Omega}\psi ds \int_{\partial\Omega}\phi ds = \int_{\partial\Omega} f\phi ds, \quad (26)$$

for all $\phi \in H^{1/2}(\partial\Omega)$. The problem of treating hypersingular integral equations has been investigated by Nedelec and his group: the idea is to transform it into an equivalent integro-differential equation with weakly singular kernel. In case of (26), Stokes' integral theorem leads to

$$\int_{\partial\Omega}\int_{\partial\Omega}\log|x-z|\frac{d\psi}{ds}(z)\frac{d\phi}{ds}(x)ds_z ds_x + \int_{\partial\Omega}\psi ds \int_{\partial\Omega}\phi ds = \int_{\partial\Omega} f\phi ds. \quad (27)$$

The complexity of this shifting operation greatly increases for more complicated kernels or higher dimensions, cf. [20], [21], [11], [2], [13].

By using a representation $x(t)$, $0 \le t \le 1$, of the curve $\partial\Omega$, the integrals of (27) are transformed into line integrals over the interval $[0,1]$, which is decomposed in portions of length $h = (N+1)^{-1}$. We are looking for a Galerkin approximate in the space H_h of periodic globally continuous and (corresponding to the given decomposition) piecewise linear functions on $[0,1]$. Clearly, $H_h \subset H^{1/2}(\partial\Omega)$ and (6)-(7) hold, with $l=2$, $m=1$.

Now some methods by [16],[17] are utilized. The logarithmic kernel is splitted

$$\log|x(t) - x(\tau)| = \log|t-\tau| + \log\left|\frac{x(t)-x(\tau)}{t-\tau}\right| \quad (28)$$

into a principal part, not depending on $x(t)$, and a regular remainder. The integrals of the principal part have been computed once for all [16], whereas the remainder is evaluated by using a mixed 2-point Gauss/3-point Kepler rule for the double integrals, e.g. Our test computations have been carried out in case of $\partial\Omega$ = unit circle, $u(x,y) = x + \text{const}$.

Fig. 2 shows the "interior superapproximation" of boundary element methods. The maximum error is computed for $|x| \le 0.7$, as a function of $N = h^{-1} - 1$. The experimental convergence orders show superconvergence for small N. Here the theoretical convergence order equals 3.0, which stems from (9) (with $s=\ell=2$, $t=-1$) in view of the smoothing property of singular boundary integrals away form the boundary [18].

N	4	9	14	19	24
max.error	1.2E-1	1.9E-2	3.8E-3	7.8E-4	2.2E-4
exp.conv.ord.	-	2.72	3.98	5.46	5.77

Fig.2: Interior superapproximation of boundary elements.

More extended boundary element computations for Neumann's problem have been carried out by [11] in $\mathbb{R}^3$.

The subsequent figures show some parameter studies of the two-level multigrid scheme (20). In Fig. 3 the constant decay of the maximum defect during the iterative process is demonstrated. There the convergence rate is given for a Galerkin ansatz with $N=9$ and $\nu=5$ smoothing steps.

step	0	1	2	3	4	5
max.defect	1.3E0	1.6E-1	1.9E-2	2.3E-3	2.8E-4	3.3E-5
conv.rate	-	0.1212	0.1212	0.1212	0.1216	0.1221

Fig.3: Convergence of the multigrid scheme.

Here $\omega=0.1$ is chosen as a damping parameter.

Fig.4 shows that in fact the convergence rate strongly depends on ω. Here the parameters are $N=9$ and $\nu=2$.

ω	0.1	0.2	0.3	0.4	0.5	0.6
conv.rate	0.428	0.097	0.007	0.158	0.550	div.

Fig.4: Convergence rate as function of ω.

For these parameters, $\omega=0.3$ appears as an optimum value.

Fig.5 confirms Theorem 1 that asserts the convergence rate as nearly independent of N (further parameters: $\omega=0.25$ and $\nu=2$).

N	4	9	19	29	49
conv.rate	0.0223	0.0214	0.0211	0.0213	0.0216

Fig.5: Convergence rate as function of N or h.

Moreover, as further experiments indicate, even the optimum ω is nearly independent of N. Consequently, in practice one should guess a satisfactory value of ω by an (inexpensive) preprocessing run with small N.

In Fig.6, the geometric decay of the convergence rate as a function of ν is established (Theorem 1). This effect turns out clearly in case of a nonoptimal $\omega=0.15$ (and $N=9$).

ν	1	2	3	4	5
conv.rate	0.481	0.232	0.113	0.056	0.028
q	0.48	0.48	0.48	0.49	0.49

Fig.6: Convergence rate as function of ν.

Finally, we remark that in case of this relatively simple example even the fast multigrid methods of the second kind are applicable (cf. [9], [22]).

REFERENCES

[1] BABUSKA I., AZIZ A.K.: "Survey lectures on the mathematical foundations of the finite element method". In A.K. AZIZ (ed.): "The mathematical foundation of the finite element method with applications to partial differential equations". New York 1972.

[2] BAMBERGER A.: "Approximation de la diffraction d'ondes élastiques: une nouvelle approche". I-III, École Polytechnique Palaiseau 1983.

[3] COLTON D., KRESS R.: "Integral equation methods in scattering theory, New York 1983.

[4] COSTABEL M.: "Boundary integral operators on Lipschitz domains". SIAM J. Math. Anal., in press.

[5] COSTABEL M., STEPHAN E.: "The normal derivative of the double layer potential on polygons and Galerkin approximation". Appl. Anal. 16 (1983), 205-228.

[6] COSTABEL M., WENDLAND W.L.: "Strong ellipticity of boundary integral operators". J. Reine Angew. Math. 372 (1986), 34-63.

[7] GIROIRE J.: "Étude de quelques problemes aux limites exterieurs et résolution par équations intégrales". Thèse de doctorat d'état. École Polytechnique Palaiseau 1987.

[8] GIROIRE J., NEDELEC J.C.: "Numerical solution of an exterior Neumann problem using a double layer potential". Math. Comp. 32 (1978), 973-990.

[9] HACKBUSCH W.: "Multigrid Methods". Berlin 1985.

[10] HACKBUSCH W., NOWAK Z.P.: "The panel clustering technique for the boundary element method". In: C.A. Brebbia, W.L. Wendland, G. Kuhn (eds.): "Boundary elements IX", vol. 1, Berlin 1987, pp. 464-474.

[11] HA DUONG T.: "A finite element method for the double layer potential solutions of the Neumann exterior problem". Math. Meth. Appl. Sci. 2 (1980), 191-208.

[12] HEBEKER F.K: "Efficient boundary element methods for 3D exterior viscous flows". Num. Meth. PDE. 2 (1986), 273-297.

[13] HEBEKER F.K.: "An integral equation of the first kind for a free boundary value problem of the stationary Stokes' equations". Math. Meth. Appl. Sci. 9 (1987), 550-575.

[14] HEBEKER F.K.: "On the numerical treatment of viscous flows against bodies with corners and edges by boundary element and multigrid methods". Numer. Math. 52 (1988), 81-99.

[15] HSIAO G.C.: "On the stability of boundary element methods for integral equations of the first kind". In C.A. Brebbia, W.L. Wendland, G. Kuhn (eds.): "Boundary elements IX", vol. 1., Berlin 1987, pp. 177-191.

[16] HSIAO G.C., KOPP P., WENDLAND W.L.:"A Galerkin collocation method for some integral equations of the first kind". Computing 25 (1980), 89-130.

[17] HSIAO G.C., KOPP P., WENDLAND W.L.: "Some applications of a Galerkin-collocation method for boundary integral equations of the first kind". Math. Meth. Appl. Sci. 6 (1984), 280-325.

[18] HSIAO G.C., WENDLAND W.L.: "The Aubin-Nitsche lemma for integral equations". J. Integral Eqs. 3 (1981), 299-315.

[19] NEČAS J.: "Les méthodes directes en théorie des équations elliptiques". Paris 1967.

[20] NEDELEC J.C.: "Approximation des équations intégrales en mécanique et en physique". Lecture Notes, École Polytechnique Palaiseau 1977.

[21] NEDELEC J.C.: "Integral equations with non integrable kernels". Integral Eqs. Oper. Th. 5 (1982), 562-572.

[22] SCHIPPERS H.: "Multigrid methods for boundary integral equations". Numer. Math. 46 (1985), 351-363.

[23] TAYLOR M.E.: "Pseudodifferential operators". Princeton NJ 1981.

[24] VOEVODIN V.V., TYRTYSHNIKOV E.E.: "Toeplitz matrices and their applications". In R. Glowinski, J.L. Lions (eds.): "Computing methods in applied sciences and engineering", vol. 6. Amsterdam 1984, pp. 75-85.

EFFECTIVE PRECONDITIONING FOR SPECTRAL MULTIGRID METHODS

by
Wilhelm Heinrichs

Mathematisches Institut der Universität Düsseldorf
4000 Düsseldorf 1, Federal Republic of Germany

Abstract

Spectral methods employ global polynomials for the approximation of elliptic problems. They give very accurate approximations for smooth solutions with relatively few degrees of freedom. On the other hand, the matrices involed are full and yield high condition numbers, growing as $O(N^4)$ (for polynomials of degree $\leq N$ in each variable). Nevertheless the spectral systems can be efficiently solved using spectral multigrid (SMG) methods. Utilizing FFT's only $O(N^2 \ln N)$ operations are necessary for the evaluation of the spectral residual. We investigate effective preconditioners for SMG based on line relaxation techniques and show the robustness of minimal residual relaxation. We also present numerical results for L-shaped regions where the Schwarz alternating procedure has been used.

1. INTRODUCTION

Spectral methods [10] give very accurate approximations for smooth solutions of elliptic problems with relatively few degrees of freedom. The matrices involved are full (and nonsymmetric) and efficient iterative methods for the solution of the spectral systems are necessary. Above all pseudospectral (or collocation) methods can be implemented very efficiently with Fast Fourier Transforms.

Zang et al. [13,14] introduced multigrid techniques for the fast solution spectral problems. Brandt et al. [2] have significantly improved spectral multigrid methods for periodic elliptic problems. Streett et al. [11] investigated combined Dirichlet and periodic problems and applied them to the transonic flow. They used alternate direction implicit (ADI) methods for relaxation. We achieved some improvements for both Dirichlet and combined Dirichlet-Fourier problems [8], [9]. This was done by using certain (alternating) line relaxation techniques (see also [1], [12]). Furthermore we show that minimal residual relaxation [4] speeds up the convergence and also yields good results for problems far from Poisson's equation. It is also shown that these techniques can easily be adoped to problems in non-rectangular domains. Here we use a Schwarz alternating procedure [3]. Numerical results are presented for L-shaped regions.

2. PSEUDOSPECTRAL DISCRETIZATION

We consider the elliptic problem

$$Lu = -(au_x)_x - (bu_y)_y = f \tag{2.1}$$

on the region $\Omega = (-1,1)^2$ with the Dirichlet boundary condition $u = g$ on $\partial\Omega$. Hereby a,b,f denote given functions. For the discretization products of Chebyshev polynomials $t_p(x)\, t_q(y) = \cos(p \arccos x) \cos(q \arccos y)$ for $p,q = 0,1,\ldots,N$ are employed. The collocation points are

$$(x_i, y_i) = \left(\cos \frac{i\pi}{N}, \cos \frac{j\pi}{N}\right) \quad \text{for} \quad i,j = 0,1,\ldots,N.$$

We introduce the following grids

$$\bar{\Omega}_N = \{(x_i,y_j): i,j=0,\ldots,N\}, \quad \Omega_N = \Omega \cap \bar{\Omega}_N, \quad \partial\Omega_N = \partial\Omega \cap \bar{\Omega}_N.$$

By $(i,j) \in I_{\bar{\Omega}_N}$ (resp. I_{Ω_N} or $I_{\partial\Omega_N}$) we mean that $(x_i,y_j) \in \bar{\Omega}_N$

(resp. Ω_N or $\partial\Omega_N$). $G(\bar{\Omega}_N)$ (resp. $G(\Omega_N)$ or $G(\partial\Omega_N)$) denote the set of grid functions defined on $\bar{\Omega}_N$ (resp. Ω_N or $\partial\Omega_N$). For the components of the grid functions $v_N \in G(\bar{\Omega}_N)$ we use the abbreviation $v_N^{i,j} = v_N(x_i,y_j)$ $((i,j)\in I_{\bar{\Omega}_N})$. Let the fixed grid functions $f_N \in G(\Omega_N)$ be defined by

$$f_N^{i,j}=f(x_i,y_j) \quad ((i,j)\in I_{\Omega_N}), \quad g_N^{i,j}=g(x_i,y_j) \quad ((i,j)\in I_{\partial\Omega_N}).$$

The pseudospectral discretization of (2.1) requires that (2.1) holds at the collocation points. It leads to a discrete problem of the form

$$L_{sp}u_N = f_N \quad \text{on} \quad \Omega_N \quad , \quad u_N = g_N \quad \text{on} \quad \partial\Omega_N \ ,$$

where L_{sp} denotes the well known spectral matrix as introduced in [13].

3. THE FINITE DIFFERENCE DISCRETIZATION

For the defect correction we need a finite difference discretization of the operator L. We introduce the five point difference star

$$L_{FD}^{i,j} = \begin{bmatrix} 0 & \gamma_{0,1}^{i,j} & 0 \\ \gamma_{-1,0}^{i,j} & \gamma_{0,0}^{i,j} & \gamma_{1,0}^{i,j} \\ 0 & \gamma_{0,-1}^{i,j} & 0 \end{bmatrix} \quad ((i,j)\in I_{\Omega_N}) \qquad (3.1)$$

with $\gamma_{k,l}^{i,j} = -\beta\beta_{k,l}^{i,j}$, $\beta = 1/(2\sin(\pi/2N)\sin(\pi/N))$

and
$$\beta_{0,1}^{i,j} = b(x_i,(y_j+y_{j+1})/2)/(\sin((j+1/2)\pi/N)\sin(j\pi/N)),$$
$$\beta_{0,-1}^{i,j} = b(x_i,(y_j+y_{j-1})/2)/(\sin((j-1/2)\pi/N)\sin(j\pi/N)),$$
$$\beta_{1,0}^{i,j} = a((x_i+x_{i+1})/2,y_j)/(\sin((i+1/2)\pi/N)\sin(i\pi/N)),$$
$$\beta_{-1,0}^{i,j} = a((x_i+x_{i-1})/2,y_j)/(\sin((i-1/2)\pi/N)\sin(i\pi/N)),$$
$$\beta_{0,0}^{i,j} = -(\beta_{-1,0}^{i,j} + \beta_{1,0}^{i,j} + \beta_{0,1}^{i,j} + \beta_{0,-1}^{i,j}).$$

Let $L_{FD}^N : G(\bar{\Omega}_N) \to G(\Omega_N)$ denote the corresponding difference operator. The above discretization is the usual five point star for the Chebyshev-Lobatto points where addition theorems are used to get numerically stable formulas. Although the collocation points are not equidistant it still yields a second order scheme. (see [7])

4. MULTIGRID METHOD

We present a spectral multigrid method which consists of a relaxation scheme as in section 4.1 and the commonly used transfer operators (see [13]). The grid transfer (restriction or interpolation) consist of setting the higher Chebyshev coefficients to zero or appending additional zero coefficients as appropriate, and transforming back to the new grid.

4.1 Relaxation scheme

Central to the multigrid method is the relaxation scheme used to smooth the error on each grid. We use a Richardson (or Euler) scheme combined with defect correction. If some approximation $u_N \in G(\bar{\Omega}_N)$, $u_N = g_N$ $(\partial\Omega_N)$ of u_N is given, the calculation of a new approximation $\bar{u}_N \in G(\bar{\Omega}_N)$, $\bar{u}_N = g_N$ $(\partial\Omega_N)$ proceeds as follows:

(1) Defect computation: $d_N = f_N - L_{sp}u_N$.

(2) Defect correction: Compute an approximation $\bar{v}_N$ to the exact

solution of

$$L_{FD}^{N} v_N = d_N \quad (\Omega_N), \quad v_N = 0 \quad (\partial\Omega_N). \tag{4.1}$$

(3) Richardson step: $\bar{u}_N = u_N + \omega\bar{v}_N$ with a suitable parameter ω.

As the defect correction requires homogeneous boundary conditions the relaxation transfers the boundary values of the start approximation to the new approximation. Hence the boundary values of the start approximation should be set exactly. Relaxations for spectral problems are only efficient if fast (Fourier) transforms are available. For this purpose the Richardson scheme in contrast to other methods (as, e.g., Gauss-Seidel relaxation) is well suited. The necessity of a defect correction is enforced by the high condition number of the spectral matrix which increases like $O(N^4)$. For exact preconditioning with the finite difference we obtain a reduced eigenvalue spectrum of $[1,\pi^2/4]$ (see [6]). Defect corrections with finite element methods have been examined by Deville and Mund [5]. Since exact preconditioning is too costly, one or more steps of an iterative method are commonly used. We see from the difference star (3.1) that the Chebyshev-nodes introduce an anisotropic behaviour. Hence we prefer one step of alternating Zebra relaxation (AZR) (see Brandt [1]) for defect correction. It consists of relaxing along lines of constant x by solving

$$\gamma_{-1,0}^{i,j}\,\bar{v}_{i-1,j}^{N} + \gamma_{0,0}^{i,j}\,\bar{v}_{i,j}^{N} + \gamma_{1,0}^{i,j}\,\bar{v}_{i+1,j}^{N} =$$

$$= r_{i,j}^{N} - \gamma_{0,-1}^{i,j}\,v_{i,j-1}^{N} - \gamma_{0,1}^{i,j}\,v_{i,j+1}^{N}$$

for the grid function $\bar{v}^N$. v^N denotes the old approximation. It is followed by an analogons sweep along lines of constant y. By solving first for the odd and then for the even lines we attain to AZR. Brandt et al. [2] also recommended this kind of defect correction for spectral Fourier problems with anisotropicity. In [8] we also tested alternating incomplete LU(ILU)-decompositions as preconditioners. But it needs twice the amount of pre-computations and the convergence factors are not much better. We further have to find suitable relaxation parameters ω. Dependent on this choice we distingnish between stationary Richardson (SR), nonstationary Richardson (NSR) and minimal residual Richardson (MRR) relaxation. For the SR relaxation the parameter ω is the same for all sweeps. It is known that the optimal parameter ω_{opt} and smoothing rate μ_{opt} are

$$\omega_{opt} = \frac{8}{\pi^2+4} \doteq 0.5768\ , \quad \mu_{opt} = \frac{\pi^2-4}{\pi^2+4} \doteq 0.4232.$$

For the NSR relaxation [13] with k different parameters the optimal ones are

$$\omega_{l,opt} = \left(\frac{\pi^2+4}{8} + \frac{\pi^2-4}{8}\cos\left(\frac{2l-1}{2k}\pi\right)\right)^{-1} (l=1,\dots,k)$$

with smoothing rate

$$\mu_{k,opt} = \left| t_k\left(\frac{\pi^2+4}{\pi^2-4}\right)\right|^{-1/k} \quad \text{where } t_k(x) = \cos k \arccos x\ .$$

In the applications we choose $k = 3$ with $\mu_{3,opt} \doteq 0.2797$ since the additional work for greater k is not paying.

Canuto et al. [4] have already tested the MRR relaxation as an iterative method. We examine its smoothing properties in connection with spectral multigrid methods. The parameter ω is chosen in order to minimize the residual $\bar{d}_N = f_N - L_{sp}\bar{u}_N$ of the new approximation $\bar{u}_N$ in the norm $\|.\|$, defined by

$$\|v_N\| = (v_N,v_N)^{1/2} \text{with } (v_N,w_N) = N^{-2} \sum_{i,j=1}^{N-1} v_N^{i,j} w_N^{i,j} .$$

If a start approximation u_N with the residual $d_N = f_N - L_{sp}u_N$ is given, one iteration step proceeds as follows:

(1) Defect computation: Compute the correction $\bar{v}_N$ as in (4.1)

(2) Parameter computation: $\omega = (d_N,w_N)/(w_N,w_N)$ with $w_N = L_{sp}\bar{v}_N$

(3) Richardson step: Compute the new approximation $\bar{u}_N$ and residual $\bar{d}_N$:

$$\bar{u}_N = u_N + \omega\bar{v}_N \ , \quad \bar{d}_N = d_N - \omega w_N \ .$$

For this type of relaxation the additional work of computing the adaptive parameters becomes abvious. The computation of w_N can be used for the evaluation of the new residual $\bar{d}_N$. Since after the coarse grid correction the residual of the last step is not needed we recommend only smoothing steps before coarse grid correction.

5. SOME NUMERICAL RESULTS

In order to estimate the convergence properties we compute the convergence factor ρ of the multigrid operator by the power method. The convergence factor per work can be defined by $\rho_w = \rho^{1/w}$. Hereby the work unit is the amount of work on the finest grid. We use a V-cycle with four grids, given as 4x4, 8x8, 16x16 and 32x32. If $\upsilon_1(\upsilon_2)$ relaxations are used in the downward (upward) direction we obtain $W = (1+1/4 + 1/16 + 1/64)(\upsilon_1+\upsilon_2)$. We compare SR, NSR and MRR relaxation where we employ $\upsilon_1=2$, $\upsilon_2=0$ for SR, MRR and $\upsilon_1=3$, $\upsilon_2=0$ for NSR. We tested isotropic and anisotropic examples with coefficients

$$a(x,y) = b(x,y) = 1 + \varepsilon\, e^{x+y} \ , \quad \varepsilon > 0 \tag{5.1}$$

and

$$a(x,y) = 1 + \varepsilon\, e^{x} \ , \ b(x,y) = 1 + \varepsilon\, e^{y} \ , \quad \varepsilon > 0. \tag{5.2}$$

Table I: ρ_w for examples (5.1), (5.2)

(5.1)

ε	0.	0.1	0.3	1.
SR	0.4811	0.5164	0.5922	0.6850
NSR	0.3933	0.4144	0.5191	0.6385
MRR	0.3927	0.4026	0.4335	0.4564

(5.2)

ε	0.1	0.3	1.
SR	0.5132	0.5592	0.6459
NSR	0.3936	0.4654	0.5884
MRR	0.4025	0.4044	0.4080

The results in table I show the effectiveness of our preconditioning. Furthermore the robustness of MRR for increasing ε becomes obvious. We also examined the Helmholtz equation

$$-\Delta u(x,y) + \gamma u(x,y) = f(x,y) \tag{5.3}$$

with a parameter $\gamma \geq 0$. The spectral and finite difference discretization is done in a straightforward manner. SR and NSR relaxation yield nearly unchanged convergence factors whereas MRR yields improved factors for large γ(see table II).

Table II: ρ_w for (5.3) and MRR relaxation

γ	0.	1.	10^4	10^5	10^7
ρ_w	0.3927	0.3708	0.3522	0.1967	0.0074

For γ about 10^5 the term γu gets dominant in $-\Delta u + \gamma u$ and the rates of MRR are dramatically better. Finally, we give some results for two examples which show the high accuracy of spectral methods:

Example 1: $-\Delta u(x,y) = 32\pi^2 \sin 4\pi x \cdot \sin 4\pi y$

Example 2: $-\Delta u(x,y) = 1$

with homogeneous boundary conditions. The solution of the first example is analytic whereas the second one has singularities in the four corners. The solution behaves like $O(r^2 \ln r)$ $(r \to 0)$ where r is the distance from the corner. We calculate the absolute error EM in the maximum norm. IT denotes the number of V-cycles needed in order to reach this accuracy.

Table III:

Example 1:

N	EM	IT	ρ_w
16	5.25E-3	2	0.4564
32	2.17E-12	7	0.4743

Example 2:

N	EM	IT	ρ_w
16	7.47E-7	4	0.3432
32	5.51E-8	5	0.3595

The high accuracy for both examples is very impressive. It becomes obvious that only a few V-cycles are needed to reach this accuracy. We further tested two examples on L-shaped domains:

Example 3: $-\Delta u_1(x,y) = 32 \sin 4\pi x \sin 4\pi y$
on $\Omega_L^1 = (0,2) \times (0,1) \cup (0,1) \times (0,2)$
with $u(x,y) = 0$ on $\partial\Omega_L^1$

and

Example 4: $-\Delta u_2(x,y) = 0$
on $\Omega_L^2 = (-1,0) \times (-1,1) \cup (-1,1) \times (0,1)$
with $u(r,\zeta) = r^{4/3} \sin \frac{4}{3}\zeta$ on $\partial\Omega_L^2$
in polar coordinates (r,ζ).

The exact solution for the examples 3 and 4 are $u(x,y) = \sin 4\pi x \cdot \sin 4\pi y$ and $u(r,\zeta) = r^{4/3} \sin \frac{4}{3}\zeta$. The first solution is smooth whereas the second one has a singularity in $r = \emptyset$. We use the Schwarz procedure in an outer loop where the problems in rectangular domains are approximately solved by one V-cycle of SMG. In the Schwarz procedure the values of the approximation on the pseudo-boundaries (in $\Omega_L^{1,2}$) are transferred by a Clenshaw-recursion since FFT's are no more available. This takes a computational effort of about $O(N^2)$ arithmetic operations which is below the total amount of $O(N^2 \ln N)$. Because of the high overlapping size (a third of the whole domain) the convergence is determined by the SMG convergence (see table IV).

Table IV:

Example 3:

N	EM	IT	ρ_w
16	1.52E-3	2	0.3554
32	5.29E-13	8	0.3832

Example 4:

N	EM	IT	ρ_w
16	2.01E-3	3	0.4699
32	8.12E-4	4	0.5095

The numerical results onesmore show the high accuracy of spectral methods and substantiate the usefulness of SMG.

REFERENCES

1. Brandt, A.: Multi-level adaptive solutions to boundary value problems. Math. Comp. 31(1977), 333.
2. Brandt, A., Fulton, S.R., Taylor, G.D.: Improved spectral multigrid methods for periodic elliptic problems. J. Comp. Phys. 58 (1985), 96.
3. Canuto, C., Funaro, D.: The Schwarz algorithm for spectral methods.To appear in Siam J. on Numerical Anal.
4. Canuto, C., Quateroni, A.: Preconditioned minimal residual methods for Chebyshev spectral calculations. J. Comp. Phys. 60 (1985), 315.
5. Deville, M., Mund, E.: Chebyshev pseudospectral solution of second order elliptic equations with finite element preconditioning. J. Comp. Phys. 60(1985), 527.
6. Haldenwang, P., Labrosse, G., Abbondi, S., Deville, M.: Chebyshev 3-D spectral and 2-D pseudospectral solvers for the Helmholtz equation. J. Comp. Phys. 55(1984), 115.
7. Heinrichs, W.: Kollokationsverfahren und Mehrgittermethoden bei elliptischen Randwertaufgaben. GMD-Bericht Nr. 168, R. Oldenbourg Verlag, München/Wien 1987.
8. Heinrichs, W.: Line relaxation for spectral multigrid methods. To appear in J. Comp. Phys..
9. Heinrichs, W.: Multigrid methods for combined finite difference and Fourier problems. To appear in J. Comp. Phys..
10. Orszag, S.A.: Spectral methods for problems in complex geometries. J. Comp. Phys. 37(1980), 70.
11. Street, C.L., Zang, T.A., Hussaini, M.Y.: Spectral multigrid methods with applications to transonic flow. J. Comp. Phys. 57 (1985), 43.
12. Stüben, K., Trottenberg, U.: Multigrid methods: fundamental algorithms, model problem analysis and applications, in Multigrid methods, Proceeding of the conference at Köln-Porz, 1981,(Hackbusch, Trottenberg (eds.)), Lecture Notes in Mathematics 960.
13. Zang, T.A., Wong, Y.S., Hussaini, M.Y.: Spectral multigrid methods for elliptic equations I. J. Comp. Phys. 48(1982), 485.
14. Zang, T.A., Wong, Y.S., Hussaini, M.Y.: Spectral multigrid methods for elliptic equations II. J. Comp. Phys. 54(1984), 489.

NUMERICAL SOLUTION OF TRANSONIC POTENTIAL FLOW IN 2D COMPRESSOR CASCADES USING MULTI-GRID TECHNIQUES

M. HUNĚK, K. KOZEL, M. VAVŘINCOVÁ
o.p.ČKD-Prague, Compressors, Klečákova 1947, CS-19200 Prague
Department of Computational Techniques and Informations, Czech Technical University Prague, Suchbátarova 4, CS-16607 Prague 6

SUMMARY

The paper deals with numerical solution of transonic potential flow in two-dimensional compressor cascades. The governing equation is full potential equation in non-conservative form. The work considers a weak formulation of the problem; for numerical solution modified Jameson's rotated difference scheme is used and to simplify periodical conditions we used the locally disturbance form of the governing equation. The resulting algebraic system of difference equations is solved by:

1/ SLOR method,
2/ line relaxation method in three grid-levels,
3/ FAS and CS-algorithm using full weighting of non--conservative residue.

Some applications of these several methods applied to transonic cascade flows are compared to other numerical and experimental results.

1. FORMULATION

The problem of transonic flows in 2D cascades is periodical and we consider the solution in domain Ω / fig. 1.1 /. We consider fulfilling governing equation

$$(\rho \Phi_x)_x + (\rho \Phi_y)_y = 0, \quad [x,y] \in \Omega \cup \overline{AF_1} \cup \overline{F_2B}. \tag{1.1}$$

in the integral form $\oint_{\partial D} \rho \Phi_x dy - \rho \Phi_y dx = 0, \quad D \subset \Omega \cup V_\varepsilon$ (1.2)

/ V_ε is some suitable vicinity of $\overline{AF_1} \cup \overline{F_2B}$; $\Phi = w_\infty \tilde{\Phi}$, $\tilde{\Phi}$ - velocity potential, $\rho = \rho(\Phi_x^2 + \Phi_y^2)$ is given known function / and next periodical and boundary conditions:

1/ $\Phi_x = \cos\alpha$, $\Phi_y = \sin\alpha$, α-angle of attack, $[x,y] \in \overline{AD}$, (1.3)
2/ $\Phi_x(E) = \Phi_x(E')$, $\Phi_y(E) = \Phi_y(E')$, $\overline{EE'} \parallel \overline{F_1F_1'}$, $E \in \overline{AF_1} \cup \overline{F_2B}$, ,
$E' \in \overline{DF_1'} \cup \overline{F_2'C}$, (1.4)
3/ $\Phi_n = 0$, $[x,y] \in \partial P \cup \partial P'$ /n is outer normal to ∂P, $\partial P'$/, (1.5)
4/ $\Phi_x = K_1(\gamma)$, $\Phi_y = K_2(\gamma)$, $[x,y] \in \overline{BC}$ (1.6)

K_1, K_2 are constants depending on velocity circulation γ along profile. Relations $K_i = K_i(\gamma)$; i=1,2 are derived using (1.2) for $\partial D = \partial \Omega$ and non-vorticity conditions $\Phi_{xy} = \Phi_{yx}$ in integral form and (1.6).

Remark 1.1: Along upstream boundary $\overline{AD}$ we consider homogenous parallel flows with $M=M_\infty$, conditions (1.4) are periodical conditions for velocity components. Nonpermeability conditions (1.5) are considered along profile surface and we assume homogenous downstream flowfield with $\Phi_x=K_1(\gamma)$, $\Phi_y=K_2(\gamma)$, where K_1, K_2 are constants / unknown in advance / depending on the solution of the problem.

2. NUMERICAL SOLUTION

The problem is solved by finite difference method using modified Jameson's rotated difference scheme. Consider local transformation of full potential equation to (s,n) coordinate system, s-streamline, n-normal,

$$(1-M^2)\Phi_{ss} + \Phi_{nn} = 0. \qquad (2.1)$$

For numerical solution (u,v) coordinate system is used / see fig. 1.1 / and for suitable realization of periodical conditions the local disturbance potential φ:

$$\varphi_u = \Phi_u - 1,\ \varphi_v = \Phi_v - \cos(\beta-\alpha),\ \beta \text{ is stagger angle,} \qquad (2.2)$$

is considered. Then the governing equation is the same (2.1); φ_{ss} and φ_{nn} can be expressed

$$\begin{aligned}\varphi_{ss} &= A_1\varphi_{uu} + B_1\varphi_{uv} + C_1\varphi_{vv},\\ \varphi_{nn} &= A_2\varphi_{uu} + B_2\varphi_{uv} + C_2\varphi_{vv}.\end{aligned} \qquad (2.3)$$

Since quasilinear governing equation (1.1), (2.1) is of mixed type: elliptic for $M<1$, parabolic for $M=1$ / sonic line/ and hyperbolic for $M>1$, we use following difference approximations. Central differencing for approximation of all second derivates φ_{uu}, φ_{uv}, φ_{vv} in (2.3)

$$\begin{aligned}\varphi_{uu}|_{i,j} &= (\varphi^{n}_{i+1,j} - \varphi^{n}_{i,j} - \varphi^{n+1}_{i,j} + \varphi^{n+1}_{i-1,j})/\Delta u^2 + O(\Delta u^2),\\ \varphi_{uv}|_{i,j} &= (\varphi^{n}_{i+1,j+1} - \varphi^{n}_{i+1,j-1} - \varphi^{n+1}_{i-1,j+1} + \varphi^{n+1}_{i-1,j-1})/(4\Delta u\Delta v) + O(\Delta u\Delta v),\\ \varphi_{vv}|_{i,j} &= (\varphi^{n+1}_{i,j+1} - 2\varphi^{n+1}_{i,j} + \varphi^{n+1}_{i,j-1})/\Delta v^2 + O(\Delta v^2)\end{aligned} \qquad (2.4)$$

for $M<1$. If $M>1$, the same differencing is used for φ_{uu}, φ_{uv}, φ_{vv} in φ_{nn}, but for terms in φ_{ss} backward differencing of the first order is used; for example, assume $\varphi_u>0$, $\varphi_v>0$, then

$$\begin{aligned}\varphi_{uu}|_{i,j} &= (2\varphi^{n+1}_{i,j} - \varphi^{n}_{i,j} - 2\varphi^{n+1}_{i-1,j} + \varphi^{n}_{i-2,j})/\Delta u^2 + O(\Delta u),\\ \varphi_{uv}|_{i,j} &= (2\varphi^{n+1}_{i,j} - \varphi^{n}_{i,j} - \varphi^{n+1}_{i-1,j} + \varphi^{n+1}_{i,j-1} + \varphi^{n}_{i-1,j-1})/(\Delta u\Delta v) + O(\Delta u+\Delta v),\\ \varphi_{vv}|_{i,j} &= (2\varphi^{n+1}_{i,j} - \varphi^{n}_{i,j} - 2\varphi^{n+1}_{i,j-1} + \varphi^{n}_{i,j-2})/\Delta v^2 + O(\Delta v).\end{aligned} \qquad (2.5)$$

Upper index n, n+1 denotes number of iteration cycle.

Remark 2.1: Realization of boundary and periodical conditions

Since we consider periodical Neumann problem, instead (1.1), it is used $\varphi=0$ along $\overline{AD}$. In the grid points along $\overline{AF_1}\cup\overline{F_2B}$ it is fulfilled (2.1) in finite-difference form, along $\overline{DF_1'}$ then periodical condition $\varphi(E)=\varphi(E')$ is fulfilled and along $\overline{F_2'C}$ in the form $\varphi(E')=\varphi(E)-\gamma$, γ is circulation,

$$\gamma=\varphi(F_2)-\varphi(F_2')=\varphi(E)-\varphi(E'),\quad E\in F_2B,\ E'\in F_2'C. \qquad (2.6)$$

For computation of circulation γ we used special iteration procedure in the numerical solution. Downstream boundary conditions (1.6) are considered in the difference scheme, $K_1=K_1(\gamma^n)$, $K_2=K_2(\gamma^n)$, where γ^n is value of γ corresponding n-th iteration cycle.

Boundary conditions along the profile surface (1.5) are approximated such that φ_n is expressed by φ_u , φ_v and φ_u are approximated by central differencing and φ_v forward/ lower profile boundary / or backward / upper profile boundary / differencing second order.

Very important problem for the numerical results is realization of difference equation for last grid point lying on $\overline{AF_1}$ in vicinity of leading edge F_1 / F_1 is not node of grid /. We have to fulfil the difference approximation of (2.1) there / see fig. 2.1 /. There is used values $\varphi_{i-1,j}$, $\varphi_{i,j}$, $\varphi_{i+1,j}$ for approximation φ_{uu} , where $\varphi_{i+1,j}$ is defined by one of following ways:

1/ $\varphi_{i+1,j}=\varphi^{+}_{i+1,j}$,

2/ $\varphi_{i+1,j}=(\varphi^{+}_{i+1,j}+\varphi^{-}_{i+1,j})/2$, (2.7)

3/ $\varphi_{i+1,j}=\varphi(F_1)$, where $\varphi(F_1)$, is extrapolated from approximation of boundary condition $\Phi_n|_{F_1} = 0$,

/ $\varphi^{+}_{i+1,j}$ or $\varphi^{-}_{i+1,j}$ is value on upper or lower profile surface, respectively /.

Resulting system of algebraic difference equations is solved by line iteration method / or in older version by succesive line relaxation method - SLOR /.

3. MULTI-GRID TECHNIQUES

The resulting system of algebraic equations is nonlinear

$$L_h \varphi_h = f_h. \tag{3.1}$$

SLOR in one-level-grid is unsufficient for practical use. It is due to lower efficienty for reduction of slow error frequencies and necessity of a good first approximation of solution / for higher M_∞ especially /. Computation is realized in sequences of flow fields with increasing value M_∞. For the first M_∞ lower subcritical value must be considered. For futher flow fields we use last solution as the first approximation for new M_∞ everytime.

A/ <u>Line iteration in three grid-levels / MULPO /</u>

The first multi-grid method uses line iterations in the three grid-levels. The method uses standart coarsering G_{4h}, G_{2h} to the finest grid G_h. For H=2h, H=4h respectively, the system of equations is set up the same way like system (3.1) for the finest grid G_h:

$$L_H \varphi_H = f_H. \tag{3.2}$$

Two elementary iterations are presented here.

a/ <u>V-cycle</u>

It consists of one line iteration in grid $G_h, G_{2h}, G_{4h}, G_{2h}, G_h$. Function φ_{2H} / the solution (3.2) in G_{2H} / is transformed into grid G_H / H=h,2h / by bilinear prolongation for each transmition from coarser to finer grid. The accuracy of V-cycle is given by step of the coarsest grid. Therefore it is necessary use iterations in the finest grid G_h for finished error smoothing.

b/ S-cycle

The iteration cycle consists of four line iterations using grid G_h.

Scheme 1/ in (2.7) is used for fulfilling of governing equation in the last point lying in $\overline{AF_1}$ near leading edge. The scheme prefers a suction profile surface and it has a negative influence on velocity value in vicinity of leading edge.

The method has a high convergence rate then one grid-level method and this method converges for higher M_∞ without good initial approximation of the solution. A disadvantage of this method is necessity of using of S-cycle.

B/ FAS-algorithm / Fully Approximation Scheme /

FAS algorithm is used for increasing of accuracy of V-cycle. Generally FAS multi-grid method can be defined recursively on the basis of the FAS two-grid method. The two-grid method can be divided into next steps.

1/ Smoothing part - if φ_h^n is old approximation of solution, then new value is computed using

$$\bar{\varphi}_h^n := S_h(\varphi_h^n, L_h, f_h). \tag{3.3}$$

S_h is one iteration of algebraic system (3.1).

2/ Coarse-grid correction - residue is computed on the finest grid G_h

$$R_h = f_h - L_h \bar{\varphi}_h^n . \tag{3.4}$$

The residue is restricted on G_{2h} grid by full weighting r, then

$$R_{2h} = rR_h, \text{where } r \triangleq \frac{1}{16} \begin{bmatrix} 1 & 2 & 1 \\ 2 & 4 & 2 \\ 1 & 2 & 1 \end{bmatrix}_h^{2h} . \tag{3.5}$$

Next we compute coarser grid solution that one must hold correction equation:

$$L_{2h} \tilde{\varphi}_{2h}^n = R_{2h} + L_{2h} \bar{\varphi}_{2h}^n , \tag{3.6}$$

$\bar{\varphi}_{2h}^n$ is injection of $\bar{\varphi}_h^n$ from G_h to G_{2h}. Matrix L_{2h} is defined by $L_{2h} = rL_h p$, where p is prolongation from G_{2h} to G_h. In the fact we use the first approximation of difference scheme on coarser grid G_{2h}. Practically the solution of (3.6) is approximated by one line iteration of the system on G_{2h}. It means

$$\tilde{\varphi}_{2h}^n := S_{2h}(\bar{\varphi}_{2h}, L_{2h}, R_{2h} + L_{2h}). \tag{3.7}$$

Then we computed the correction of the solution and interpolated on G_h. New approximation can be written in the form

$$\varphi_h^{n+1} = \bar{\varphi}_h^n + p\,(\tilde{\varphi}_{2h}^n - \bar{\varphi}_{2h}^n) \; ; \tag{3.8}$$

where prolongation p must hold $p = cr^*$ / r^* is adjoint operator to r, c is scale /. For approximation of governing equation in the last point in $\widehat{AF_1}$ in front of leading edge scheme 2/ in

(2.7) is used.

C/ CS-algorithm / Correction Scheme /

CS method is simplified linear multi-grid method explained in our case by following steps.

1/ Let φ_h^n is approximation of solution of the problem (2.1). Matrix $\mathcal{L}_h$ is constant for one multi-grid cycle and is defined as derivation of L_h in φ_h^n / Newton iteration /.

2/ Smoothing part: We solve system

$$\mathcal{L}_h \Delta\varphi_h = R_h, \; R_h = f_h - \mathcal{L}_h \varphi_h^n , \tag{3.9}$$

in the finest grid. Value $\Delta\varphi_h$ is approximated by one line iteration of this system

$$\Delta\varphi_h := S_h(0, \mathcal{L}_h, R_h) . \tag{3.10}$$

3/ Coarse-grid correction: We compute defect

$$d_h = R_h - \mathcal{L}_h \Delta\varphi_h . \tag{3.11}$$

The next equation is solved in the coarser grid

$$\mathcal{L}_{2h} \Delta\varphi_{2h} = d_{2h}, \text{ where } d_{2h} = r d_h , \tag{3.12}$$

and the following approximation is used in the form

$$\Delta\varphi_{2h} := S_{2h}(0, \mathcal{L}_{2h}, d_{2h}) . \tag{3.13}$$

New approximation φ_h^{n+1} is obtained

$$\varphi_h^{n+1} = \varphi_h^n + \Delta\varphi_h + p\Delta\varphi_{2h} . \tag{3.14}$$

The two-grid method can be also recursively defined for three grid-levels. For fulfilling of full potential equation in the last point in $\overline{AF}_1$ near leading edge scheme 3/ (2.7) is used.

4. SOME NUMERICAL RESULTS

Transonic flows in variety of 2D cascades was computed using all versions of presented method / SLOR, MULPO, FAS, CS /. We would like to present some of mentioned results describing main features of the methods.

I/ First case of numerical results is numerical solution of transonic flows in DCA-8% cascade. Interferometric measurements of Institute of Thermomechanics of Czechoslovak Academy of Sciences were published in [3]. We can compare our numerical solution presented using lines M=const. to experimental results because white and black strips are lines with ρ=const. and difference of neighbour strips is equal $\Delta\rho$=const.. We can compare numerical and experimental results not only qualitatively / to compare shape of lines M=const. in fig.4.1 and lines ρ=const. in fig. 4.2 / but also quantitatively, because sonic line M=1 corresponds dotted line in experimental results.

II/ Second presented results are numerical results of transonic flow through compressor cascade DCA-2S. Fig. 4.3 shows comparison of numerical results using Mach number distribution achieved by our method and method [2] / for $M_\infty=0.87$ /. For this type of cascade flows we can follow convergence of SLOR, FAS and CS algorithm to the steady state in L_2-norm and C-norm / maximal norm / of residue during iteration proces / see fig. 4.4 and 4.5 /. Probably the way of fulfilling (2.1) in the last grid point near F_1 is the reason, that CS-algorithm seems to be better then FAS, where maximal value of residue is in this point.

III/ Third case of numerical results is numerical results of transonic flow through MCA-2S compressor cascade of ČKD-Prague. We can compare experimental and numerical results using:

a/ distribution of Mach number along upper part / suction side / of given profile surface / fig. 4.6 / for for $M_\infty=0.6$, $\alpha=19.64^\circ$, $\beta=58.68^\circ$ and $M_\infty=0.863$, $\alpha=19.64^\circ$, $\beta=58.68^\circ$ / fig. 4.6 /,

b/ interferometric measurements of Institute of Thermomechanics are compared to our numerical results using lines M=const. for $M_\infty=0.6$, $\alpha=19.64^\circ$, $\beta=58.68^\circ$/ fig. 4.7 and 4.9 / and for $M_\infty=0.863$, $\alpha=19.64^\circ$, $\beta=58.68^\circ$ / see fig. 4.8 and 4.10 /.

One can see good agreement between numerical and experimental results for M =0.6. In the other case we can observe good agreement in distribution of Mach number along suction side in the region without separation of boundary layer / including position and shape of shock wave /, but in the region with separated flow is not possible to expect agreement, because we consider only potential model of transonic flows.

5. CONCLUSION

Presented method of numerical solution of potential transonic cascade flows is suitable for calculation of theoretical and practical cases of two-dimensional cascades where physical model corresponds to potential flows / weak shock waves, negligible viscous effects /. Using multigrid techniques we developed suitable numerical method for the case of computation of transonic flow in compressor cascades and our computers. Convergence to steady state is controlled by residue in L_2--norm and C-norm and seems to be very good especially for CS--algorithm. Multigrid techniques also enables to realize computation of several cases of upstream Mach numbers and angle of attack α using our computers.

REFERENCES

[1] JAMESON, A.:"Numerical Computation of Transonic Flow with

Shock Waves", Symposium Transonicum II., Goettingen (1975), Berlin Springer Verlag (1976).

[2] FOŘT, J., KOZEL, K.: "Numerical Solution of Inviscid Two-Dimensional Transonic Flow Through a Cascade", Transaction of the ASME, Journal of Turbomachinary, January (1987), pp. 108-113.

[3] DVOŘÁK, R.: "On the Development and Structure of Transonic Flow in Cascades", Symposium Transonicum II., Oswatitsch, Rues, R. (eds.), Springer Verlag Berlin, pp. 297-307.

[4] HACKBUSCH, W.: "Multi-Grid Methods and Applications", Computational Mathematics 4, Springer Verlag Berlin Heidelberg New York Tokyo, (1985).

[5] BRANDT, A.: "Multi-Level Adaptive Solutions to Boundary-Value Problems", Math. Comp., 31 (1977), pp. 333-390.

[6] HUNĚK, M., KOZEL, K., VAVŘINCOVÁ, M.: "Numerical Solution of 2D Transonic Flow Problem in Compressor Cascades Using the Full Potential Equation and Multi-Grid Techniques; Numerical Solution of Boundary-Layer and Transonic Euler Equations", technical report o.p. ČKD-Prague, Compressors, KKS-TK 2.7-285, November (1987), (in Czech).

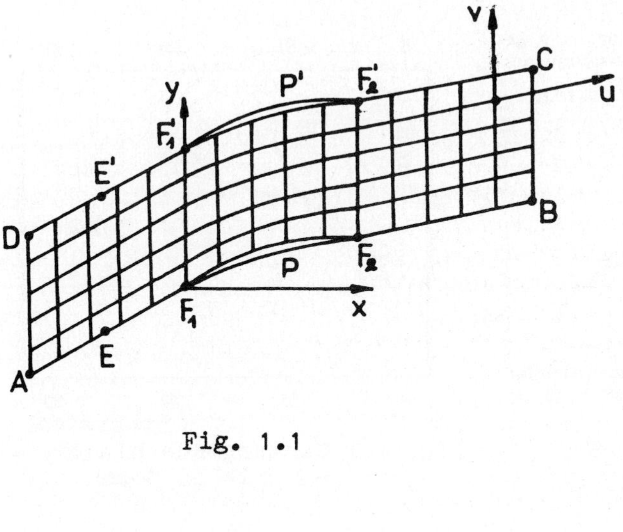

Fig. 1.1

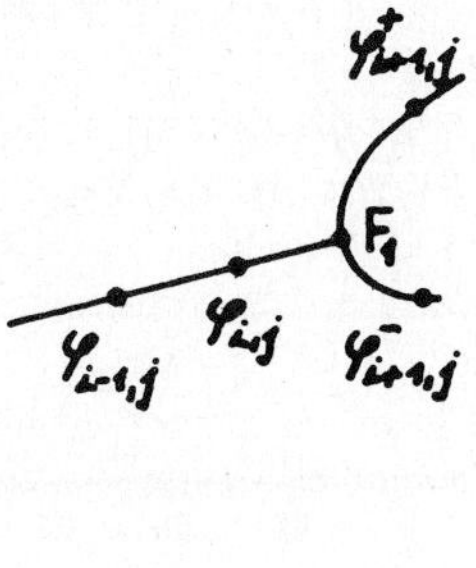

Fig. 2.1

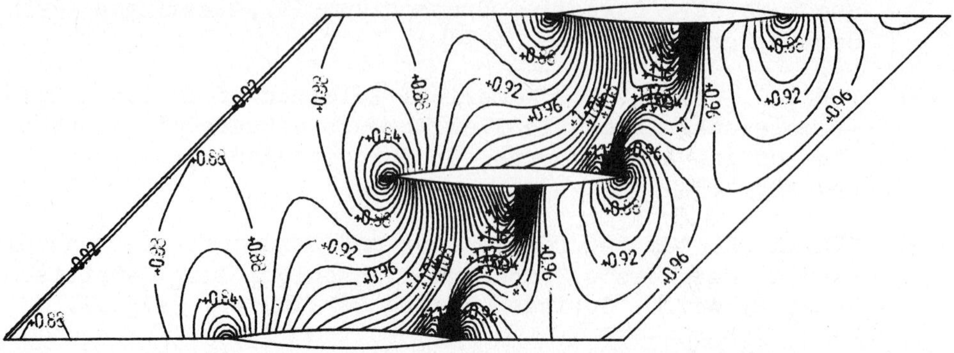

Fig. 4.1 DCA-8%, $\alpha=0.55^\circ$, $\beta=45^\circ$, $M_\infty=0.94$; computation

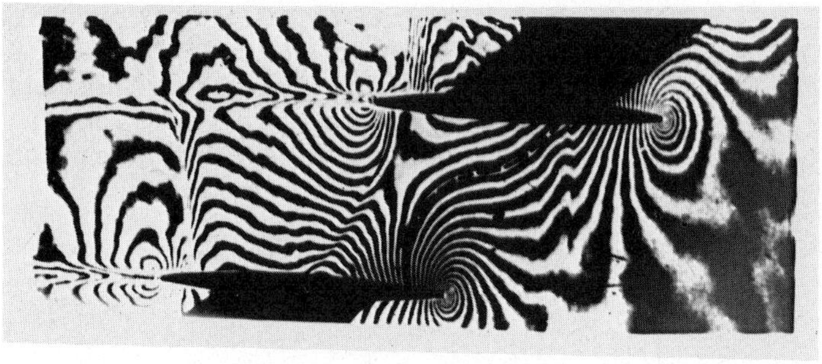

Fig. 4.2 DCA-8%, interferometric measurement

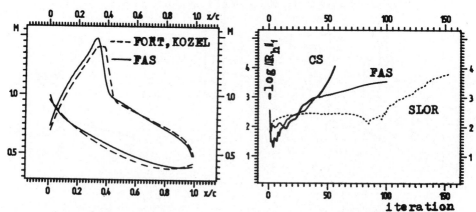

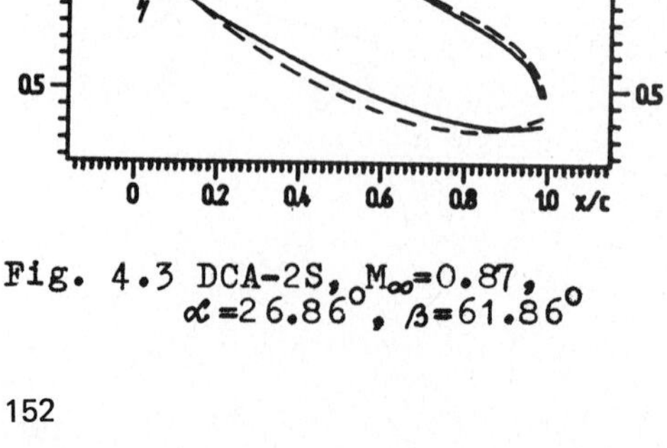

Fig. 4.3 DCA-2S, $M_\infty=0.87$, $\alpha=26.86^\circ$, $\beta=61.86^\circ$

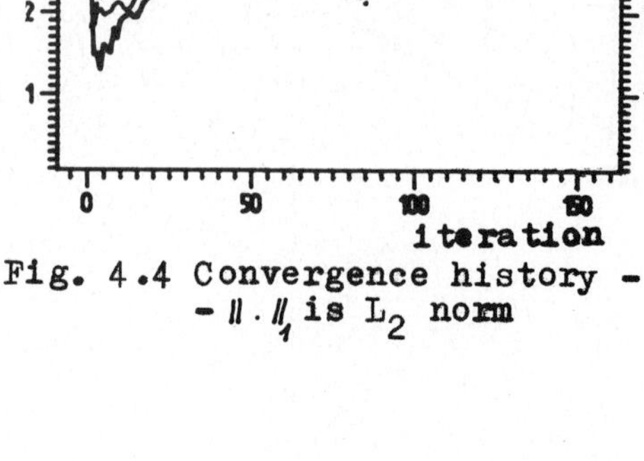

Fig. 4.4 Convergence history - $\|\cdot\|_1$ is L_2 norm

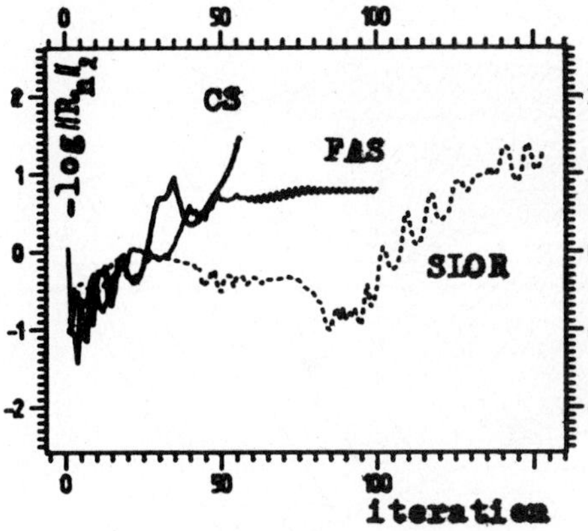

Fig. 4.5 Convergence history- $\| \cdot \|_2$ is C-norm

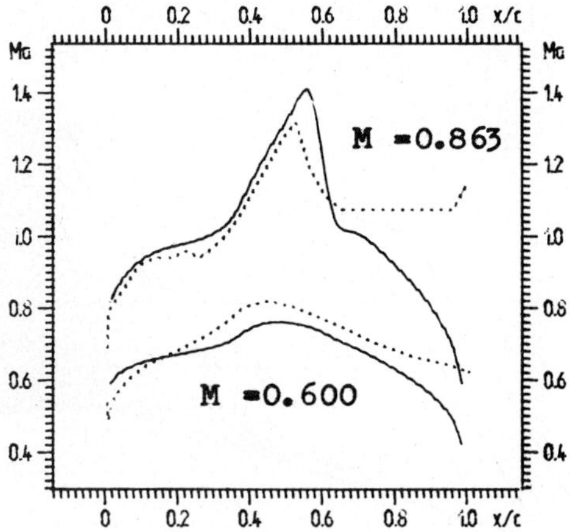

Fig. 4.6 MCA-2S, suction side ....comput., —— exper.

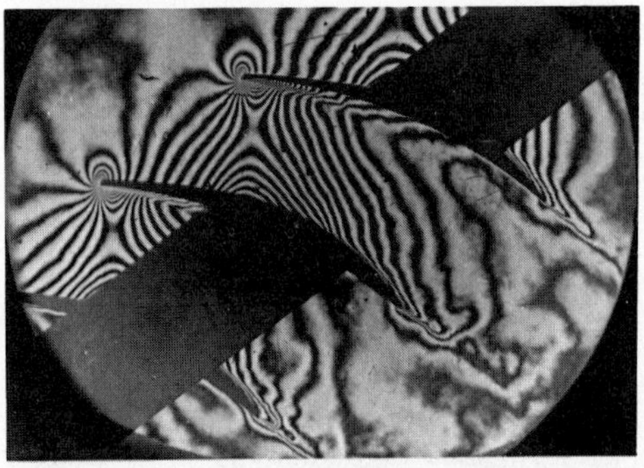

Fig. 4.7 Interferometric measurement; MCA-2S, M_∞= 0.600

Fig. 4.8 Interferometric measurement; MCA-2S, M_∞= 0.863

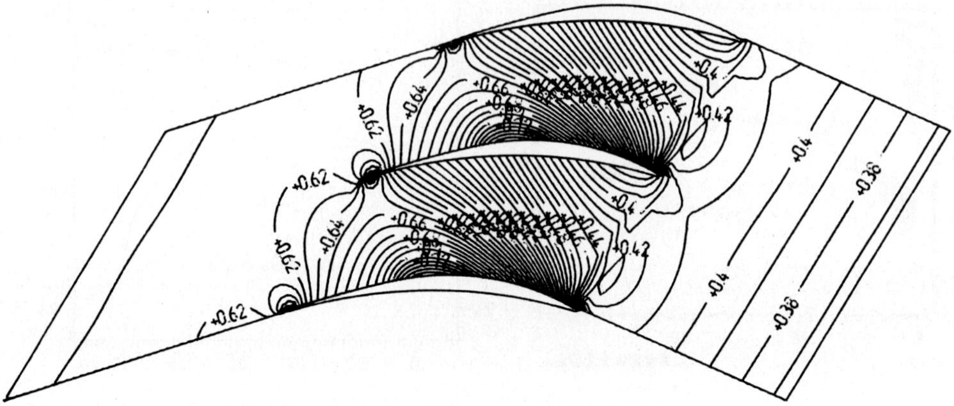

Fig. 4.9 Computation; MCA-2S, M_∞=0.600, α=19.64°, β=58.68°

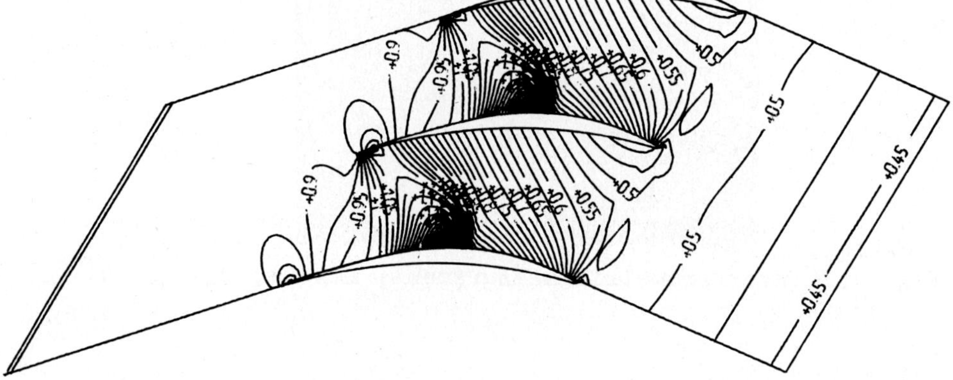

Fig. 4.10 Computation; MCA-2S, M_∞=0.863, α=19.64°, β=58.68°

LOCAL MODE SMOOTHING ANALYSIS OF VARIOUS INCOMPLETE FACTORIZATION ITERATIVE METHODS

M. Khalil
Laboratoire d'Analyse Numerique, Univ. Paul Sabatier,
118, Route de Narbonne, 31062 Toulouse-cedex, France

SUMMARY

By local mode analysis an infinite domain we can assess the smoothing efficiency of iterative methods based on incomplete factorizations. Our analysis consists of an analytic and numerical study of a simple two-dimensional problem: $-\epsilon u_{xx} - u_{yy} = f$ discretized by finite differences. This analysis is done for pointwise and blockwise incomplete factorizations. This study show how this analysis is helpful to obtain simple expressions for the smoothing factors and to introduce new smoothers.

INTRODUCTION

The most common smoothers in multigrid are stationary iterative methods such as Jacobi, Gauss-Seidel and iterative methods based on incomplete factorizations of the coefficient matrix.
Consider the system:

$$A\phi = f. \tag{0.1}$$

Let a splitting of A be represented as:

$$A = C - R, \tag{0.2}$$

where C is an approximation of A which is nonsingular, and R is the error matrix. It is assumed here that C is obtained by pointwise or blockwise incomplete factorization.
From (1.2) we obtain a stationary iterative method of the form:

$$\phi^{(m+1)} = (I - C^{-1}A)\phi^{(m)} + C^{-1}f. \tag{0.3}$$

We define the iteration matrix S by:

$$S = I - C^{-1}A. \tag{0.4}$$

Let $\epsilon^{(m)} = \phi - \phi^{(m)}$, (0.5)

denote the error. We have

$$\epsilon^{(m+1)} = S\epsilon^{(m)} \tag{0.6}$$

Local mode analysis applied to (0.6) has been used to evaluate the smoothing behaviour of methods such as Jacobi and Gauss-Seidel [4,7,8] and some pointwise incomplete factorizations [6,7] analytically or blockwise incomplete factorization numerically [7,8]. Here we will introduce an analysis for incomplete factorization by points or blocks of the 2D anisotropic diffusion problem. We will show how this analysis helps us to introduce a new efficient smoother in 2D.

Let M_h be a finite difference matrix with stencil $[M_{h,\nu}]$. We may write:

$$M_h\psi(x) = \sum_{\nu\in J_{M_h}} M_{h,\nu}\psi(x+\nu h), \tag{1.1}$$

where J_{M_h} is the non-zero pattern of M_h:

$$J_M = \left\{\nu = (\nu_1,\nu_2) \in \mathbb{Z}^2 : M_{h,\nu} \neq 0\right\}. \tag{1.2}$$

Let us take [3] $\psi_{h,\theta}(x) = e^{ij\theta_s + ik\theta_t}$, with $x = (jh,kh)$, $h = 1/(N+1)$, $\theta_s = s\pi h$, $\theta_t = t\pi h$; $s,t = -N,\ldots,N$. Assuming the computational grid the unbounded and $[M_{h,\nu}]$ to be independent of position, we have:

$$M_h\psi_{h,\theta}(x) = \hat{M}_h(\theta)\psi_{h,\theta}(x), \tag{1.3}$$

where $\theta = (\theta_s,\theta_t)$ and

$$\hat{M}_h(\theta) = \sum_{\nu\in J_M} M_{h,\nu}\exp(i\theta\cdot\nu), \qquad \theta\cdot\nu = \theta_s\nu_1 + \theta_t\nu_2. \tag{1.4}$$

If we take $\epsilon^{(m)} = c_m\psi_{h,\theta}$, by substitution in (0.6) we obtain $c_{m+1} = \hat{S}c_m$.
The applification factor of the error is then given by:

$$\rho(\theta) = \left|\frac{c_m + 1}{c_m}\right| = |\hat{S}(\theta)| = \left|1 - \frac{\hat{A}(\theta)}{\hat{C}(\theta)}\right|, \tag{1.5}$$

and the smoothing factor by:

$$\bar{\rho} = \max_{\theta\in\Theta} \rho(\theta), \tag{1.6}$$

where

$$\Theta = \left\{\theta \,/\, \frac{\pi N}{2(N+1)} \leq \max\,(|\theta_s|,|\theta_t|) \leq \frac{N\pi}{N+1}\right\}, \tag{1.7}$$

N even fixed.

LOCAL MODE SMOOTHING ANALYSIS OF INCOMPLETE FACTORIZATIONS

It is assumed that the domain is infinite and that the matrix A arises as a discretization of an equation with constant coefficients, so that A is an infinite Toeplitz matrix [6]. To derive the smoothing factor for an incomplete factorization smoothing, we shall employ the limit matrix [5,8,10] which arises in the recursion of the incomplete matrix factorization, thus obtaining the incomplete factorization of A. It may be expected that the analysis in this case is representative for the local behaviour in the interior of a finite domain.
We consider the problem:

$$-\epsilon u_{xx} - u_{yy} = f. \tag{2.1}$$

A five point discretization in 2D produces the following stencil:

$$[A] = \begin{bmatrix} & -1 & \\ -\epsilon & 2(\epsilon + 1) & -\epsilon \\ & -1 & \end{bmatrix}. \tag{2.2}$$

A can be written as:

$$A = \ell_2 + \ell_1 + d + u_1 + u_2. \tag{2.3}$$

POINTWISE INCOMPLETE FACTORIZATION

We define C, approximation of A by:

$$\begin{aligned} C &= (\ell_2 + \ell_1 + \delta)\delta^{-1}(\delta + u_1 + u_2) \\ &= A + \delta - d + \ell_1\delta^{-1}u_1 + \ell_2\delta^{-1}u_2 + \ell_1\delta^{-1}u_2 + \ell_2\delta^{-1}u_1, \end{aligned} \tag{2.4}$$

with δ a diagonal matrix.
We have many ways to define δ and then R. Two examples are:

a) $\delta = d$, this case gives us symmetric point Gauss-Seidel (SPGS) (cf. Appendix 1).

b) $\delta = d - \ell_1\delta^{-1}u_1 - \ell_2\delta^{-1}u_2$. In this case, known as DKR [5], (Dupont-Kendall-Rachford) decomposition, R is given by:

$$R = \ell_1\delta^{-1}u_2 + \ell_2\delta^{-1}u_1. \tag{2.5}$$

The error terms can be taken into account in an approximate way by modification of δ. This can be done for example by requiring C to have the same row sums as A:

$$Ae = Ce, \qquad e = (1,\ldots,1)^T. \tag{2.6}$$

This condition gives for δ:

c) $\delta = d - \ell_1\delta^{-1}u_1 - \ell_2\delta^{-1}u_2 - \text{Rowsum}(R)$.

The corresponding iterative method will be referred to as DKRM(odified). We propose to introduce a relaxation parameter ω and define δ by:

$$\text{d) } \delta = d - \ell_1\delta^{-1}u_1 - \ell_2\delta^{-1}u_2 - \omega\,\text{Rowsum}(R). \tag{2.7}$$

The corresponding iterative method will be designated by DKRM(ω).
Note that DKRM(0) = DKR, DKRM(1) = DKRM.

SMOOTHING ANALYSIS OF DKRM(ω)

With A an infinite Toeplitz matrix, d, ℓ_1, ℓ_2, u_1, u_2 in (2.7) can be interpreted as constants. Equation (2.7) can be written as:

$$\delta_{ij} = d - \ell_1u_1/\delta_{i-1,j} - \ell_2u_2/\delta_{i,j-1} - \omega(\ell_1u_2/\delta_{i-1,j} + \ell_2u_1/\delta_{i,j-1}). \tag{2.8}$$

As i,j go to the infinity, δ_{ij} goes to the limit value δ satisfying

$$\delta = 2(\epsilon + 1) - \frac{\epsilon^2 + 2\omega\epsilon + 1}{\delta}. \tag{2.9}$$

As A is an M-matrix, $\delta > 0$ [9]. We will require $\delta \le d$, hence:

$$\omega \ge -\frac{\epsilon^2 + 1}{2\epsilon} = \omega_{min}. \tag{2.10}$$

Remark: for $\omega = \omega_{min}$, $\delta = d$ (SPGS).
The largest root of (2.9) (which gives the smallest R) is given by:

$$\delta = \epsilon + 1 + \sqrt{2\epsilon(1 - \omega)}\ , \tag{2.11}$$

which is real iff $\omega \le 1$.
We recapitulate: equation (2.9) has a real solution δ such that $0 < \delta \le d$ iff $\omega \in D = [\omega_{min}, 1]$.
Using the definition (1.5), we obtain the amplification factor ρ_1 associated with DKRM(ω):

$$\rho_1(\omega,\theta) = \left| \frac{\delta^2 - 2\delta(\epsilon + 1) + C_1}{\delta^2 - 2\delta(\epsilon \cos\theta_s + \cos\theta_t) + C_1} \right|, \tag{2.12}$$

with $C_1 = 2\epsilon\cos(\theta_s - \theta_t) + \epsilon^2 + 1$.

LINEWISE INCOMPLETE FACTORIZATION (ILF)

The matrix A given by (2.3) can be written as a block tridiagonal matrix:

$$A = \ell_2 + D + u_2, \tag{2.13}$$

with

$$D = \ell_1 + d + u_2. \tag{2.14}$$

We define C by:

$$\begin{aligned} C &= (\ell_2 + \Delta)\Delta^{-1}(\Delta + u_2) \\ &= A + \Delta - D + \ell_2\Delta^{-1}u_2. \end{aligned} \tag{2.15}$$

We can define Δ by:

$$\Delta = D - \ell_2\Delta^{-1}u_2, \tag{2.16}$$

which gives $R = 0$ (exact decomposition). However, Δ^{-1} is almost full. To have C with the same sparsity as A we define Δ by:

$$\Delta = D - \ell_2(\Delta^{-1})^{(p)}u_2, \tag{2.17}$$

with $p = 0$ ($\Delta = D$), $p = 1$ or $p = 3$ (which corresponds respectively to a diagonal and tridiagonal approximation of the exact inverse). Note that in that case $R = \ell_2(\Delta^{-1} - (\Delta^{-1})^{(p)})u_2$.
We study here the case $p = 3$, the case $p = 1$ is deduced easily.
Let Δ be given by the stencil:

$$[\Delta] = [\alpha, \beta, \alpha]. \tag{2.18}$$

Let T ($\simeq (\Delta^{-1})^{(3)}$) be such that:

$$[T] = [\tau_2, \tau_1, \tau_2]. \tag{2.19}$$

To compute the entries of T we use:

$$\alpha\tau_1 + \beta\tau_2 = 0, \qquad 2\alpha\tau_2 + \beta\tau_1 = 1. \tag{2.20}$$

We deduce:

$$[T] = \left[-\frac{\alpha}{\beta^2 - 2\alpha^2}, \frac{\beta}{\beta^2 - 2\alpha^2}, -\frac{\alpha}{\beta^2 - 2\alpha^2} \right]. \tag{2.21}$$

We can obtain Δ (limit matrix) by using the following algorithm (which converges in few iterations):

Algorithm:

c- Initialisation of α,β:

$\alpha = \alpha_0$

$\beta = \beta_0$

c- Iteration:

IT = 0

10 IT = IT + 1

$\alpha_1 = \alpha$

$\beta_1 = \beta$

$$\alpha = -\epsilon + \frac{\alpha_1}{\beta_1^2 - 2\alpha^2}$$

$$\beta = 2(\epsilon + 1) - \frac{\beta_1}{\beta_1^2 - 2\alpha^2}$$

IF ($|\alpha_1 - \alpha| >$ TOL or $|\beta_1 - \beta| >$ TOL) GOTO 10

END

(TOL = 10^{-6}).

We have

$$\hat{A}(\theta) = 2(\epsilon + 1) - 2\epsilon \cos\theta_s - 2\cos\theta_t, \tag{2.22}$$

$$\hat{C}(\theta) = \frac{\delta - 2\delta\cos\theta_t + 1}{\delta}, \tag{2.23}$$

where

$$\delta = \hat{\Delta}(\theta) = 2\alpha\cos\theta_s + \beta. \tag{2.24}$$

We have for the iterative method with C given by (2.15) and Δ defined by (2.18),

$$\rho_2(\theta) = \left| \frac{\delta^2 - 2\delta\,[\epsilon(1 - \cos\theta_s) + 1] + 1}{\delta^2 - 2\delta\cos\theta_t + 1} \right|. \tag{2.25}$$

A is as before given by (2.13,14) and C by (2.15). The difference with ILF is the approximation of Δ. We define Δ here by ([1]):

$$\Delta = (\ell_1 + \delta)\delta^{-1}(\delta + u_1), \tag{2.26}$$

with δ a diagonal matrix. We have:

$$C = A + \delta - d + \ell_1\delta^{-1}u_1 + \ell_2\Delta^{-1}u_2. \tag{2.27}$$

With the constraint (2.6) we can define δ by:

$$\delta = d - \ell_1\delta^{-1}u_1 - \text{Rowsum}\,(\ell_2\Delta^{-1}u_2), \tag{2.28}$$

which gives $R = -\ell_2\Delta^{-1}u_2 + \text{Rowsum}\,(\ell_2\Delta^{-1}u_2)$, and $Re = 0$, $(e = (1,\dots,1,\dots,1)^T)$.
Remark: we can also take the sum of columns (in the case where A is symmetric the two definitions of δ are equivalent). If $r^{(o)}$ is such that $e^T r^{(o)} = 0$ then $e^T r^{(n)} = 0$ for any iteration index n (cf. Appendix 2).
The calculation of Rowsum $(\ell_2\Delta^{-1}u_2) = v$ can be done by solving the system:

$$\ell_2\Delta^{-1}u_2 e = v. \tag{2.29}$$

We propose to take, instead of (2.28) the following definition of δ (which defines NF(ω)):

$$\delta = d - \ell_1\delta^{-1}u_1 - \omega\,\text{Rowsum}\,(\ell_2 T u_2), \tag{2.30}$$

where T is an approximation of Δ^{-1}. This will allows us to have an explicit expression of Rowsum $(\ell_2 T u_2)$ without solving the system (2.29). The approximation T of Δ^{-1} can be obtained in two manners:

a) by calculating the three main diagonals of the approximate inverse as in (2.21).
b) by Neumann truncation, in the case of a diagonally dominant matrix. To first order, for example, we have,

$$T = \delta^{-1} - \delta^{-1}(\ell_1 + u_1)\delta^{-1} + \delta^{-1}u_1\delta^{-1}\ell_1\delta^{-1}, \tag{2.31}$$

which is tridiagonal.

This approximation will facilitate the analysis and allow us to obtain an analytic expression for the amplification factor. With the Neumann truncation approximation, the limit of δ given by (2.30) is a solution of:

$$\delta = d - \frac{\epsilon^2}{\delta} - \omega\frac{(\delta + \epsilon)^2}{\delta^3}, \tag{2.32}$$

which can be solved for its largest solution by a few iterations. With the calculation of the three main diagonals of the approximate inverse, δ is a solution of:

$$\delta = d - \frac{\epsilon^2}{\delta} - \omega\frac{(\delta + \epsilon)^2}{\delta^3 + (\epsilon^4/\delta)} \simeq d - \frac{\epsilon^2}{\delta} - \omega\frac{(\delta + \epsilon)^2}{\delta^3} + 0\left(\left(\frac{\epsilon}{\delta}\right)^4\right). \tag{2.33}$$

The amplification factor is given by:

$$\rho_3(\omega,\theta) = \left|1 - \frac{\hat{A}(\theta)}{\hat{C}(\theta)}\right|, \tag{2.34}$$

with $\hat{A}(\theta)$ given by (2.22) and $\hat{C}(\theta) = \dfrac{\hat{\Delta}^2(\theta) - 2\hat{\Delta}(\theta)\cos\theta_t + 1}{\hat{\Delta}(\theta)}$, where

$$\hat{\Delta}(\theta) = \frac{\epsilon^2 + \delta^2 - 2\epsilon\delta\cos\theta_s}{\delta} .$$

NUMERICAL CALCULATION OF SMOOTHING FACTORS

In order to compare the smoothing efficiencies of the iterative methods based on the three incomplete factorizations above, we calculate the smoothing factor for various values of ϵ in (2.1). For calculating $\bar{\rho}$, the maximum is taken over Θ (1.7), which is the set of wave numbers that occur on an (N+1)(N+1) periodically extended grid. We have taken N = 64 and N = 128. The results are summarized in the following tables:

TABLE 1 Smoothing factors of DKRM(ω) for N = 64.

ω \ ϵ	10^{-6}	10^{-4}	10^{-2}	1	10^2	10^4	10^6
ω_{min}	1	1	0.962	0.256	0.962	1	1
ω_{opt}	0.366	0.345	0.303	0.160	0.303	0.345	0.352
(•)	(-0.55)	(-0.5)	(-0.45)	(-0.05)	(-0.45)	(-0.5)	(-0.5)
-0.5	0.406	0.345	0.312	0.205	0.312	0.345	0.352
0	0.940	0.972	0.767	0.172	0.767	0.971	0.999
1	>> 1	>> 1	>> 1	1	>> 1	>> 1	>> 1

TABLE 2 Smoothing factors of DKRM(ω) for N = 128.

ω \ ϵ	10^{-6}	10^{-4}	10^{-2}	1	10^2	10^4	10^6
ω_{min}	1	1	0.962	0.253	0.962	1	1
ω_{opt}	0.362	0.338	0.304	0.158	0.304	0.338	0.366
(•)	(-0.55)	(-0.5)	(-0.45)	(-0.1)	(-0.45)	(-0.5)	(-0.55)
-0.5	0.395	0.338	0.308	0.203	0.304	0.338	0.353
0	0.940	0.972	0.767	0.172	0.767	0.971	0.906
1	>> 1	>> 1	>> 1	1	>> 1	>> 1	>> 1

TABLE 3 Smoothing factors of ILF for N = 64.

ρ \ ϵ	10^{-6}	10^{-4}	10^{-2}	1	10^{2}	10^{4}	10^{6}
0	1	1	0.962	0.204	0.204	0.204	0.204
1	0.415	0.935	0.597	0.124	0.202	0.204	0.204
3	0.358	0.163	0.151	0.071	0.199	0.204	0.204

TABLE 4 Smoothing factors of NF(ω) for N = 64.

ω \ ϵ	10^{-6}	10^{-4}	10^{-2}	1	10^{2}	10^{4}	10^{6}
$\omega = 1$	> 1	> 1	> 1	0.156	0.197	0.204	0.195
ω_{opt} (•)	1	0.801 (0.999)	0.252 (0.976)	0.091 (0.850)	-	-	-

COMPARISON AND CONCLUSION

From Tables 1 and 2 we see that we can improve efficiently the smoothing factor of DKR by introducing the relaxation factor ω. For $\omega = -0.5$ we have a good smoothing factor for different values and directions of anisotropy. This smoother needs to store just one supplementary diagonal. From Table 3 we see that ILF(p) is efficient when p = 3 for all values of ϵ taken here. For $\epsilon >> 1$ ILF(1) is equivalent to ILF(3) and both of them are equivalent to SLGS(ILF(0)) when ϵ is large. This can be seen easily from the equation (2.17). Note that ILF(1) can be also a good smoother for some values of $\epsilon << 1$. ILF(3) is better than DKRM(-0.5) but it needs more storage (three diagonals) and more calculation. From Table 4 we see that NF(1) has a nice smoothing factor if the ordering corresponds to the strong anisotropy. NF(1) needs less storage than ILF(3) but unfortunately its performance depends on the ordering.

This method of local mode analysis is useful for studying the smoothing properties of iterative methods based on various incomplete factorizations. This analysis helps us to exhibit a new simple smoother (DKRM(-0.5)) for the 2D-anisotropic diffusion problem, and to appreciate easily the smoothing properties of ILF and NF(ω) for such problems.

ACKNOWLEDGEMENTS

This work has been done while I was a guest at the Delft University of Technology. I wish to thank Prof. P. Wesseling for his help, advice and inspiring ideas. I wish to thank also Prof. H.A. van der Vorst for the discussions that I had with him.

APPENDIX 1

Let $A = \ell_1 + \ell_2 + d + u_1 + u_2$. A can be written as $A = L + D + U$ where $L = \ell_1 + \ell_2$, $D = d$ and $U = u_1 + u_2$ or $L = \ell_2$, $D = \ell_1 + d + u_1$ and $U = u_2$. Let C be given by:

$$C = (L + \Delta)\Delta^{-1}(\Delta + U).$$

If we choose $\Delta = D$ then the iterative method with iteration matrix $S = I - C^{-1}A$ is a symmetric (point or line, depending of the choice of L, D and U) Gauss-Seidel method.

Proof: Let $\tilde{L} = L + \Delta$, $\tilde{U} = \Delta + U$ then $A = \tilde{L} + \tilde{U} - D$ and $C = \tilde{L}D^{-1}\tilde{U}$ since $\Delta = D$. As:

$$\tilde{U}^{-1}A\tilde{L}^{-1}A = \tilde{U}^{-1}A + \tilde{L}^{-1}A - \tilde{U}^{-1}D\tilde{L}^{-1}A,$$

then

$$S = (I - \tilde{U}^{-1}A)(I - \tilde{L}^{-1}A)$$

which is the iteration matrix of the symmetric Gauss-Seidel.

APPENDIX 2

Let $C\phi_0 = f$. Then we have for the residual, $r_0 = f - A\phi_0 = (C - A)\phi_0$. If $e^TC = e^TA$ then $e^Tr_0 = 0$.
We know that: $\phi^{(n+1)} = \phi^{(n)} + C^{-1}r^{(n)}$ where $r^{(n)} = b - A\phi^{(n)}$, then

$$A\phi^{(n+1)} = A\phi^{(n)} + AC^{-1}r^{(n)}$$

and then

$$r^{(n+1)} = r^{(n)} - AC^{-1}r^{(n)}$$

$$= RC^{-1}r^{(n)} \quad (\text{as } I - AC^{-1} = RC^{-1})$$

so that

$$e^Tr^{(n+1)} = (e^TR)C^{-1}r^{(n)} = 0$$

since $e^TR = 0$. Then the sum of the components of the residual is zero at any iteration.

REFERENCES

[1] APPLEYARD, J.R., CHESHIRE, I.M. and POLLARD, R.K.: "Special techniques for fully implicit simulators", Proc. European Symposium on Enhanced Oil Recovery, Bournemouth, England (1981) pp. 395-408.

[2] AXELSSON, O., BRINKKEMPER, S. and IL'IN, V.P.: "On some versions of incomplete block-matrix factorization iterative methods", Linear Algebra and Application, **59** (1984) pp. 3-15.

[3] BJÖRCK, Å and DAHLQUIST, G.: "Numerical Methods", Prentice-Hall, Inc., Englewood Cliffs, N.J., 1974.

[4] BRANDT, A.: "Multilevel adaptive solutions to boundary value problems", Math. Comp., **31** (1977) pp. 333-390.

[5] CHAN, T. and ELMAN, H.: "Fourier analysis of iterative methods for elliptic problems", Technical Report 1763, January 1987, Yale University, New Haven.

[6] HEMKER, P.: "The incomplete LV decomposition as a relaxation method in multigrid method algorithms", Proc. BAIL I Conference. In: J.J.H. Miller (ed), Dublin, June 1980.

[7] KETTLER, R.: "Analysis and computation of relaxation schemes in robust multigrid and preconditioned conjugate gradient". In: W. Hackbusch, U. Trottenberg (eds), Multigrid Methods, Letter Notes in Mathematics **960**, Springer-Verlag, Berlin, 1982.

[8] KETTLER, R.: "Linear multigrid methods for numerical reservoir simulation", Ph.D. Thesis, Delft University of Technology, Dept. of Math. and Inf., December 1987.

[9] MEIJERINK, J.A. and VAN DER VORST, H.A.: "An iterative method for linear systems of which the coefficient matrix is a symmetric matrix", Math. Comp., **31** (1977) pp. 148-162.

[10] WESSELING, P.: "A robust and efficient multigrid method", Proc. of Conf. on Multigrid Method (1981), Letter Notes in Mathematics **960**, Springer-Verlag, Berlin, 1982.

Multigrid and Defect Correction for the Steady Navier-Stokes Equations

Barry Koren
Centre for Mathematics and Computer Science
P.O. Box 4079, 1009 AB Amsterdam, The Netherlands

Theoretical and experimental convergence results are presented for multigrid and iterative defect correction applied to finite volume discretizations of the steady, 2D, compressible Navier-Stokes equations. Iterative defect correction is introduced for circumventing the difficulty in finding a solution of discretized equations with a second- or higher-order accurate convective part. As a smoothing technique, use is made of point Gauss-Seidel relaxation with inside the latter, Newton iteration as a basic solution method. The multigrid echnique appears to be very efficient for smooth as well as non-smooth problems. Iterative defect correction appears to be very efficient for smooth problems only, though still reasonably efficient for non-smooth problems.

1980 Mathematics Subject Classification: 65B05, 65N10, 65N20, 65N30, 76N10.
Keywords & Phrases: multigrid, defect correction, Navier-Stokes equations.
Note: This work was supported by the European Space Agency (ESA), via Avions Marcel Dassault - Bréguet Aviation (AMD-BA).

1. Introduction

1.1. Navier-Stokes equations

The Navier-Stokes equations considered are:

$$\frac{\partial}{\partial x}\begin{bmatrix}\rho u\\ \rho u^2+p\\ \rho uv\\ \rho u(e+\frac{p}{\rho})\end{bmatrix}+\frac{\partial}{\partial y}\begin{bmatrix}\rho v\\ \rho uv\\ \rho v^2+p\\ \rho v(e+\frac{p}{\rho})\end{bmatrix}- \tag{1.1a}$$

$$\frac{1}{Re}\left[\frac{\partial}{\partial x}\begin{bmatrix}0\\ \tau_{xx}\\ \tau_{xy}\\ \tau_{xx}u+\tau_{xy}v+\frac{1}{\gamma-1}\frac{1}{Pr}\frac{\partial(c^2)}{\partial x}\end{bmatrix}+\frac{\partial}{\partial y}\begin{bmatrix}0\\ \tau_{xy}\\ \tau_{yy}\\ \tau_{yy}v+\tau_{xy}u+\frac{1}{\gamma-1}\frac{1}{Pr}\frac{\partial(c^2)}{\partial y}\end{bmatrix}\right]=0,$$

with

$$\begin{aligned}\tau_{xx}&=\frac{4}{3}\frac{\partial u}{\partial x}-\frac{2}{3}\frac{\partial v}{\partial y},\\ \tau_{xy}&=\frac{\partial u}{\partial y}+\frac{\partial v}{\partial x},\\ \tau_{yy}&=\frac{4}{3}\frac{\partial v}{\partial y}-\frac{2}{3}\frac{\partial u}{\partial x}.\end{aligned} \tag{1.1b}$$

For a detailed description of the various quantities used, assumptions made and so on, we refer to any standard textbook. Suffice to say here that these are the full Navier-Stokes equations with as main assumptions made: zero bulk viscosity and constant diffusion coefficients. (So, the flow is assumed to be laminar and its diffusion coefficients are assumed to be temperature independent.)

1.2. Discretization method

For a detailed description of the discretization method, we refer to [8,9]. Here, a brief summary is given of its main characteristics only.

Since we also want to be able to compute Euler flow solutions (with possibly occurring discontinuities), the Navier-Stokes equations (1.1) are discretized in integral form. A straightforward and simple discretization of the integral form is obtained by subdividing the integration region into finite volumes, and by requiring that the integral form holds for each finite volume separately. This discretization requires an evaluation of a convective and diffusive flux vector at each volume wall.

1.2.1. Evaluation of convective fluxes. Based on experience with the Euler equations [3,4,5,6,7,11], for the evaluation of the convective fluxes we prefer an upwind approach. In here, the convective flux vector is assumed to be constant along each volume wall, and to be determined by a uniformly constant left and right state only. For the 1D Riemann problem thus obtained, an approximate Riemann solver is applied.

The choice of the left and right state, to be used as entries for the approximate Riemann solver, determines the accuracy of the convective discretization. First-order accuracy is simply obtained by taking the left and right state equal to that in the corresponding adjacent volume [5]. Higher-order accuracy is obtained by applying low-degree piecewise polynomial functions, using two or three adjacent volume states for the left and right state separately [3]. The higher-order accurate polynomial function used is van Leer's κ-function [13]. This function is general in the sense that it contains a variable $\kappa \in [-1,1]$ that can be used for chosing any higher-order approximation ranging from central ($\kappa=1$) to fully one-sided upwind ($\kappa=-1$). A survey of some characteristic κ-values and their corresponding properties in the case of Euler flow computations has been given in [7]. As an optimal value for κ in the case of Navier-Stokes flows, we found by error analysis: $\kappa=1/3$ [10]. For this specific κ-value, we also constructed a new (monotonicity preserving) limiter [10].

For the approximate Riemann solver, we considered two possibilities which both have continuous differentiability (a prerequisite for applying Newton's method), namely OSHER's [14] and van LEER's [12] scheme. Theoretical analysis has shown that Osher's scheme is to be preferred above van Leer's scheme [10]. This has been confirmed by computations [10].

1.2.2. Evaluation of diffusive fluxes. For the evaluation of the diffusive fluxes, use is made of the standard central technique as outlined in [15]. For the necessary computation at each volume wall of $\nabla u, \nabla v$ and ∇c^2, the technique uses (at inner volume walls) a shifted volume overlying the volume wall considered.

2. CONVERGENCE OF MULTIGRID

The same multigrid method which has been used with success for the first-order discretized Euler equations [5] is taken as a point of departure for both the first- and second-order discretized Navier-Stokes equations. The method makes use of symmetric point Gauss-Seidel relaxation as a smoothing technique. In here, one or more Newton steps are performed for the collective relaxation of the four state vector components in each finite volume. (Usually, the tolerance for the Newton iteration is so large that in a substantial majority of all cells, only one Newton step is performed.) For the first-order discretized Euler equations, point Gauss-Seidel relaxation turned out to be a good smoother, thus enabling a good multigrid acceleration. However, for higher-order discretized Euler equations the good smoothing property is lost. Obviously, this will also be the case for Navier-Stokes flows with high Reynolds number. We do not anticipate to this by looking for some remedy already, but we investigate at first how smoothing evolves with increasingly dominating convection. The complete multigrid method as developed for Euler (see [11] for a detailed description) is carried over to Navier-Stokes, with as the only a priori change, a replacement of the piecewise constant correction prolongation by a bilinear prolongation [8], thus satisfying the rule that the sum of orders of the prolongation and restriction should exceed the order of the differential equation ($m_p + m_r > 2m$) [1]. Notice that as a consequence the Galerkin property [5] is definitely lost.

2.1. Investigation method

Both theoretical and experimental convergence results are presented; the theoretical results being obtained by local mode analysis, the experimental results being obtained by considering two standard flow problems; a smooth and a non-smooth problem.

2.1.1. Smoothing analysis. For the smoothing analysis we consider the linear and scalar convection diffusion equation

$$\frac{\partial u}{\partial x} + \frac{\partial u}{\partial y} - \epsilon\left(\frac{\partial^2 u}{\partial x^2} + \frac{\partial^2 u}{\partial x \partial y} + \frac{\partial^2 u}{\partial y^2}\right) = 0. \tag{2.1}$$

For the integral form to be considered for each finite volume $\Omega_{j,k}, j=1,2,...,J,\ k=1,2,...,K$, we take:

$$\oint_{\partial\Omega_{j,k}} (un_x + un_y)ds - \epsilon \oint_{\partial\Omega_{j,k}} \left(\frac{\partial u}{\partial x}n_x + \frac{\partial u}{\partial x}n_y + \frac{\partial u}{\partial y}n_y\right)ds = 0,\ \forall_{j,k}, \tag{2.2}$$

with $\partial\Omega_{j,k}$ the boundary of $\Omega_{j,k}$. The two parts of the discretization to be modelled are (i) an upwind treatment of convection, either first- or higher-order accurate (non-limited $\kappa = 1/3$), and (ii) a central second-order accurate treatment of diffusion. Assuming a finite volume grid with equidistant walls parallel to the x- and y-axis ($\Delta x = \Delta y = h$, fig. 2.1), the evaluation of convective flux terms yields:

$$\begin{aligned} \oint_{\partial\Omega_{j,k}} un_x ds &= (u^l_{j+½,k} - u^l_{j-½,k})h, \\ \oint_{\partial\Omega_{j,k}} un_y ds &= (u^l_{j,k+½} - u^l_{j,k-½})h, \end{aligned} \tag{2.3a}$$

with

$$\begin{aligned} u^l_{j+½,k} &= \alpha_1 u_{j-1,k} + \alpha_2 u_{j,k} + \alpha_3 u_{j+1,k}, \\ u^l_{j,k+½} &= \alpha_1 u_{j,k-1} + \alpha_2 u_{j,k} + \alpha_3 u_{j,k+1}, \end{aligned} \tag{2.3b}$$

and similar expressions for $u^l_{j-½,k}$ and $u^l_{j,k-½}$ (the coefficients α_i still free).

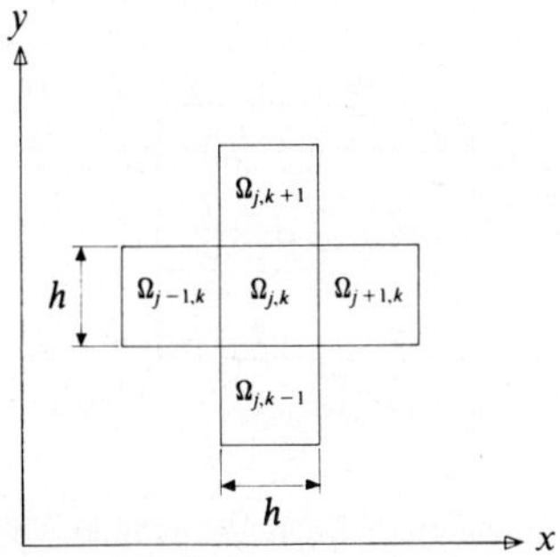

Fig. 2.1. Model volume $\Omega_{j,k}$ with neighbours

For the diffusive terms we get:

$$\begin{aligned} \oint_{\partial\Omega_{j,k}} \frac{\partial u}{\partial x} n_x ds &= \left\{\left(\frac{\partial u}{\partial x}\right)_{j+½,k} - \left(\frac{\partial u}{\partial x}\right)_{j-½,k}\right\}h, \\ \oint_{\partial\Omega_{j,k}} \frac{\partial u}{\partial x} n_y ds &= \left\{\left(\frac{\partial u}{\partial x}\right)_{j,k+½} - \left(\frac{\partial u}{\partial x}\right)_{j,k-½}\right\}h, \\ \oint_{\partial\Omega_{j,k}} \frac{\partial u}{\partial y} n_y ds &= \left\{\left(\frac{\partial u}{\partial y}\right)_{j,k+½} - \left(\frac{\partial u}{\partial y}\right)_{j,k-½}\right\}h, \end{aligned} \tag{2.4a}$$

with

$$\begin{aligned} \left(\frac{\partial u}{\partial x}\right)_{j+½,k} &= \frac{1}{h^2}\oint_{\partial\Omega_{j+½,k}} un_x ds = \frac{1}{h}(u_{j+1,k} - u_{j,k}), \\ \left(\frac{\partial u}{\partial x}\right)_{j,k+½} &= \frac{1}{h^2}\oint_{\partial\Omega_{j,k+½}} un_x ds = \frac{1}{h}\Big\{\frac{1}{4}(u_{j,k} + u_{j+1,k} + u_{j,k+1} + u_{j+1,k+1}) - \\ &\qquad \frac{1}{4}(u_{j-1,k} + u_{j,k} + u_{j-1,k+1} + u_{j,k+1})\Big\} = \\ &= \frac{1}{4h}(u_{j+1,k} + u_{j+1,k+1} - u_{j-1,k} - u_{j-1,k+1}), \\ \left(\frac{\partial u}{\partial y}\right)_{j,k+½} &= \frac{1}{h^2}\oint_{\partial\Omega_{j,k+½}} un_y ds = \frac{1}{h}(u_{j,k+1} - u_{j,k}), \end{aligned} \tag{2.4b}$$

and similar expressions for $(\frac{\partial u}{\partial x})_{j-1/2,k}$, $(\frac{\partial u}{\partial x})_{j,k-1/2}$ and $(\frac{\partial u}{\partial y})_{j,k-1/2}$. In (2.4b), $\partial\Omega_{j+1/2,k}$ and $\partial\Omega_{j,k+1/2}$ denote the boundary of shifted volume $\Omega_{j+1/2,k}$ respectively $\Omega_{j,k+1/2}$ (fig. 2.2).

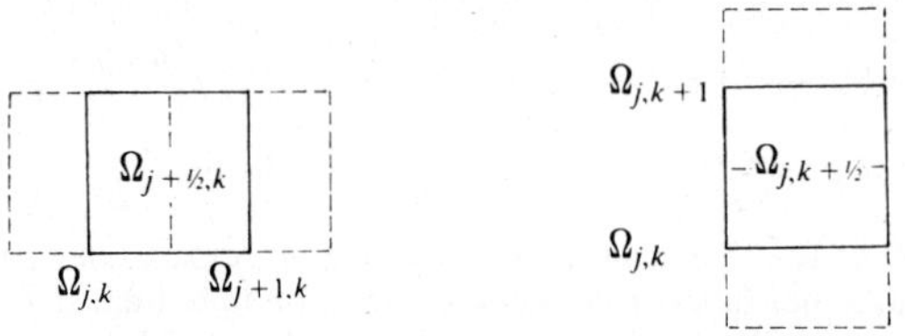

a. $\Omega_{j+1/2,k}$ b. $\Omega_{j,k+1/2}$

Fig. 2.2. Shifted volumes

With the previous flux evaluations we get for each finite volume $\Omega_{j,k}$ the algebraic equation:

$$\begin{aligned}
&\frac{1}{4}\frac{\epsilon}{h}u_{j-1,k+1}+(\alpha_3-\frac{\epsilon}{h})u_{j,k+1}-\frac{1}{4}\frac{\epsilon}{h}u_{j+1,k+1}-\\
&\alpha_1 u_{j-2,k}+(\alpha_1-\alpha_2-\frac{\epsilon}{h})u_{j-1,k}+(2\alpha_2-2\alpha_3+4\frac{\epsilon}{h})u_{j,k}+(\alpha_3-\frac{\epsilon}{h})u_{j+1,k}-\\
&\frac{1}{4}\frac{\epsilon}{h}u_{j-1,k-1}+(\alpha_1-\alpha_2-\frac{\epsilon}{h})u_{j,k-1}+\frac{1}{4}\frac{\epsilon}{h}u_{j+1,k-1}-\\
&\alpha_1 u_{j,k-2}=0,
\end{aligned} \tag{2.5}$$

with corresponding stencil:

	$j-2$	$j-1$	j	$j+1$
$k+1$		$\frac{1}{4}\frac{\epsilon}{h}$	$\alpha_3-\frac{\epsilon}{h}$	$-\frac{1}{4}\frac{\epsilon}{h}$
k	$-\alpha_1$	$\alpha_1-\alpha_2-\frac{\epsilon}{h}$	$2\alpha_2-2\alpha_3+4\frac{\epsilon}{h}$	$\alpha_3-\frac{\epsilon}{h}$
$k-1$		$-\frac{1}{4}\frac{\epsilon}{h}$	$\alpha_1-\alpha_2-\frac{\epsilon}{h}$	$\frac{1}{4}\frac{\epsilon}{h}$
$k-2$			$-\alpha_1$	

(2.6)

The parameter ϵ/h in (2.5) and (2.6) models the inverse of the mesh Reynolds number. Of course, for the model grid considered, the finite volume discretization boils down to a finite difference discretization for which the above stencil can be given directly. The purpose of the previous finite volume derivation merely is to illustrate for a model problem, the way of evaluating the intricate convective and diffusive fluxes arising for (1.1).

For point Gauss-Seidel relaxation applied to (2.5), four basically different sweep directions can be considered: downwind, upwind and twice crosswind. Introducing the counter n for the number of sweeps performed, these four possibilities can be illustrated as has been done in fig. 2.3. (In fig. 2.3, $u_{j,k}^{n+1}$ denotes the $(n+1)$th iterand of $u_{j,k}$.)

a. Downwind

	$j-1$	j	$j+1$
$k+1$	u^n	u^n	u^n
k	u^{n+1}	u^{n+1}	u^n
$k-1$	u^{n+1}	u^{n+1}	u^{n+1}

b. Upwind

	$j-1$	j	$j+1$
$k+1$	u^{n+1}	u^{n+1}	u^{n+1}
k	u^n	u^{n+1}	u^{n+1}
$k-1$	u^n	u^n	u^n

c. Crosswind

	$j-1$	j	$j+1$
$k+1$	u^n	u^n	u^n
k	u^n	u^{n+1}	u^{n+1}
$k-1$	u^{n+1}	u^{n+1}	u^{n+1}

d. Crosswind

	$j-1$	j	$j+1$
$k+1$	u^{n+1}	u^{n+1}	u^{n+1}
k	u^{n+1}	u^{n+1}	u^n
$k-1$	u^n	u^n	u^n

Fig. 2.3. Basic sweep directions, convection direction: 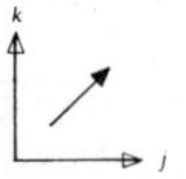

For local mode analysis, we introduce in the standard way: (i) the iteration error

$$\delta^n_{j,k} = u^*_{j,k} - u^n_{j,k}, \tag{2.7}$$

with $u^*_{j,k}$ the converged numerical solution in $\Omega_{j,k}$, and: (ii) the Fourier form

$$\delta^n_{j,k} = D\mu^n e^{i(\omega_1 j + \omega_2 k)h}, \tag{2.8}$$

with D some constant, μ the amplification factor, and ω_1 and ω_2 the error frequency in j- respectively k-direction. The frequencies to be considered are: $\pi/2h \leqslant |\omega_1|, |\omega_2| \leqslant \pi/h$. By introducing (2.8) into (2.5), defining $\theta_1 = \omega_1 h, \theta_2 = \omega_2 h$, and considering for instance the downwind relaxation sweep, we get:

$$\begin{array}{c|c|c|c|c|}
e^{i\theta_2} & & \frac{1}{4}\frac{\epsilon}{h} & \alpha_3 - \frac{\epsilon}{h} & -\frac{1}{4}\frac{\epsilon}{h} \\
1 & -\alpha_1\mu & (\alpha_1 - \alpha_2 - \frac{\epsilon}{h})\mu & (2\alpha_2 - 2\alpha_3 + 4\frac{\epsilon}{h})\mu & \alpha_3 - \frac{\epsilon}{h} \\
e^{-i\theta_2} & & -\frac{1}{4}\frac{\epsilon}{h}\mu & (\alpha_1 - \alpha_2 - \frac{\epsilon}{h})\mu & \frac{1}{4}\frac{\epsilon}{h}\mu \\
e^{-2i\theta_2} & & & -\alpha_1\mu & \\
 & e^{-2i\theta_1} & e^{-i\theta_1} & 1 & e^{i\theta_1}
\end{array} \tag{2.9}$$

with $\pi/2 \leqslant |\theta_1|, |\theta_2| \leqslant \pi$.

Results for this smoothing analysis are given in section 2.2.

2.1.2. Experiments. The smooth flow problem considered is a subsonic flat plate flow at $Re = 100$, for which we can use the Blasius solution [16] as a reference solution. The non-smooth problem considered is a supersonic flat plate flow at $Re = 2.96\ 10^5$, with an oblique shock wave impinging upon the flat plate boundary layer. This problem has been taken from [2]. For both flow problems, use is made of: $\gamma = 1.4$ and $Pr = 0.71$.

Geometry and boundary conditions as applied for the subsonic flat plate flow are given in fig. 2.4. As far as convection is concerned, the eastern boundary is considered to be an outflow boundary. For diffusion, the northern, southern and eastern boundary are assumed to be far-field boundaries with zero diffusion. For this subsonic problem we only apply grids composed of square finite volumes. The coarsest grid in all multigrid computations is the 4×2-grid given in fig. 2.4. For details about boundary conditions and so on, we refer to [8].

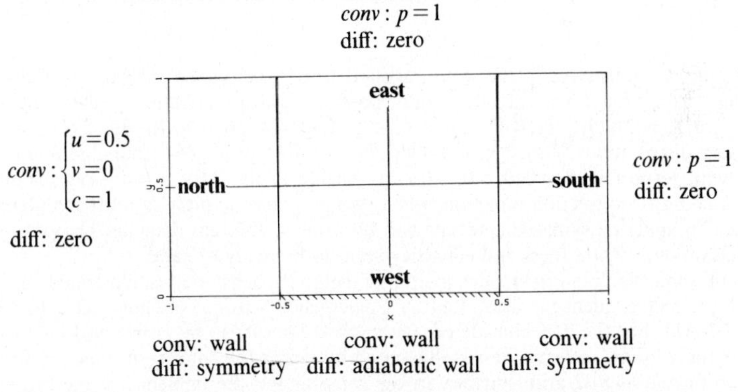

Fig. 2.4. Geometry, boundary conditions and coarsest grid subsonic flat plate flow (conv: convection, diff: diffusion)

Geometry and boundary conditions for the supersonic flat plate flow are indicated globally in fig. 2.5. For details see again [8]. In all multigrid computations, the coarsest grid applied is the 5×2-grid given in fig. 2.5. The grid has been optimized for convection by introducing a stretching in j-direction, and in particular by aligning it with the impinging shock wave [9]. A grid adaptation for diffusion has been realized by introducing stretching in k-direction only. Notice that this problem essentially differs from the previous problem, both in flow and in grid.

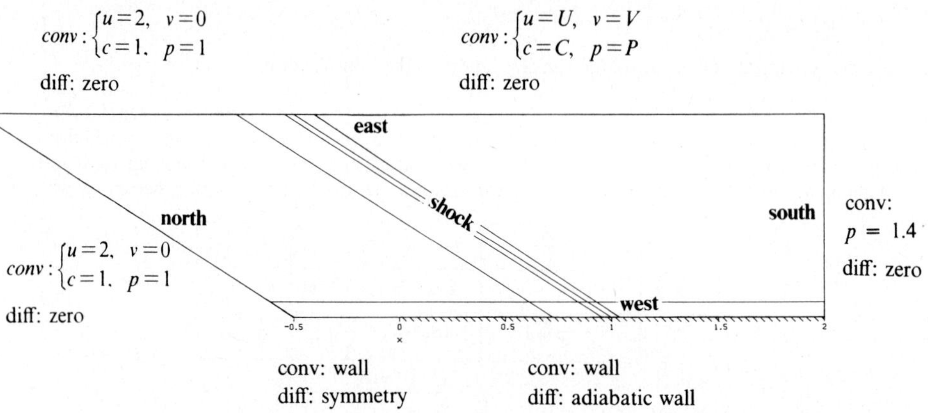

Fig. 2.5. Geometry, boundary conditions and coarsest grid supersonic flat plate flow (conv: convection, diff: diffusion)

2.2. Results

2.2.1. First-order discretized equations. For the first-order accurate model discretization we have $\alpha_1=\alpha_3=0, \alpha_2=1$. With this the general 11-point stencil (2.6) reduces to the following 9-point stencil

	$j-1$	j	$j+1$
$k+1$	$\frac{1}{4}\frac{\epsilon}{h}$	$-\frac{\epsilon}{h}$	$-\frac{1}{4}\frac{\epsilon}{h}$
k	$-(1+\frac{\epsilon}{h})$	$2+4\frac{\epsilon}{h}$	$-\frac{\epsilon}{h}$
$k-1$	$-\frac{1}{4}\frac{\epsilon}{h}$	$-(1+\frac{\epsilon}{h})$	$\frac{1}{4}\frac{\epsilon}{h}$

(2.10)

Introducing the iteration error in the way suggested before we get the smoothing results given in fig. 2.6. In fig. 2.6a, for each of the four possible sweep directions, the smoothing factor $\mu_s=\sup|\mu(\theta_1, \theta_2)|$, $\pi/2 \leq |\theta_1|$, $|\theta_2| \leq \pi$ is given as a function of ϵ/h. In fig. 2.6b, for $\epsilon/h=1$, the corresponding distributions $|\mu(\theta_1,\theta_2)|$, $\pi/2 \leq |\theta_1|,|\theta_2| \leq \pi$ are given. (All four distributions are point-symmetric with respect to $\theta_1=0, \theta_2=0$.) Clearly visible in fig. 2.6a is the good smoothing for any value of ϵ/h and any convection direction, when sweeping alternatingly in all four different directions (for instance by applying symmetric sweeps and by using a different diagonal sweep direction in pre- and post-relaxation). Robustness and efficiency seem to be ready to hand.

For the subsonic flat plate flow, the multigrid method's behaviour is illustrated in fig. 2.7a. The measure of grid independence is illustrated by convergence histories obtained on a 16×8-, a 32×16- and a 64×32-grid. For the flow considered, the method appears to be nearly grid independent. In the same figure, the multigrid effectiveness is illustrated by giving the convergence history for a single grid computation on the 64×32-grid. Further, in the same figure, the influence of the higher-order accuracy of the prolongation is illustrated by giving also the convergence history for a strategy with $m_p=1$ (so violating the rule $m_p+m_r>2m$ [1]). In agreement with [18], for this convection dominated flow, the positive influence of the second-order prolongation already appears to be negligible. Using the Blasius solution as a reference, in [8] it is shown that only a single FAS-cycle is sufficient for converging to discretization error accuracy.

For the supersonic flat plate flow, results are shown in fig. 2.7b for a 20×8-, a 40×16- and a 80×32-grid. Here we used the first-order prolongation only. Despite of the slight deterioration with respect to the subsonic flow, the multigrid properties are still acceptable.

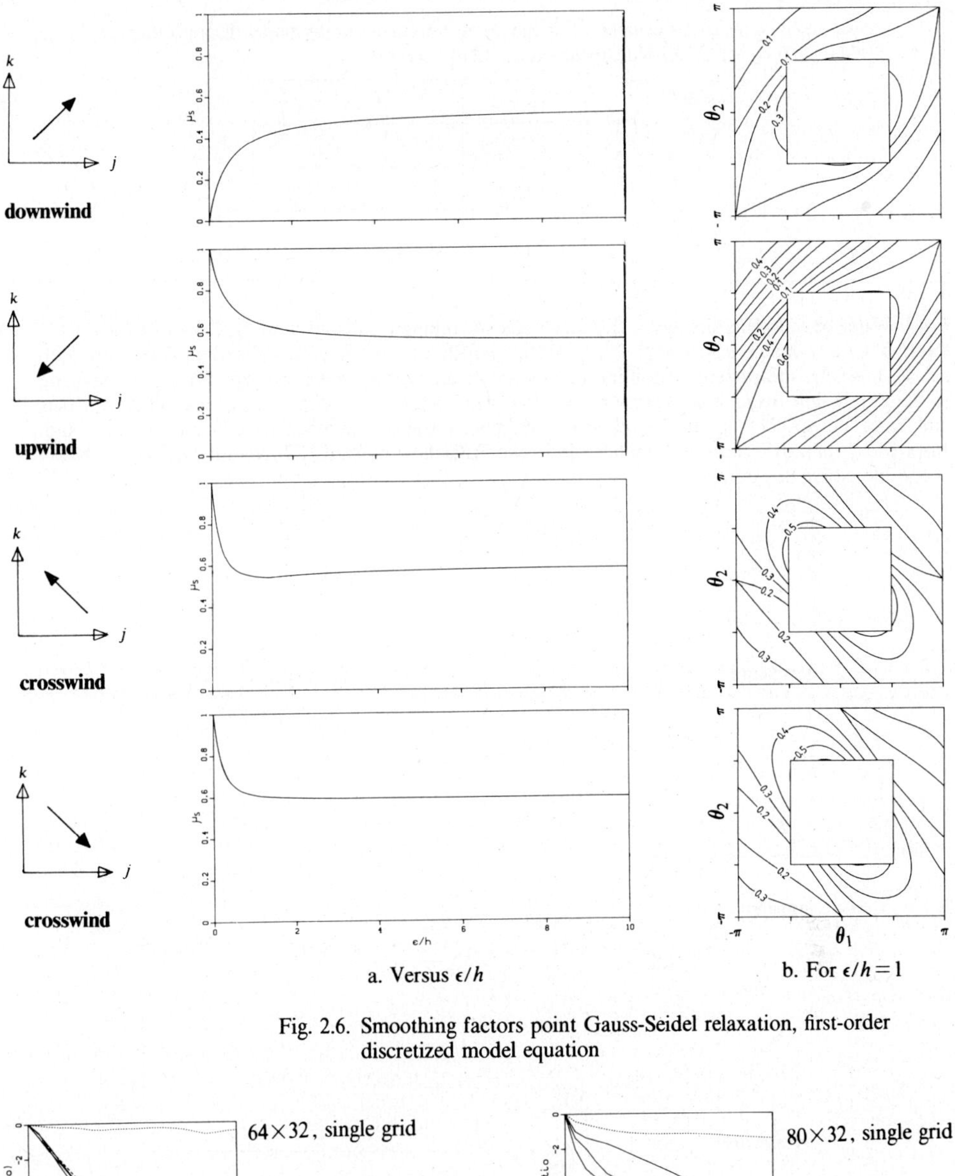

a. Versus ϵ/h

b. For $\epsilon/h=1$

Fig. 2.6. Smoothing factors point Gauss-Seidel relaxation, first-order discretized model equation

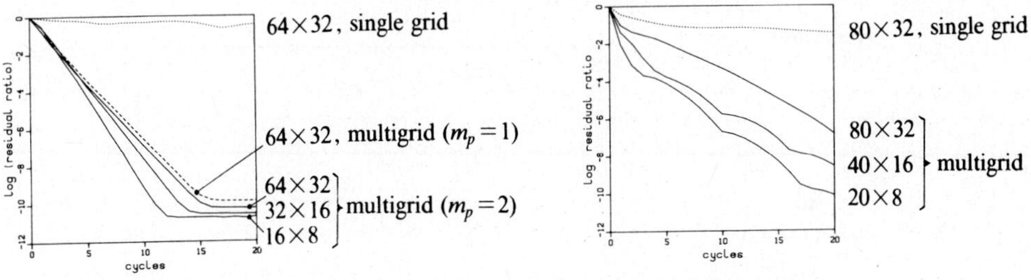

a. Subsonic flat plate flow

b. Supersonic flat plate flow

Fig. 2.7. Multigrid behaviour first-order discretized Navier-Stokes equations

2.2.2. Second-order discretized equations. For the second-order accurate model discretization we have: $\alpha_1 = -1/6, \alpha_2 = 5/6, \alpha_3 = 1/3$ [9]. With these values, (2.6) becomes

$$\begin{array}{c|c|c|c|c|}
 & j-2 & j-1 & j & j+1 \\
k+1 & & \frac{1}{4}\frac{\epsilon}{h} & \frac{1}{3}-\frac{\epsilon}{h} & -\frac{1}{4}\frac{\epsilon}{h} \\
k & \frac{1}{6} & -(1+\frac{\epsilon}{h}) & 1+4\frac{\epsilon}{h} & \frac{1}{3}-\frac{\epsilon}{h} \\
k-1 & & -\frac{1}{4}\frac{\epsilon}{h} & -(1+\frac{\epsilon}{h}) & \frac{1}{4}\frac{\epsilon}{h} \\
k-2 & & & \frac{1}{6} &
\end{array} \qquad (2.11)$$

For the four basic sweep directions, this yields the smoothing results given in fig. 2.8. Only for $\epsilon/h > 1$ there is some valuable smoothing. For problems which are locally convection dominated, say with $\epsilon/h \ll 1$ locally, the present smoothing factors are unacceptable, except perhaps for those belonging to the purely downwind sweep. Since purely downwind relaxation sweeps are not feasible in practice, and since no specific alternation of sweep directions suffices, another remedy has to be found. Inspired by its rather successful application in the Euler flow method [3,7], iterative defect correction is considered for this purpose.

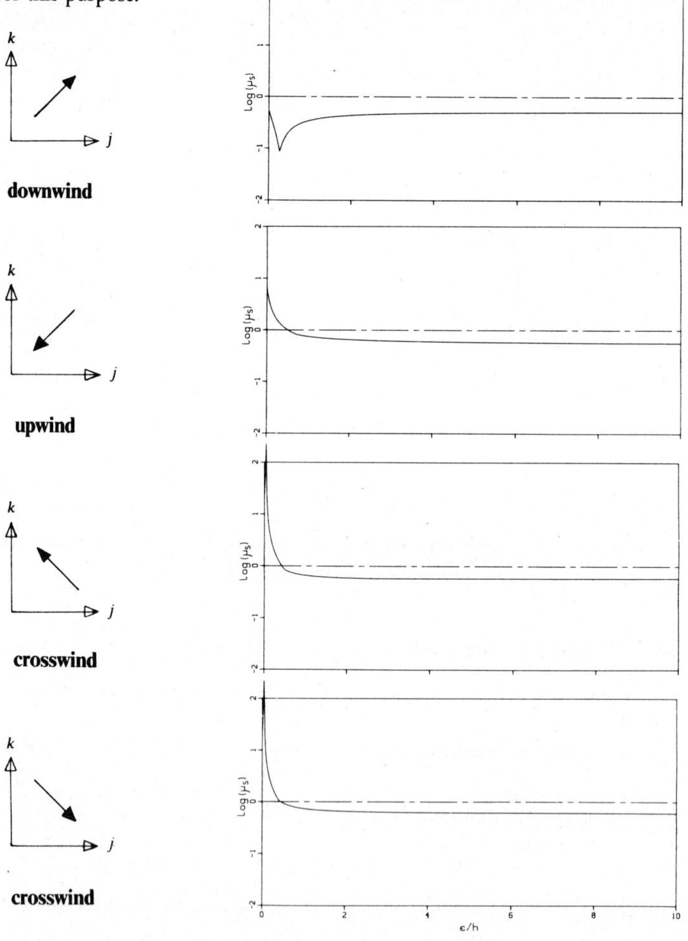

Fig. 2.8. Smoothing factors point Gauss-Seidel relaxation, second-order discretized model equation

3. Convergence of iterative defect correction

The iterative defect correction (IDeC-) method can be written as:

$$\begin{aligned}\tilde{F}_h(q_h^1) &= 0,\\ \tilde{F}_h(q_h^{n+1}) &= \tilde{F}_h(q_h^n)-\omega F_h(q_h^n),\ \ n=1,2,...,N.\end{aligned} \tag{3.1}$$

with the superscript n denoting the iteration counter and ω a possible damping factor. (A standard value for ω is $\omega=1$.) The two discrete operators considered are: the higher-order accurate operator F_h which for the model problem is defined by (2.11), and: the approximate operator $\tilde{F}_h$, the operator to be inverted. A requirement to be fulfilled by $\tilde{F}_h$ as seen in section 2.2.2, is that it must have a first-order accurate convective part only. The choice of the diffusive part is still free. Two in this sense extreme possibilities are already available: (i) the operator without diffusive terms as used in the Euler work, and (ii) the operator with second-order accurate diffusion as just considered in section 2.2.1. The advantage of the first approximate operator is its greater simplicity. For the second operator this is its closer resemblance to the target operator F_h. It complies with the theory that for sufficiently smooth problems, the solution will be second-order accurate after a single IDeC-cycle only [1]. As an intermediate alternative we also consider the approximate operator which neglects the cross derivatives. This operator will combine, in some sense, simplicity and good resemblance.

As in section 2, both theoretical and experimental results are presented. The theoretical results are obtained by local mode analysis for the same model problem as in section 2, and the experimental results are obtained by considering the same two flow problems as in section 2. Local mode analysis is made for both the outer and inner iteration (convergence respectively smoothing analysis).

3.1. Theoretical results

Concisely written, the three approximate operators to be considered are: (i) the first-order accurate convection operator

$$\begin{array}{c|c|c|} & j-1 & j \\ \hline k & -1 & 2 \\ \hline k-1 & & -1 \\ \hline \end{array} \tag{3.2}$$

(ii) the zeroth-order accurate convection-diffusion operator

$$\begin{array}{c|c|c|c|} & j-1 & j & j+1 \\ \hline k+1 & & -\frac{\epsilon}{h} & \\ \hline k & -(1+\frac{\epsilon}{h}) & 2+4\frac{\epsilon}{h} & -\frac{\epsilon}{h} \\ \hline k-1 & & -(1+\frac{\epsilon}{h}) & \\ \hline \end{array} \tag{3.3}$$

and, (iii) the first-order accurate convection-diffusion operator (2.10).

For the linear model problem (2.1), iteration (3.1) can be rewritten as

$$\begin{aligned}\tilde{F}_h(u_h^1) &= 0,\\ \tilde{F}_h(u_h^{n+1}) &= (\tilde{F}_h-\omega F_h)(u_h^n),\ \ n=1,2,...,N.\end{aligned} \tag{3.4}$$

Introducing as before the iteration error (2.7) in its Fourier form (2.8), we can write for the convergence factor μ:

$$\mu(\theta_1,\theta_2) = 1-\omega F_h(\theta_1,\theta_2)\tilde{F}_h^{-1}(\theta_1,\theta_2),\ 0\leqslant|\theta_1|,|\theta_2|\leqslant\pi. \tag{3.5}$$

For $\omega=1$, convergence results are given in fig. 3.1. In fig. 3.1a, for each of the three approximate operators (3.2), (3.3) and (2.10), the convergence factor $\mu_c=\sup|\mu(\theta_1,\theta_2)|$, $\omega=1$, $0\leqslant|\theta_1|,|\theta_2|\leqslant\pi$ is given as a function of ϵ/h. In fig. 3.1b, for $\epsilon/h=4/9$, $\epsilon/h=1$ and $\epsilon/h=\infty$, the corresponding distributions of $|\mu(\theta_1,\theta_2)|$, $\omega=1$, $0\leqslant|\theta_1|$, $|\theta_2|\leqslant\pi$ are given. (Again, all distributions are point-symmetric with respect to $\theta_1=0$, $\theta_2=0$.) From fig. 3.1a it appears that for small values of ϵ/h, the approximate operator (3.2) yields the best convergence rate. However, as was to be expected, its convergence starts to deteriorate (from $\epsilon/h=4/9$) and finally turns into divergence. Even for high-Reynolds number flows, local regions with diffusion dominating convection may arise. Therefore, approximate operator (3.2) has to be rejected. As far as the convergence rate of the two remaining operators is concerned, the 9-point operator (2.10) clearly is to be preferred above the 5-point alternative (3.3).

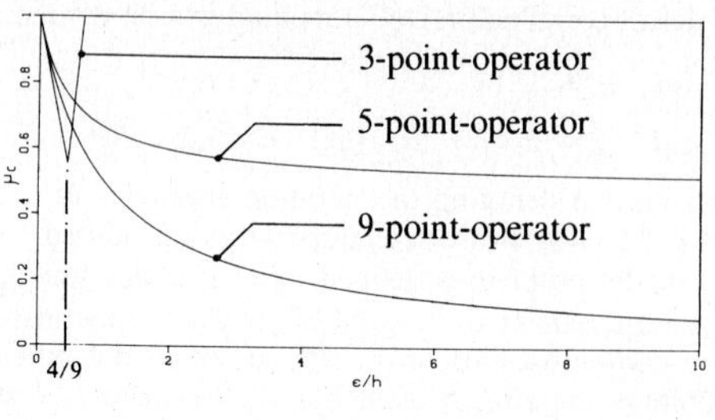

a. Versus ϵ/h

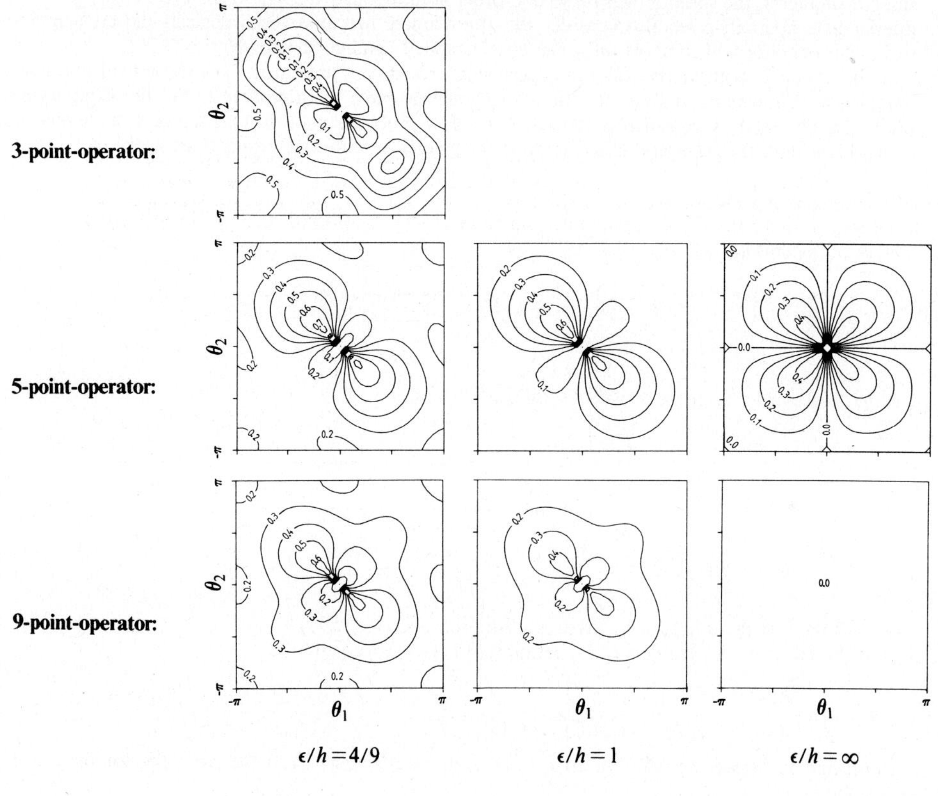

b. For $\epsilon/h=4/9$, $\epsilon/h=1$ and $\epsilon/h=\infty$

Fig. 3.1. Convergence factors iterative defect correction, second-order discretized model equation

However, the 5-pointer might behave better in the inner iteration (Gauss-Seidel accelerated by multigrid). In fig. 3.2, for the four basic sweep directions, its smoothing factors μ_s are given as a function of ϵ/h. The smoothing factors which were already presented for the 9-point operator (fig. 2.6) have been added. It appears that both operators practically have the same good smoothing behaviour, the 5-pointer being only slightly better. Because of its superior behaviour in IDeC, we prefer the 9-pointer as operator to be inverted. (Its relative complexity is taken for granted.)

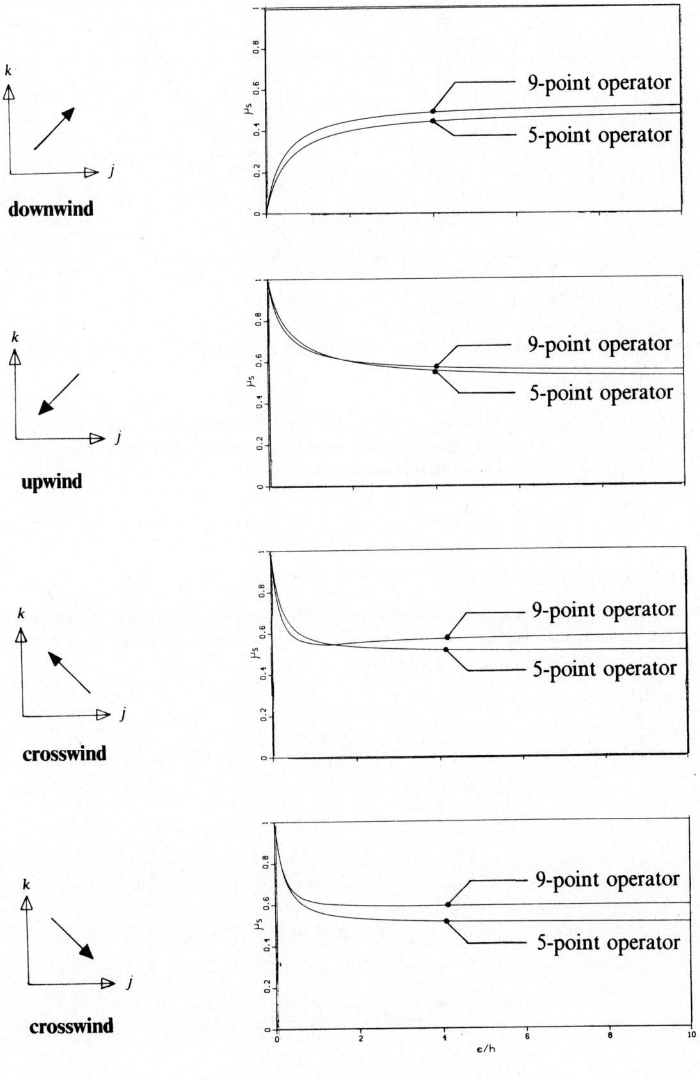

Fig. 3.2. Smoothing factors point Gauss-Seidel relaxation, zeroth- and first-order discretized model equation

3.2. Experimental results
For the subsonic flat plate flow, results are shown in fig. 3.3. Given for the 16×8-, 32×16- and 64×32-grid is the velocity profile obtained on the middle of the plate after 1 and 50 IDeC-cycles. (In all cases we performed a single FAS-cycle per IDeC-cycle only.) In agreement with theory [1], only a single IDeC-cycle appears to be sufficient for obtaining higher-order accuracy.

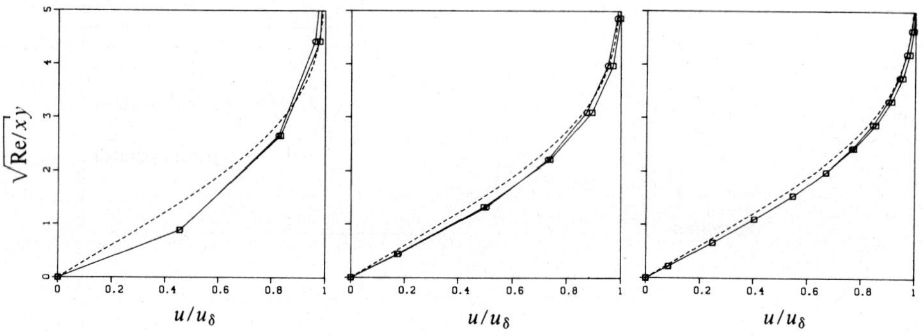

a. On 16 × 8-grid b. On 32 × 16-grid c. On 64 × 32-grid

Fig. 3.3. Velocity profiles subsonic flat plate flow, $Re=100$, $x=0.5$ (O: after 1 IDeC-cycle, □ : after 50 IDeC-cycles, ------: Blasius solution)

For the supersonic flat plate flow, the second-order accurate results are given in fig. 3.4. Here, we had to use the limiter, and further we had to take $\omega=½$. Compared with the subsonic flat plate flow, again a decrease in convergence rate is observed.

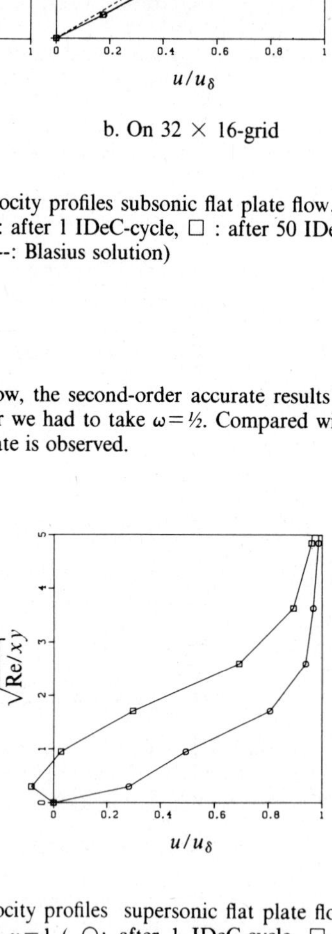

Fig. 3.4. Velocity profiles supersonic flat plate flow, $Re = 2.96\ 10^5$, $x=1$ (O: after 1 IDeC-cycle, □ : after 50 IDeC-cycles)

4. Conclusions

For the first-order discretized Navier-Stokes equations, point Gauss-Seidel relaxation accelerated by multigrid has been applied to the target equations directly. Both theory and practice show a fast convergence for smooth problems. For problems with non-smooth solutions and (consequently) non-uniform grids, practical computations show the same.

For the second-order discretized equations, iterative defect correction has been introduced, with point Gauss-Seidel and multigrid applied to the first-order discretized equations as an approximate solver. Both theory and practice show a fast convergence for smooth problems. For problems with non-smooth solutions and non-uniform grids the convergence rate is less good, though still satisfactory.

Acknowledgement

The author wants to thank dr. Hemker and prof. Wesseling for their constructive comments.

References

1. W. Hackbusch (1985). *Multigrid Methods and Applications.* Springer, Berlin.
2. R.J. Hakkinen, I. Greber, L. Trilling and S.S. Abarbanel (1958). *The Interaction of an Oblique Shock Wave with a Laminar Boundary Layer.* NASA-memorandum 2-18-59 W.
3. P.W. Hemker (1986). *Defect Correction and Higher Order Schemes for the Multi Grid Solution of the Steady Euler Equations.* Proceedings of the 2nd European Conference on Multigrid Methods, Cologne 1985, Springer, Berlin.
4. ------ and B. Koren (1986). *A Non-linear Multigrid Method for the Steady Euler Equations.* Proceedings GAMM-Workshop on the Numerical Simulation of Compressible Euler Flows, Rocquencourt 1986, Vieweg, Braunschweig.
5. ------ and S.P. Spekreijse (1986). *Multiple Grid and Osher's Scheme for the Efficient Solution of the Steady Euler Equations.* Appl. Num. Math. 2, 475-493.
6. B. Koren (1988). *Euler Flow Solutions for a Transonic Wind Tunnel Section.* Proceedings High Speed Aerodynamics II, Aachen 1987 (to appear).
7. ------ (1988). *Defect Correction and Multigrid for an Efficient and Accurate Computation of Airfoil Flows.* J. Comput. Phys. (to appear).
8. ------ (1988). *First-Order Upwind Schemes and Multigrid for the Steady Navier-Stokes Equations.* CWI report NM-R88xx (to appear).
9. ------ (1988). *Higher-Order Upwind Schemes and Defect Correction for the Steady Navier-Stokes Equations.* CWI report NM-R88yy (to appear).
10. ------ (1988). *Upwind Schemes for the Navier-Stokes Equations.* Proceedings Second International Conference on Hyperbolic Problems, Aachen, 1988 (to appear).
11. ------ and S.P. Spekreijse (1988). *Solution of the Steady Euler Equations by a Multigrid Method.* Proceedings Third Copper Mountain Conference on Multigrid Methods, Copper Mountain, 1987 (to appear).
12. B. van Leer (1982). *Flux-Vector Splitting for the Euler Equations.* Proceedings 8th International Conference on Numerical Methods in Fluid Dynamics, Aachen 1982, Springer, Berlin.
13. ------ (1985). *Upwind-Difference Methods for Aerodynamic Problems governed by the Euler Equations.* Proceedings 15th AMS - SIAM Summer Seminar on Applied Mathematics, Scripps Institution of Oceanography 1983, AMS, Providence, Rhode Island.
14. S. Osher and F. Solomon (1982). *Upwind-Difference Schemes for Hyperbolic Systems of Conservation Laws.* Math. Comp. 38, 339-374.
15. R. Peyret and T.D. Taylor (1983). *Computational Methods for Fluid Flow.* Springer, Berlin.
16. H. Schlichting (1979). *Boundary-Layer Theory.* McGraw-Hill, New York.
17. S.P. Spekreijse (1987). *Multigrid Solution of Monotone Second-Order Discretizations of Hyperbolic Conservation Laws.* Math. Comp. 49, 135-155.
18. P. Wesseling (1987). *Linear Multigrid Methods.* In: Multigrid Methods, SIAM, Philadelphia.

TOWARDS MULTIGRID ACCELERATION OF 2D COMPRESSIBLE NAVIER-STOKES FINITE VOLUME IMPLICIT SCHEMES

Y. MARX & J. PIQUET
CFD Group, ENSM (UA 1217 CNRS)
1 Rue de la Noe, 44072 Nantes, France

SUMMARY

A multigrid solution of the twodimensional compressible Navier-Stokes equations is described. The method rests on the non linear FAS multigrid procedure. Implicit upwind schemes are used as relaxation schemes. This method allows the computation of oscillation-free shocks without the need of introducing artificial viscosity. Results are presented on the GAMM double-throat nozzle configuration [1].

1. INTRODUCTION

A lot of efforts has been recently devoted to the study of highly accurate non oscillatory schemes. Most of them use some kind of upwinding associated with the so-called TVD limiters introduced in order to increase the accuracy without spurious oscillations. If threedimensional calculations on realistic configurations are looked for, procedures have to be developped in order to reduce the computational time. Some of them are presented here on the twodimensional GAMM nozzle test case. The work is outlined as follows. First, the upwind method used for solving the Navier-Stokes equations (§2) is described (§3) ; then the acceleration techniques are presented in §4 (implicit schemes) and in §5 (multigrid procedure). Finally the obtained results are discussed (§6).

2. GOVERNING EQUATIONS

The Navier-Stokes equations written in their two dimensional conservation form are

$$\frac{\partial U}{\partial t} + \frac{\partial F}{\partial x} + \frac{\partial G}{\partial y} = 0 \qquad (1)$$

with $F = F' + F''$; $G = G' + G''$ where F' and G' are the advective fluxes while F'' and G'' are the diffusive fluxes:

$$F'' = N_x \partial\tilde{U}/\partial x + M_x \partial\tilde{U}/\partial y \; ; \; G'' = N_y \partial\tilde{U}/\partial y + M_y \partial\tilde{U}/\partial x$$

$$U = \begin{Vmatrix} \rho \\ \rho u \\ \rho v \\ e \end{Vmatrix} ; \; F' = \begin{Vmatrix} \rho u \\ \rho u^2+P \\ \rho uv \\ u(e+P) \end{Vmatrix} ; \; G' = \begin{Vmatrix} \rho v \\ \rho uv \\ \rho v^2+P \\ v(e+P) \end{Vmatrix} ; \; \tilde{U} = \begin{Vmatrix} \rho \\ u \\ v \\ P \end{Vmatrix}$$

$$N_x = \begin{Vmatrix} 0 & 0 & 0 & 0 \\ 0 & \lambda+2\mu & 0 & 0 \\ 0 & 0 & \mu & 0 \\ \frac{-\gamma\mu P}{(\gamma-1)\rho^2\sigma} & (\lambda+2\mu)u & \mu v & \frac{\gamma\mu}{\rho(\gamma-1)\sigma} \end{Vmatrix} ; \; M_x = \begin{Vmatrix} 0 & 0 & 0 & 0 \\ 0 & 0 & \lambda & 0 \\ 0 & \mu & 0 & 0 \\ 0 & \lambda v & \lambda u & 0 \end{Vmatrix}$$

$$N_y = \begin{Vmatrix} 0 & 0 & 0 & 0 \\ 0 & \mu & 0 & 0 \\ 0 & 0 & \lambda+2\mu & 0 \\ \frac{-\gamma\mu P}{(\gamma-1)\rho^2\sigma} & \mu u & (\lambda+2\mu)v & \frac{\gamma\mu}{\rho(\gamma-1)\sigma} \end{Vmatrix} ; \; M_y = \begin{Vmatrix} 0 & 0 & 0 & 0 \\ 0 & 0 & \mu & 0 \\ 0 & \lambda & 0 & 0 \\ 0 & \lambda v & \lambda u & 0 \end{Vmatrix}$$

where σ is the Prandtl number, λ and μ the viscosities.

3. THE EXPLICIT STAGE

The MUSCL approach [2] is followed, it rests on a finite volume approach where the variables U are assumed to vary linearly in each cell. The piecewise constant or linear distribution L(u) -leading respectively to first order or second order accurate schemes- is computed applying the Harten & Osher [3] reconstruction. Having calculated the space evolution of the variables U, it remains to solve at the interfaces the Riemann problems arising from the piecewise distribution. In order to save computing time without losing accuracy, the Riemann problems are solved in the direction normal to the cell interfaces only approximately, following [4].

4. THE IMPLICIT CORRECTION

In order to accelerate the convergence towards the steady state, the explicit increment ΔU has to be corrected through an implicit operator (2)

$$\frac{\delta U}{\Delta t} + \frac{\partial}{\partial x}[A\delta U] + \frac{\partial}{\partial y}[B\delta U]$$

$$- \frac{\partial}{\partial x}\left[N_x \frac{\partial}{\partial x}\delta\tilde{U} + M_x \frac{\partial}{\partial y}\delta\tilde{U}\right] - \frac{\partial}{\partial y}\left[N_y \frac{\partial}{\partial y}\delta\tilde{U} + M_y \frac{\partial}{\partial y}\delta\tilde{U}\right] = \frac{\Delta U}{\Delta t} \qquad (2)$$

where $\delta U = U^{n+1} - U^n$; $\delta\tilde{U} = \tilde{U}^{n+1} - \tilde{U}^n$; $A = \partial F'^n/\partial U$; $B = \partial G'^n/\partial U$

The implicit operator is obtained after (i) a Taylor series expansion of U

$U^{n+1} = U^n + \Delta t\,[\partial U/\partial t]^{n+1} + O(\Delta t^2)$ where

$$\frac{\partial U}{\partial t}^{n+1} = - \frac{\partial F'}{\partial x}^{n+1} - \frac{\partial G'}{\partial y}^{n+1}$$

$$+ \frac{\partial}{\partial x}\left[N_x^{n+1} \frac{\partial \tilde{U}}{\partial x}^{n+1} + M_x^{n+1} \frac{\partial \tilde{U}}{\partial y}^{n+1}\right] + \frac{\partial}{\partial y}\left[N_y^{n+1} \frac{\partial \tilde{U}}{\partial y}^{n+1} + M_y^{n+1} \frac{\partial \tilde{U}}{\partial x}^{n+1}\right]$$

(ii) a linearisation of the fluxes F′ and G′, (iii) a time freezing of the M and N matrices. Different implicit methods result from different space discretisations of (2). Three of them have been tested.

4.1. The (factored) Coakley scheme

In this case, N_x and N_y are replaced by their spectral radius while M_x and M_y are set equal to zero. The 2D operator is also factorized and the system is "diagonalized" in the sense of [5]. Finally a first order upwind approximation is used for the advection terms and a second order centered approximation is used for the diffusion terms. The resulting scheme is written

$$S_x^{-1}\{1 + \Delta t \Lambda_x^+ \partial^- \cdot/\partial x + \Delta t \Lambda_x^- \partial^+/\partial x - \Delta t \nu \partial^2 \cdot/\partial x^2\} S_x \cdot$$
$$S_y^{-1}\{1 + \Delta t \Lambda_y^+ \partial^-/\partial y + \Delta t \Lambda_y^- \partial^+/\partial y - \Delta t \nu \partial^2/\partial y^2\} S_y \delta U = \Delta U \qquad (3)$$

with $\nu = \rho^{-1} \max(\mu, \lambda+2\mu, \gamma\mu/\sigma)$ and $\mathbf{A} = R^{-1}S_x^{-1}\,\boldsymbol{\Lambda}_x\, S_x R$; $\mathbf{B} = R^{-1}S_y^{-1}\,\boldsymbol{\Lambda}_y\, S_y R$. $\boldsymbol{\Lambda}_x$ and $\boldsymbol{\Lambda}_y$ are the diagonal matrices of the eigenvalues of **A** and **B**, R is the matrix which transforms the conservative increments to the convective increments, S_x and S_y are the matrices transforming the convective increments to the characteristic increments δU^x and δU^y in the x and y direction. The Coakley scheme leads to a scalar tridiagonal system, however an implicit treatment of the boundaries may introduce a coupling between some of the variables.

4.2. The unfactored Coakley scheme.

An unfactored version of the Coakley scheme has been developped [6]; it differs from the previous scheme mainly by the fact that no factorization is performed. The implicit operator to solve is then:

$$\{1 + \Delta t \mathbf{\Lambda}_x^{+} \partial^{-} \cdot / \partial x + \Delta t \mathbf{\Lambda}_x^{-} \partial^{+} \cdot / \partial x - \Delta t \nu \partial^2 \cdot / \partial x^2 + \\ + \Delta t S^{-1} \mathbf{\Lambda}_y^{+} S \partial^{-} \cdot / \partial y + \Delta t S^{-1} \mathbf{\Lambda}_y^{-} S \partial^{+} \cdot / \partial y - \Delta t \nu \partial^2 \cdot / \partial y^2\} \ \delta U^x = \Delta U \qquad (5)$$

with $S = S_y S_x^{-1}$. The form of the S and S^{-1} matrices forces now a coupling between all the variables but the entropy. As the upwinding leads to a diagonally dominant matrix, a Gauss-Seidel relaxation method is efficient.

4.3. The full unfactored scheme.

The full unfactored scheme is derived directly from (2) by bringing out of the space derivatives the matrices $R^{-1}S_x^{-1}\mathbf{\Lambda}_x S_x$, $R^{-1}S_y^{-1}\mathbf{\Lambda}_y S_y$, M_x, M_y, N_x, N_y. As for the Coakley schemes, a first order upwind approximation is used for the advection terms and a centered approximation for the diffusive terms. The result is given in the appendix.

With this scheme, all the variables are coupled and relaxation iterations are performed to calculate δU. In the three cases, the boundary conditions are treated implicitly.

5. THE MULTIGRID PROCEDURE

The use of implicit schemes on a single grid leads to a spectacular improvement in the convergence towards the steady state. The computational time can be further reduced if the implicit schemes are used in a multigrid procedure.

5.1. The FAS and FMG algorithms.

For the non linear system $NU = \tau$, let us assume that an approximate solution U_k^n is known on a grid G_k where k=1 is the finest grid.

(i) The errors are smoothed out applying m_1 prerelaxation iterations.

(ii) The solutions $U_k^{n+m_1}$ and the residual $N_k U_k^{n+m_1} - \tau_k$ are restricted on a coarser grid G_{k+1} $(\tau_1 \equiv 0)$.

(iii) ω FAS iterations are performed on the grid G_{k+1} to obtain an updated value U_{k+1} to the system

$$N_{k+1} U_{k+1} = N_{k+1} [I_k^{k+1} U_k^{n+m_1}] - I_k^{k+1} [N_k U_k^{n+m_1} - \tau_k] = \tau_{k+1}$$

where $\omega = 1$ gives a V-cycle and $\omega = 2$ gives a W-cycle.

(iv) The fine grid solution is corrected using

$$U_k^{n+m_1} = U_k^{n+m_1} + I_{k+1}^k [U_{k+1} - I_k^{k+1} U_k^{n+m_1}]$$

(v) The high frequencies introduced by the prolongation operator I_{k+1}^k are damped using m_2 postrelaxation iterations.

In the FAS algorithm, the initial guess is U_1^0, the procedure starts therefore with a crude approximation on the finest grid.

A different strategy is to use the so-called FMG algorithm. A "converged" solution of $NU = 0$ is first calculated on the coarsest grid G_k. Then a prolongation gives an initial estimate on G_{k-1} which is improved with FAS iterations between the grids G_k and G_{k-1} : the improved approximation is prolongated to G_{k-2}, and so on, until the finest grid is reached. The FMG stage is a way of providing on the finest grid a good initial approximation. To complete the description of the multigrid procedure, the relaxation scheme and the transfers have to be specified.

5.2. *The relaxation scheme*

The implicit schemes described in §4 are used as the relaxation schemes. With the symbolic notation $L\delta U = \Delta U = -\Delta t\ NU$ for the implicit step, one relaxation iteration with the implicit scheme will be on a grid G_k:

$$L_k^n\ \delta U_k^n = -\Delta t\ [N_k^n U_k^n - \tau_k]\ ;\ U_k^{n+1} = U_k^n + \delta U_k^n. \qquad (6)$$

The success of the MG procedure depends strongly on the high frequency damping characteristics of the relaxation schemes. They are now analyzed on model problems.

5.3. *Analysis of the damping characteristics of the relaxation schemes*

5.3.1. Implicit factored Coakley scheme.

Let us consider the twodimensional linear advection-diffusion equation

$$\frac{\partial u}{\partial t} + c_x \frac{\partial u}{\partial x} + c_y \frac{\partial u}{\partial y} = d_x \frac{\partial^2 u}{\partial x^2} + d_y \frac{\partial^2 u}{\partial y^2}\ ;\ c_x > 0\ ;\ c_y > 0 \qquad (7)$$

For high Courant numbers C_x, C_y and diffusion numbers D_x, D_y (high Δt), the amplification factor $g(\zeta,\eta)$ resulting from a Von Neumann analysis has a unit value for $(\zeta,\eta)=(\pi,\pi)$: the high frequencies are not damped (fig. 1a). Using the procedure introduced in [7], it is possible to force $|\ g(\pi,\pi)\ | = 0$ if a suitable modification of dissipation terms is introduced at the implicit step.

$$D_{xi} = (\alpha+\beta)/4 - 1/2 - C_x/2\ ;\ D_{yi} = (\alpha-\beta)/4 - C_y \qquad (8)$$

with $\beta = 1 + (C_x - C_y) + 2(D_x - D_y)\ ;\ \alpha^2 = \beta^2 + 4C_y + 8D_y - 1$.

As figs. 1b,c indicate, the high frequencies are effectively slightly better damped with the modified implicit scheme, but the scheme is no more unconditionally stable. We can conclude that the implicitly factored scheme is not a well suited relaxation scheme for a multigrid process.

5.3.2. Unfactored Coakley scheme.

The model problem used for the unfactored scheme is the following:

$$\begin{aligned}\frac{\partial u}{\partial t} + c_{x1} \frac{\partial u}{\partial x} + c_{y1} \frac{\partial u}{\partial y} + a_{y1} \frac{\partial v}{\partial y} - d_{x1} \frac{\partial^2 u}{\partial x^2} - d_{y1} \frac{\partial^2 u}{\partial y^2} &= 0 \\ \frac{\partial v}{\partial t} - a_{x2} \frac{\partial v}{\partial x} + a_{y2} \frac{\partial v}{\partial y} + c_{y2} \frac{\partial u}{\partial y} - d_{x2} \frac{\partial^2 v}{\partial x^2} - d_{y2} \frac{\partial^2 v}{\partial y^2} &= 0\end{aligned} \qquad (9)$$

with positive coefficients.

If the implicit matrix of the first order unfactored Coakley scheme is "solved" with two line Gauss Seidel relaxations (one in each direction), the amplification factor of the scheme is given in fig. 2. This scheme shows good damping properties on the model problem.

5.4. *Restriction and prolongation operators*

For a finite volume method, a natural choice for I_h^{2h} and I_h^{2h} is to take an operator which conserves the cell averages (fig. 3).

$$I_h^{2h}\ \theta_h = I_h^{2h}\ \theta_h = [\upsilon_{2h}]^{-1} \sum_{\upsilon_h \in \upsilon_{2h}} \upsilon_h\ \theta_h\ .$$

The coarse-to-fine grid transfer is performed using a bilinear interpolation as indicated in fig. 4 . At the boundaries, the values are neither restricted nor prolongated on each grid ; the boundary values are computed so that the boundary conditions are satisfied.

6. RESULTS

The aforementionned techniques are tested on the GAMM double throat nozzle. Results obtained with the first order scheme are firstly described with and without multiple grids.

6.1. Single grid results

As already mentionned, implicit schemes appear to increase efficiently the convergence rate (fig. 5). With the factored scheme, an obvious change in the slope of the convergence curve occurs at high CFLs : the highest values of the residues are generated by the wall and the exit boundary and they are of high frequency. They are therefore not efficiently damped by the factored Coakley scheme when large Δt are used. A low Δt should be utilized when the high frequencies dominate the error. A natural procedure is to perform a Fourier analysis of the residual and to relate Δt to the range of the dominating frequencies. This procedure improves the convergence of the factored Coakley scheme (fig. 6) but the relation between the time step and the dominating frequencies might be problem or mesh-dependent. The rate of convergence of the unfactored Coakley scheme is higher than the rate of convergence of the factored scheme (figs. 5 & 7). But as with the unfactored scheme, δU is computed with two sweeps -one in each direction- of a non vectorizable line Gauss-Seidel relaxation, the unfactored scheme is more time-costly than the factored scheme. Moreover, the efficiency of the unfactored scheme depends on the spectral radius ν. This sensitivity to the viscous coefficients can be avoided if the fully implicit unfactored scheme is used. A mesh-independent convergence then results (fig. 8). An initial approximation δU for the unfactored scheme is provided by the factored scheme and allows a correct treatment of the advection terms, but the treatment of the diffusive terms remains to be improved. The zebra relaxation -which is vectorizable and which damps efficiently the high frequencies- is used for this purpose. Even and odd points are updated only once and, as a result, one is left with a full unfactored scheme which is paid by a 10 to 35 % increase -the finer the mesh, the lower the increase- of the computing time on VP200 due to the resolution of the unfactored implicit operator.

6.2. Multigrid results, first order

The convergence with a MG procedure using the factored Coakley scheme as a relaxation scheme cannot be increased, as predicted by the analysis of the danping properties of this scheme. More surprising is the failure of the unfactored Coakley scheme as it appears to be contradicted by the analysis performed on the model problem. This is probably because the implicit step is not the exact jacobian of the explicit step for the considered test case ; moreover, the most important part of the domain is diffusion-controlled, the Coakley schemes therefore retain at the implicit step too crude an approximation of the most important terms ; this argument did prompt us to develop the full unfactored scheme.

When the full unfactored scheme is used as a relaxation scheme in the MG procedure, the cpu time can be saved by a factor of 3 (fig. 9) with a FAS strategy and by a factor about 4 with the FMG strategy (fig. 10). The multigrid procedure appears insensitive to the retained cycles -V or W- and to the number, 2, 3 or 4, of grids (fig. 11). Moreover, the influence of the CFL choices on each grid decreases when the number of grids increases (figs. 12). The best convergence rate is obtained when a high CFL is used on the coarsest grid. In order to keep the efficiency of the MG procedure, the CFLs must not be too high on the finer grids. In fact, the high CFLs are important for the damping of low frequencies and this is accomplished only on the coarsest grids. So it is natural to find that the convergence is rather insensitive to the value of the CFL on the finest grid. It is also not necessary to iterate too much on the finest grids as their aim is only to allow the damping of the high frequencies. With the FMG algorithm, it is worthwhile to converge on the coarsest grid where the computations are cheap. Only one or two FAS cycles are then performed on the intermediate grids.

From the various multigrid curves, it can be seen that the convergence slows down when a residual $O(10^{-4}-10^{-5})$ is reached. Then, the difference between the truncation error of the true jacobian of the explicit step and the "false" used jacobian becomes important. As they correspond to differences in the local approximations, they are high-frequency controlled and they cannot be efficiently damped by the "full" unfactored scheme. This "conflict" in truncation errors explains why the convergence slows down. Nevertheless, a mesh - independent convergence is found before the slow-down (Table 1) ; on each grid , the fields are initialized with an interpolation of a "converged solution" on a coarser grid.

6.3 Second order results

Second order steady states are calculated with a defect correction type method : a second order accurate scheme is used at the explicit step, but a first order scheme is retained at the implicit step. Using this procedure, a limit cycle behavior is usually obtained. With the full unfactored scheme, it is possible to lower the level of the residuals by underrelaxing (.5) the increments (fig. 13). For Re=100 (no shock), when the grid is not too fine, convergence to machine zero can be reached (fig. 14) and the obtained results correspond to a 1/10000 difference in the mass-flow rate. First and second order Coakley schemes behave in a similar way and show a change in the slope of the convergence curve (fig. 15). No improvement is obtained with underrelaxation. For high Re cases (with shocks), limit cycles corresponding to the non linear behavior of the slope limiter in the vicinity of the shock are always found. Nevertheless the defect correction procedure leads to a significant improvement of the accuracy as results similar to those of [8] are obtained whereas no shock is obtained with a first order scheme. Here again, the FMG defect correction allows a great reduction in the computing time (fig. 16).

7. CONCLUSION

A MG procedure where implicit schemes are used as the relaxation schemes has been proved to be efficient for the 2D compressible Navier-Stokes equations, if the implicit operator does not differ too much from the true jacobian of the explicit approximation. However, limit cycles related to the non linearity of the slope limiter yield the efficiency of the procedure difficult to assess in cases with shocks. Results obtained are nevertheless in good agreement with those obtained by other methods and a significant reduction in the computing time is found with the FMG procedure.

Acknowledgments Partial financial support of DRET (contract 86.107) is gratefully acknowledged. Computations have been performed on the VP200 (CIRCE) with a dotation of 5h by the DS-SPI and on the Cray2 of the CCVR with a dotation of 20h by its Scientific Committee.

REFERENCES

1 M.O. Bristeau, R. Glowinski, J. Périaux and H. Viviand, Eds. "Numerical Simulation of Compressible Navier-Stokes Equations", *Proc. Gamm Workshop, Notes on Numerical Fluid Dynamics*, Vol. 18, Vieweg-Verlag, 1986.
2 B. Van Leer, "Flux Vector Splitting for the Euler Equations", in *Lecture Notes in Physics*, Vol. 170, pp. 507-511, Springer Verlag, 1982.

3 A. Harten and S. Osher, "Uniformly High-Order Accurate Non Oscillatory Schemes I", *NASA CR 175768*, june 1985.

4 S. Osher, "Shock Modelling in Aeronautics", in *Numerical Methods for Fluid Dynamics*, pp. 179-217, Morton & Baines Eds., Academic Press, 1982.

5 D.S. Chaussee and T.H. Pulliam, "A Diagonal Form of an Implicit Approximate Factorization Algorithm with Application to a Two Dimensional Inlet", *AIAA Paper 80-0067*, 1980.

6 Y. Marx & J. Piquet, "Two-dimensional compressible Navier-Stokes Finite Volume Computations by means of Implicit Schemes", to appear in *Int. J. Num. Meth. Fluids*, 1988.

7 A. Jameson and S. Yoon, "Multigrid Solution of the Euler Equations using Implicit Schemes", *AIAA Paper 85-0293*, 1985.

8 Y. Marx, "Computations for Viscous Compressible Flows in a Double Throat Nozzle", *in* [1], pp. 273-290.

FIGURES

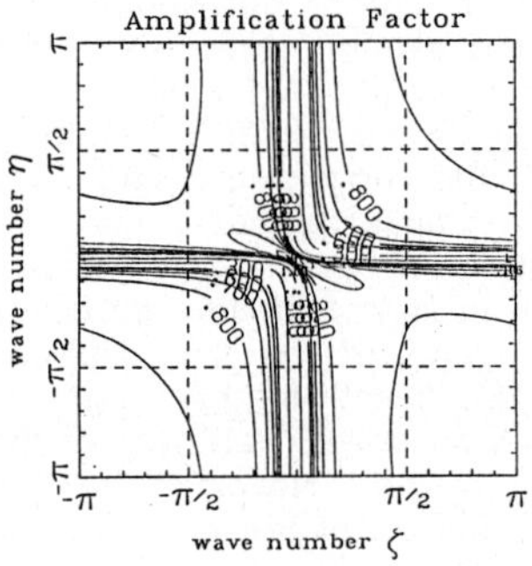

Fig.1.a

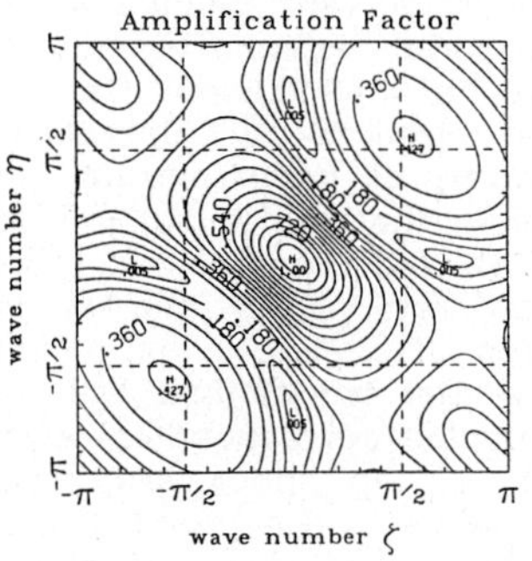

Fig.1.b

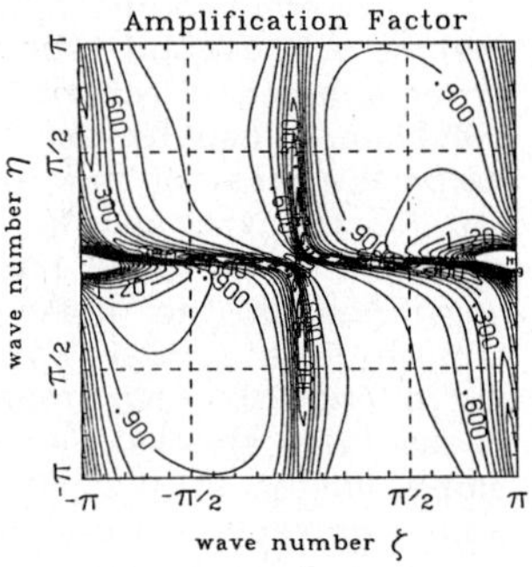

Fig.1.c

Fig.1. Amplification factor for the factored scheme

1.a Initial scheme : $C_x=C_y=10.$; $D_x=D_y=1.$

1.b Scheme with "optimal" implicit dissipation : $C_x=C_y=1.$; $D_x=D_y=0.1$

1.c : $C_x=C_y=10.$; $D_x=D_y=1.$

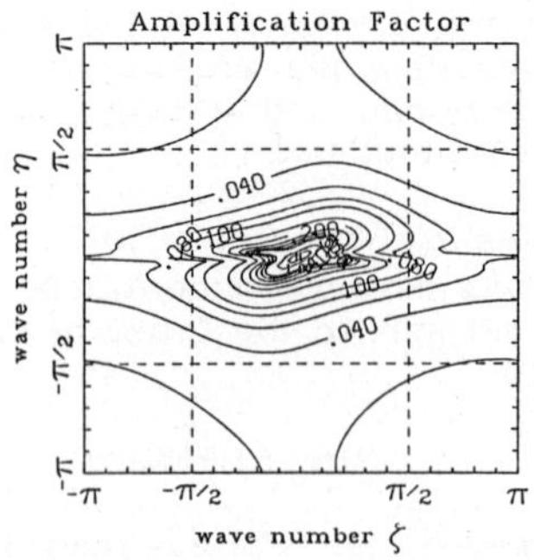

Fig.2. Amplification factor for the unfactored scheme.
$C_{x1}=10.$; $C_{y1}=100.$; $A_{y1}=30.$; $A_{y2}=50.$; $A_{x2}=20.$
$D_{x1}=0.7$; $D_{y1}=0.5$; $C_{y2}=40.$; $D_{x2}=1.1$; $D_{y2}=0.8$

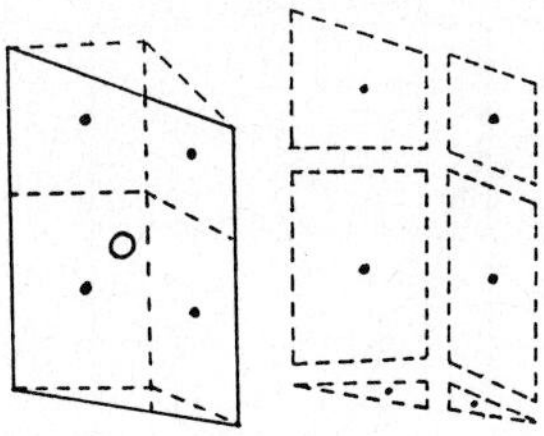

Fig.3. Restriction operator

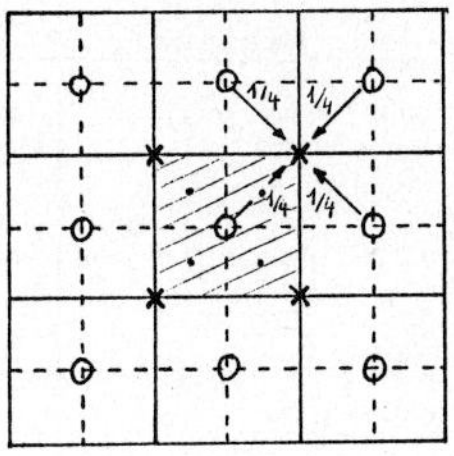

Fig.4. Prolongation operator;
o : Coarse grid points
• : fine grid points
x : nodal grid points for the bilinear interpolation.

CONVERGENCE HISTORIES, 1st ORDER SCHEMES

GAMM Double throat nozzle test case, Re=100.

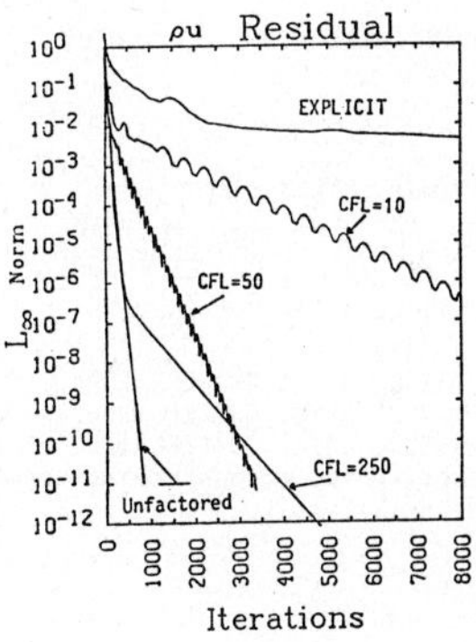

Fig.5.

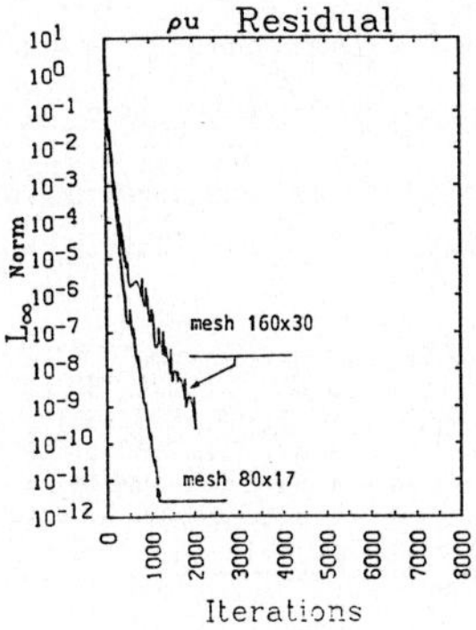

Fig.6.

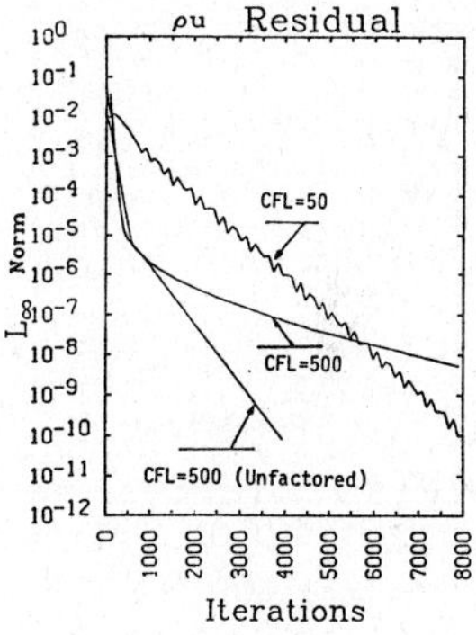

Fig.7.

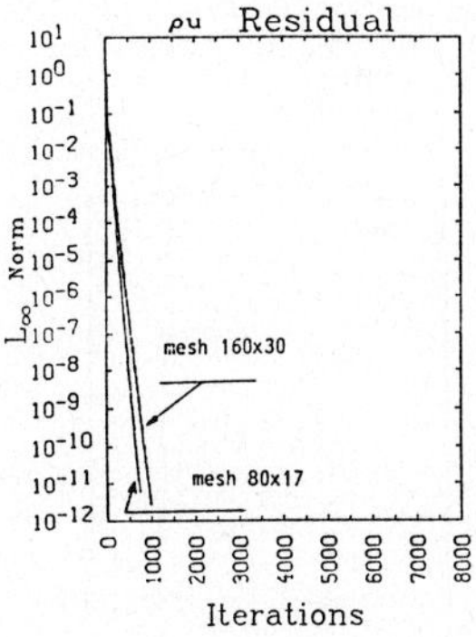

Fig.8.

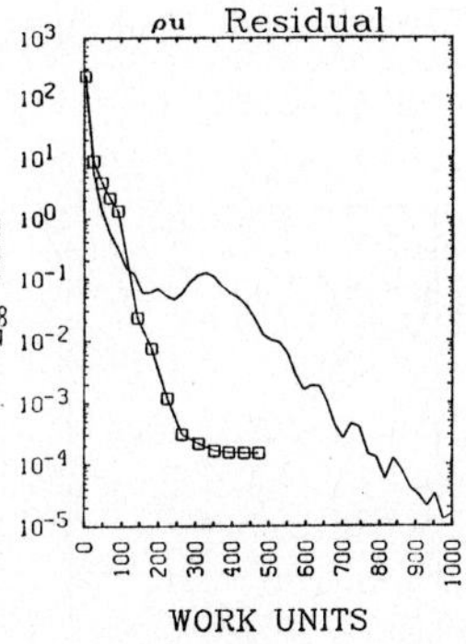

Fig.9.

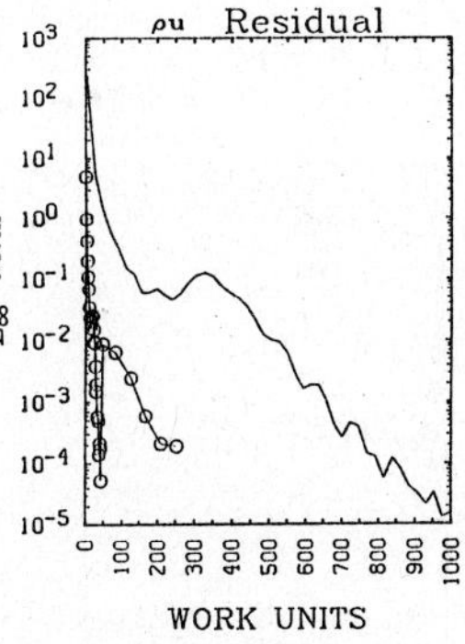

Fig.10.

Fig.5. Coakley schemes; 80x17 mesh

Fig.6. Factored Coakley scheme; Variable CFL number.

Fig.7. Coakley schemes; 160x30 mesh.

Fig.8. Full unfactored implicit scheme; CFL=250 80x17 mesh; CFL=500 160x30 mesh

Fig.9. FAS 3 grids process □ ; — single grid CFL=4500; 321x65 mesh.

Fig.10. FMG 3 grids process O ; — single grid CFL=4500; 321x65 mesh.

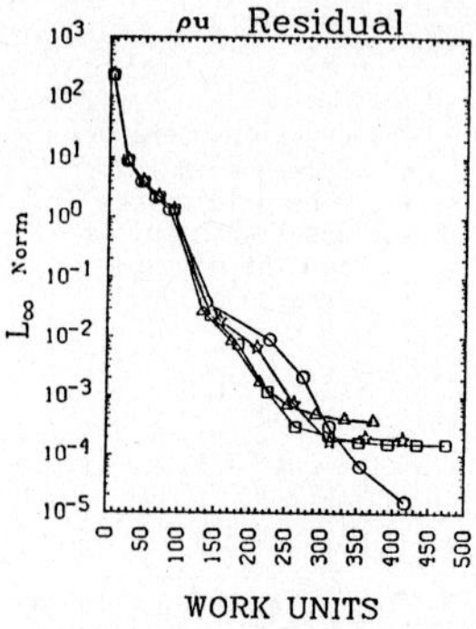

Fig.11.

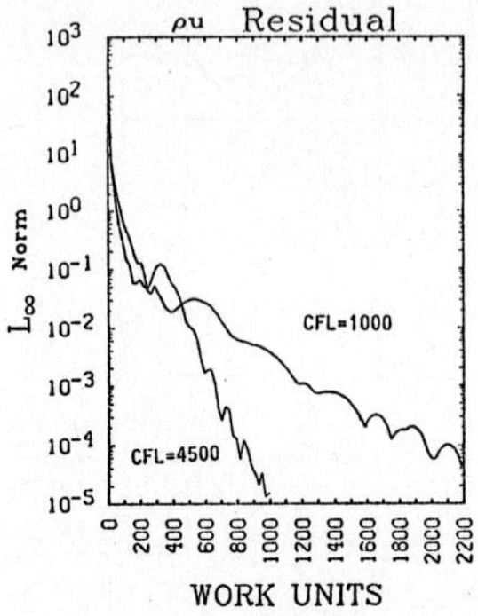

Fig.12.a

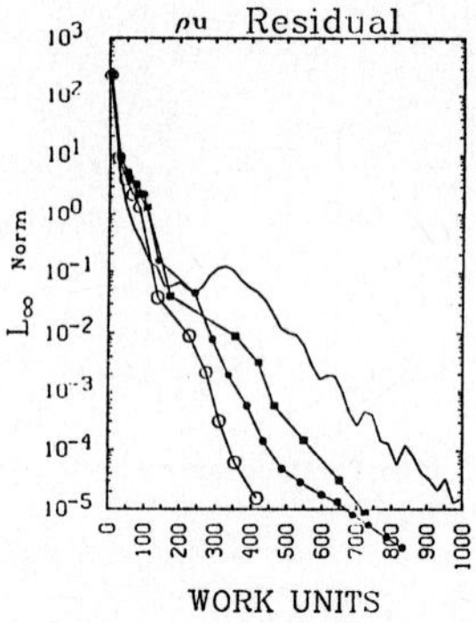

Fig.12.b

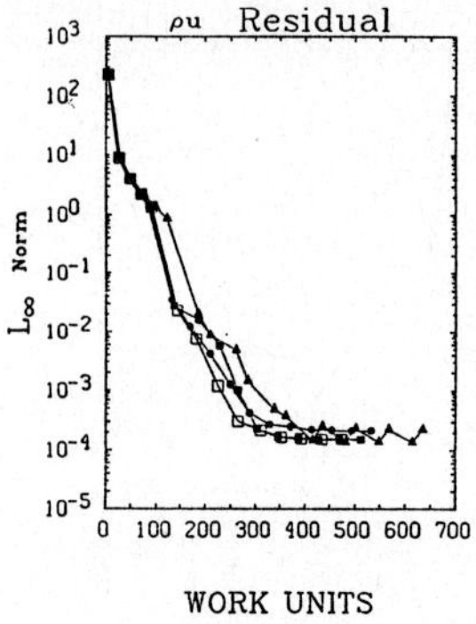

Fig.12.c

Fig.11. FAS process; ○ 2 grids, □ 3 grids, Δ 4 grids : V cycle; ☆ 3 grids W cycle.

Fig.12.a Full unfactored implicit scheme; 321x65 mesh.

Fig.12.b 2 grids V cycle ; ○ G1: CFL=500, 5 it. G2: CFL=1500.
• G1: CFL=150, 5 it. G2: CFL=1500.
▪ G1: CFL=500, 5 it. G2: CFL=500.

Fig.12.c 3 grids V cycle;
□ G1: CFL=500, 5 it. G2: CFL=250, 5 it. G3: CFL=500.
▪ G1: CFL=500, 5 it. G2: CFL=250, 5 it. G3: CFL=250.
• G1: CFL=250, 5 it. G2: CFL=100, 5 it. G3: CFL=500.
▲ Same CFL numbers as in □ but 10 iterations are performed before each fine to coarse transfers.

CONVERGENCE HISTORIES, 2nd ORDER SCHEMES

GAMM Double throat nozzle test case, Re=100.

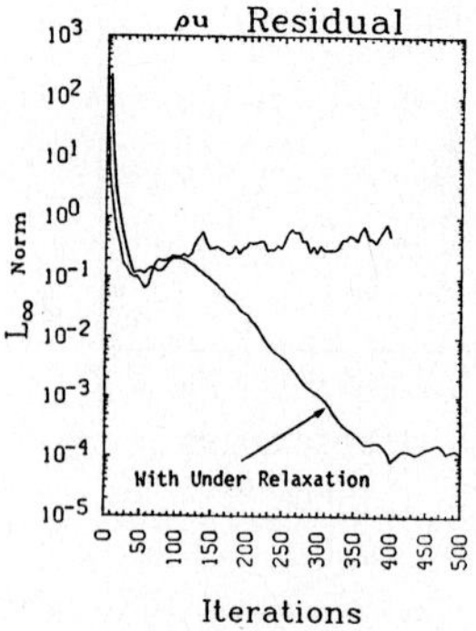

Fig.13.

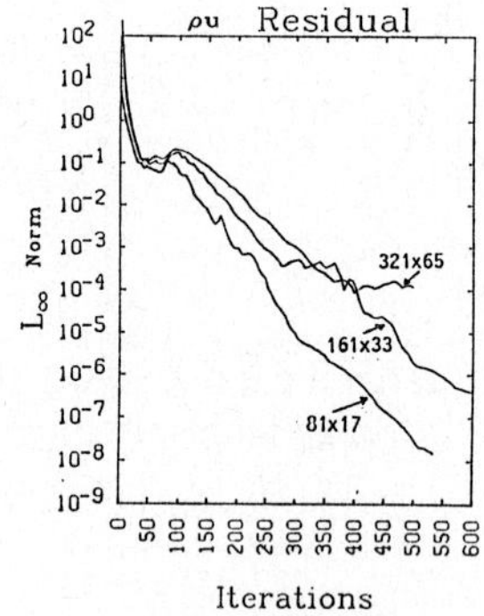

Fig.14.

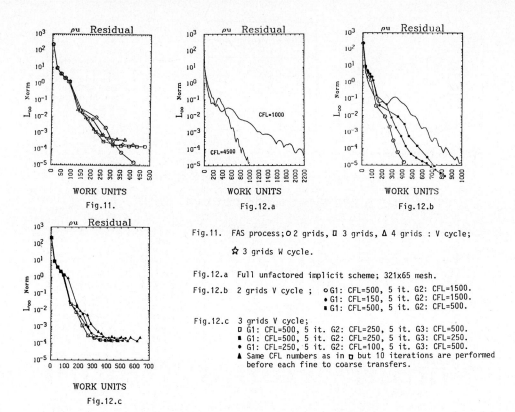

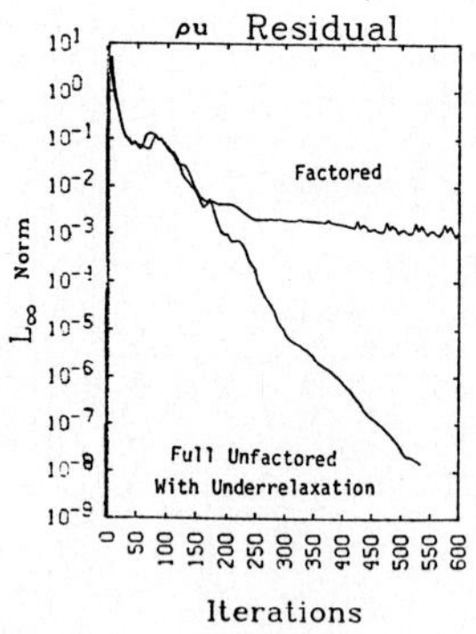

Fig.15.

Fig.13. Full unfactored implicit scheme with an without underrelaxation; 321x65 mesh; CFL=4500.

Fig.14. Full unfactored implicit scheme with underrelaxation; 321x65 mesh CFL=4500; 161x33 mesh CFL=1500; 81x17 mesh CFL=500.

Fig.15. Factored Coakley scheme and Full unfactored implicit scheme; 81x17 mesh CFL=500.

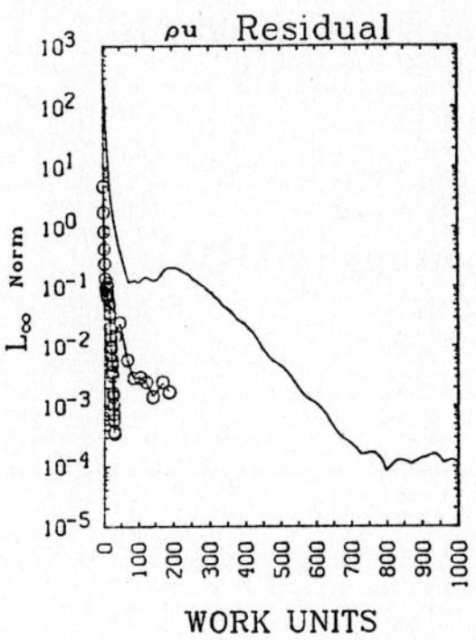

Fig.16. FMG result, O 3 grids; —single grid; 321x65 mesh; CFL=4500.

TABLE 1

iteration	residual	
	161x33 grid	321x65 grid
0	$1.42\ 10^{-2}$	$1.41\ 10^{-2}$
1	$6.33\ 10^{-2}$	$6.03\ 10^{-2}$
2	$1.51\ 10^{-3}$	$1.83\ 10^{-3}$
3	$3.86\ 10^{-4}$	$4.39\ 10^{-4}$
4	$9.28\ 10^{-5}$	$1.47\ 10^{-4}$

APPENDIX

$$-\frac{\Delta t}{\Delta x_{i-\frac{1}{2}}}\left[\tilde{A}^{+}_{i-\frac{1}{2}}+\frac{\tilde{N}_{xi}}{\Delta x_i}\right]_j \delta\tilde{U}_{i,j}-\frac{\Delta t}{\Delta x_{i+\frac{1}{2}}}\left[-\tilde{A}^{-}_{i+\frac{1}{2}}+\frac{\tilde{N}_{xi}}{\Delta x_i}\right]_j \delta\tilde{U}_{i+1,j}$$

$$-\frac{\Delta t}{\Delta y_{j-\frac{1}{2}}}\left[\tilde{B}^{+}_{j-\frac{1}{2}}+\frac{\tilde{N}_{yj}}{\Delta y_j}\right]_i \delta\tilde{U}_{i,j-1}-\frac{\Delta t}{\Delta y_{j+\frac{1}{2}}}\left[-\tilde{B}^{-}_{j+\frac{1}{2}}+\frac{\tilde{N}_{yj}}{\Delta y_j}\right]_i \delta\tilde{U}_{i,j+1}$$

$$+\left\{\mathbb{1}+\frac{\Delta t}{\Delta x_{i+\frac{1}{2}}}\left[\tilde{A}^{+}_{i-\frac{1}{2}}+\frac{\tilde{N}_{xi}}{\Delta x_i}\right]_j+\frac{\Delta t}{\Delta x_{i+\frac{1}{2}}}\left[-\tilde{A}^{-}_{i+\frac{1}{2}}+\frac{\tilde{N}_{xi}}{\Delta x_i}\right]_j+\frac{\Delta t}{\Delta y_{j-\frac{1}{2}}}\left[\tilde{B}^{+}_{j-\frac{1}{2}}+\frac{\tilde{N}_{yj}}{\Delta y_j}\right]_i+\right.$$

$$\left.+\frac{\Delta t}{\Delta y_{j+\frac{1}{2}}}\left[-\tilde{B}^{-}_{j+\frac{1}{2}}+\frac{\tilde{N}_{yj}}{\Delta y_j}\right]_i\right\}\delta\tilde{U}_{i,j}=\Delta\tilde{U}_{i,j}+$$

$$+\frac{\Delta t}{2\Delta x_i}\left(\tilde{M}_x+\tilde{M}_y\right)\left[\frac{\delta\tilde{U}_{i+1,j+1}-\delta\tilde{U}_{i+1,j-1}}{\Delta y_{i+1,j+\frac{1}{2}}+\Delta y_{i+1,j-\frac{1}{2}}}-\frac{\delta\tilde{U}_{i-1,j+1}-\delta\tilde{U}_{i-1,j+1}}{\Delta y_{i-1,j+\frac{1}{2}}+\Delta y_{i-1,j-\frac{1}{2}}}\right]$$

Multigrid with ILU-smoothing: systematic tests and improvements

Klaus-Dieter Oertel
Klaus Stüben
Gesellschaft für Mathematik und Datenverarbeitung (GMD)
Institut für methodische Grundlagen
Postfach 1240
D-5205 St.Augustin 1

Summary

Robustness is an important requirement for multigrid methods. In standard multigrid methods, the smoothing process is usually the most crucial component. In particular, the robustness of a multigrid method and that of the underlying smoother are closely related. ILU is generally considered to provide a particularly robust smoothing process. Although often more robust than standard relaxation methods, ILU is not really robust in a strict sense. This can be seen by applying a corresponding multigrid method to certain "limit cases" of typical model problems.

Two possibilities are presented which substantially improve the robustness of multigrid methods with ILU-smoothing. One is based on a modification of the ILU-decomposition, while the other uses alternating ILU-smoothing. A smoothing process which turns out to be robust for all kinds of anisotropic model problems is obtained by combining both techniques: the alternating modified ILU-smoothing.

1. Multigrid with ILU-smoothing

A well-known method which can in principle be used for the iterative solution of sparse matrix equations $Au = b$, is based on ILU-factorization, i.e., on an incomplete factorization of A in lower and upper triangular matrices L and U with prescribed sparsity structure

$$A = LU - R \tag{1}$$

where R denotes the corresponding (sparse) "error matrix". For reasons of uniqueness one additionally has to require a normalization condition. In the following we always assume $\operatorname{diag}(L) \equiv 1$. One step of ILU-iteration proceeds as follows

$$LUu^{new} = Ru^{old} + b \quad \text{or} \quad \begin{cases} Lv &= Ru^{old} + b \\ Uu^{new} &= v \end{cases} \quad . \tag{2}$$

Generally, ILU-iteration is known to be a quite inefficient solver by itself. On the other hand, ILU has very good smoothing properties for a wide range of problems and, consequently, can be incorporated as a smoothing process into multigrid methods (see, e.g., [3–6,8,11–15]). In fact, ILU is considered to be one of the most *robust* smoothing processes. However, if robustness is defined in a strict way, ILU is not really robust. This is demonstrated in [10] by means of a systematic investigation of multigrid methods with ILU-smoothing for the elliptic problem

$$\begin{aligned} \mathcal{L}u = a_{11}\,u_{xx} + a_{12}\,u_{xy} + a_{22}\,u_{yy} + b_1\,u_x + b_2\,u_y + c = f \quad & (\Omega) \\ u = g \quad & (\partial\Omega). \end{aligned} \tag{3}$$

Possibilities are shown, however, which substantially improve the robustness. In this paper, we present results from [10] for the special case $b_i = c = 0$, a_{ij} constant and $\Omega = (0,1) \times (0,1)$.

$\mathcal{L}$ is discretized by means of finite differences on a grid Ω_h with meshsize h. The derivatives u_{xx} and u_{yy} are approximated by central differences and the mixed derivative by the "left-oriented" 7-point discretization

$$u_{xy} \leftarrow \frac{1}{2h^2} \begin{bmatrix} -1 & 1 & \\ 1 & -2 & 1 \\ & 1 & -1 \end{bmatrix}_h u_h. \tag{4}$$

Disregarding the special case $a_{12} = 0$, this leads to a 7-point difference approximation for the operator $\mathcal{L}$. After elimination of the boundary condition, the resulting grid problem is denoted by

$$A_h u_h = b_h \quad (\Omega_h). \tag{5}$$

In order to use ILU for smoothing, one first has to decide on a concrete decomposition of the difference operator A_h. In grid terminology, this requires a decision on an *ordering* of the grid points as well as on the *non-zero structure* of the "lower" and "upper" decomposition operators L_h and U_h (which correspond to L and U in (1)). (In this context, non-zero always means *possibly* non-zero.)

In order to distinguish different orderings, we refer to each ordering by two letters, both of which are either N, S, E or W (standing for north, south, etc.). While the first one describes the ordering of grid points within the rows/columns, the second one refers to the ordering of the rows/columns themselves. For instance, EN means that the grid points are ordered in rows from west to *east*, and the rows themselves are ordered from south to *north*.

As non-zero structure for the ILU-decomposition we use the most natural and smallest reasonable one, namely the one which is induced by the 7-point structure of A_h. In case of EN-ordering, e.g., the structure of the corresponding grid operators is given by

$$L_h \mathrel{\hat{=}} \begin{bmatrix} \cdot & \cdot & \\ * & * & \cdot \\ & * & * \end{bmatrix}_h, \qquad U_h \mathrel{\hat{=}} \begin{bmatrix} * & * & \\ \cdot & * & * \\ & \cdot & \cdot \end{bmatrix}_h, \qquad R_h \mathrel{\hat{=}} \begin{bmatrix} * & \cdot & \cdot & & \\ & \cdot & \cdot & \cdot & \\ & & \cdot & \cdot & * \end{bmatrix}_h \tag{6}$$

where the $*$'s denote (possibly) non-zero entries. This (7-point) decomposition will be denoted by ILU^{EN}.

Besides "plain" ILU, we will also consider "alternating" ILU for smoothing, one step of which consists of two steps of plain ILU with "perpendicular" ordering of grid points. Numerical results have indicated that, e.g., the best counterpart for ILU^{EN} is ILU^{SW}, which has the following structure

$$L_h \mathrel{\hat{=}} \begin{bmatrix} \cdot & * & \\ \cdot & * & * \\ & \cdot & * \end{bmatrix}_h, \qquad U_h \mathrel{\hat{=}} \begin{bmatrix} * & \cdot & \\ * & * & \cdot \\ & * & \cdot \end{bmatrix}_h, \qquad R_h \mathrel{\hat{=}} \begin{bmatrix} * & & & & \\ \cdot & \cdot & & & \\ \cdot & \cdot & \cdot & & \\ & & \cdot & \cdot & \\ & & & & * \end{bmatrix}_h . \tag{7}$$

One alternating ILU sweep, consisting of one step of ILU^{SW} followed by one step of ILU^{EN}, will be denoted by $\mathrm{ILU}^{SW\text{-}EN}$.

For the remaining multigrid components, we use standard coarsening (doubling the meshsize) and

$$I_h^{2h} := \frac{1}{8}\begin{bmatrix} 1 & 1 & \\ 1 & 2 & 1 \\ & 1 & 1 \end{bmatrix}_h^{2h}, \qquad I_{2h}^h := \frac{1}{2}\left]\begin{matrix} 1 & 1 & \\ 1 & 2 & 1 \\ & 1 & 1 \end{matrix}\right[_{2h}^{h} \tag{8}$$

for restriction and interpolation, respectively. (The terminology used for I_{2h}^h means that values at coarse-grid points are "distributed" to their fine-grid neighbors with weights according to the entries of the above interpolation star.) The coarse-grid operators are defined by finite differences in the same way as on the finest grid.

2. Remarks on the numerical results

For our numerical tests, the following common model operators are chosen:

$$\mathcal{L}^{\varepsilon} u = -\varepsilon\, u_{xx} - u_{yy}, \ \varepsilon > 0; \tag{9}$$

$$\mathcal{L}^{\tau} u = -\Delta u + \tau\, u_{xy}, \ |\tau| < 2; \tag{10}$$

$$\mathcal{L}^{\alpha,\varepsilon} u = -(s^2 + \varepsilon c^2)\, u_{xx} + 2(1-\varepsilon) cs\, u_{xy} - (c^2 + \varepsilon s^2)\, u_{yy}, \tag{11}$$

$$s = \sin\alpha, \ c = \cos\alpha, \ \varepsilon > 0.$$

The conditions on ε and τ ensure ellipticity. The operator $\mathcal{L}^{\alpha,\varepsilon}$ is obtained by rotating $\mathcal{L}^{\varepsilon}$ by an angle of α.

In the tables shown below, we always consider the convergence factors ρ_h of the multigrid iteration using V-(1,0)-cycles (one pre-smoothing, no post-smoothing). ρ_h is computed by a v. Mises iteration (with at least 25 iterations), applied to the homogeneous Dirichlet problem on the unit square. Except for Table 1, the meshsize of the finest grid is always $h = \frac{1}{128}$. The coarsest grid consists of one point only.

We want to point out that in [10] V-cycles with more than one smoothing step and W-cycles are also investigated. For the main effects we are going to discuss below, however, it is sufficient to consider the V-(1,0)-cycle. In particular, this simplest multigrid cycle will turn out to give robust convergence for all of the above model problems if only the ILU-smoothing is chosen properly.

At places we make comparisons with predictions by local mode analysis which, generally, provides a very useful tool in developing and tuning multigrid methods [1]. This is, in particular, true also in case of smoothing by ILU. Here, however, local mode analysis has to be used with care because ILU is, in a sense, "less local" than standard smoothing methods like Gauss-Seidel relaxation.

The ILU-smoothing operator is, in grid terminology, given by $S_h = (A_h + R_h)^{-1} R_h$. Although, for (9)–(11), A_h is a constant grid operator, R_h (and hence also S_h) changes from grid point to grid point. Thus, mode analysis cannot be applied directly. Instead of applying mode analysis to a given ILU-decomposition on a real grid, however, one easily can apply it to a corresponding "asymptotic" decomposition (on an infinite grid, cf. [3,8,12,13]). Neglecting the influence of boundary conditions is especially justified if A_h corresponds to a M-matrix (see [12]). This is also confirmed by numerical experience.

3. Results with ILU

In Table 1 we consider the multigrid method with plain ILU^{EN}-smoothing for the operators $\mathcal{L}^{\varepsilon}$ and $\mathcal{L}^{\tau}$ and for different meshsizes on the finest grid. For both operators, the convergence factors obviously deteriorate for certain parameter values and for decreasing h.

In case of $\mathcal{L}^{\varepsilon}$, ($h$-independent) local mode analysis predicts the following behavior of the smoothing factor μ^* and the two-grid convergence factor ρ^*: $\mu^*, \rho^* \to 1$ ($\varepsilon \to \infty$) and $\mu^*, \rho^* \to 3 - 2\sqrt{2} \approx 0.172$ ($\varepsilon \to 0$) (cf. [8,11,12]). At first glance, this is not reflected by the convergence factors ρ_h shown in Table 1a. Actually, the results given there indicate the following: For every fixed h, there are parameters $\overline{\varepsilon}_h > 1$ and $\underline{\varepsilon}_h < 1$ such that ρ_h increases for $\varepsilon \nearrow \overline{\varepsilon}_h$ and $\varepsilon \searrow \underline{\varepsilon}_h$, respectively. Moreover, for $h \to 0$, $\overline{\varepsilon}_h$ increases and $\underline{\varepsilon}_h$ decreases. For $\varepsilon \in (\underline{\varepsilon}_h, \overline{\varepsilon}_h)$ and small h, the convergence factors ρ_h turn out to be quite realistically predicted by local mode analysis. In particular, $\rho_h(\overline{\varepsilon}_h)$ approaches 1 and $\rho_h(\underline{\varepsilon}_h)$ tends to approximately 0.17 if $h \to 0$.

For $\varepsilon < \underline{\varepsilon}_h$ and $\varepsilon > \overline{\varepsilon}_h$, however, the prediction of local mode analysis is too pessimistic, which is due to the fact that, for *fixed* meshsize and $\varepsilon \to \infty$ or $\varepsilon \to 0$, ILU^{EN} approaches more closely a direct solver. Consequently, the resulting fast ILU-convergence "supersedes" the multigrid convergence expected [6,10]. This is *not* reflected by local mode analysis because it is a boundary effect which vanishes for fixed ε and $h \to 0$.

Table 1: ILU^{EN}: multigrid convergence factors ρ_h

a)	$\mathcal{L}^{\varepsilon}$			
$\varepsilon \backslash h$	1/256	1/128	1/64	1/32
10000	0.534	0.233	0.069	0.017
1000	0.782	0.645	0.383	0.135
100	0.593	0.582	0.537	0.402
10	0.268	0.267	0.266	0.256
2	0.126	0.126	0.125	0.121
1	0.127	0.127	0.126	0.124
0.5	0.141	0.141	0.139	0.137
0.1	0.162	0.161	0.160	0.150
0.01	0.166	0.161	0.134	0.077
0.001	0.151	0.104	0.035	0.005
0.0001	0.056	0.011	0.001	$<10^{-4}$

b)	$\mathcal{L}^{\tau}$		
$\tau \backslash h$	1/128	1/64	1/32
1.99	0.651	0.601	0.438
1.95	0.447	0.443	0.398
1.9	0.352	0.350	0.331
1.7	0.202	0.202	0.198
1.0	0.168	0.167	0.164
0.0	0.126	0.125	0.124
−1.0	0.099	0.090	0.073
−1.7	0.144	0.103	0.055
−1.9	0.145	0.146	0.139
−1.95	0.467	0.251	0.192
−1.99	>1	>1	0.252

According to the results shown in Table 1b for $\mathcal{L}^{\tau}$, we have to distinguish between the two cases $\tau \to 2$ and $\tau \to -2$. For $\tau \to 2$, the multigrid behavior is similiar to the one shown in Table 1a for large ε: From local mode analysis, one expects the convergence factors to approach 1. In practice, however, convergence is better because, for fixed h and $\tau \to 2$, ILU^{EN} again approaches a direct solver. The resulting effect can be seen if τ is chosen sufficiently close to 2. For instance, for $h = \frac{1}{128}$, one obtains the convergence factors $\rho_h = 0.651$ and $\rho_h = 0.227$ if $\tau = 1.99$ and $\tau = 1.9999$, respectively.

The situation is quite different if τ approaches -2. Here multigrid diverges for small meshsizes. This is in contrast to the local mode analysis which can be shown to predict a two-grid convergence factor of around 0.17 if τ is chosen close to -2. However, we here cannot expect local mode analysis to necessarily provide realistic predictions because A_h no longer corresponds to a M-matrix (cf. Section 2). For the same reason we, generally, cannot expect ILU-iteration to converge [9]. In fact for τ close to -2 strong divergence of ILU^{EN} occurs which also destroys the multigrid convergence.

A close examination shows that ILU^{EN} still has good smoothing properties and that ILU-divergence occurs in the low frequency range. This can be seen if local mode analysis is not applied to the asymptotic decomposition (as sketched in Section 2) but rather to the *real* grid operator S_h with "frozen" coefficients [10]. This way one can investigate worst cases. It turns out that the behavior of S_h close to the "origin" of the decomposition (i.e., the lower left corner of the grid) causes the most troubles. This can also be verified numerically by using *alternating* ILU for smoothing, where the corresponding partial ILU steps have origins in *opposite* corners of the grid. For instance, if $ILU^{SW\text{-}EN}$ is used for smoothing, the corresponding multigrid method converges perfectly (for more details, see Section 5).

Note that our multigrid method with plain ILU^{EN}-smoothing essentially corresponds to the methods investigated in [6,14]. The main difference lies in the coarse-grid operator, for which there the Galerkin approach $A_{2h} = I_h^{2h} A_h I_{2h}^h$ is used. This difference, however, does not influence the multigrid behavior essentially. The numerical results for $\mathcal{L}^{\varepsilon}$ and $\mathcal{L}^{\tau}$ presented in [6,14] are comparable to ours. There, however, fewer test cases are considered (in particular, larger meshsizes). As a consequence, some effects did not show up in the same way as above.

Due to the behavior of ILU^{EN} in extreme situations, our opinion about the robustness of multigrid with ILU is less optimistic than in [6,14]. This opinion is reinforced by the multigrid behavior in case of the operator $\mathcal{L}^{\alpha,\varepsilon}$. As a demonstration we give convergence factors for the case $\varepsilon = 0.001$ in Table 2c (in the column denoted by $\delta = 0$). For nearly all values of α shown, multigrid either diverges or converges very poorly. In [5] it has already been pointed out that multigrid with ILU^{EN} behaves unsatisfactorily for $\varepsilon = 0.01$ and $\alpha \in (52.5^\circ, 82.5^\circ)$*. Note that $\mathcal{L}^{\alpha,\varepsilon}$ with $\alpha = 45^\circ, 135^\circ$ and $\varepsilon = 0.001$ corresponds to $\mathcal{L}^{\tau}$ with $\tau \approx \pm 1.996$. Consequently, the above discussion on $\mathcal{L}^{\tau}$ applies also here.

4. Modified ILU

The numerical results discussed above show that multigrid with ILU-smoothing is not unconditionally robust. The robustness can substantially be improved, however, by a proper modification of the decomposition.

In case of the usual ILU-decomposition, the error matrix R contains zeros in all positions which correspond to the non-zero structure of L and U, especially along the main diagonal. With some parameter δ we define a modified ILU-decomposition

$$A = L_\delta U_\delta - R_\delta \tag{12}$$

for which the diagonal of R_δ is no longer zero: Assuming that the incomplete factorization is computed row by row, we modify the entries r_{ii} and u_{ii} immediately before row number $i+1$ is computed in the following way

$$r_{ii} \leftarrow \delta \sum_{j \neq i} |r_{ij}|, \qquad u_{ii} \leftarrow u_{ii} + r_{ii}. \tag{13}$$

Reasonable values for δ are given by $\delta \in [0,1]$, where $\delta = 0$ corresponds to the non-modified decomposition. Below, we will consider the two choices $\delta = 0.5$ and $\delta = 1$, which will be refered to as the "semi-modified" and the "fully-modified" case, respectively. For $\delta = 1$, (13) coincides with a modification introduced in [7]. During the

* Note that our α corresponds to $90^\circ - \alpha \pmod{180^\circ}$ in [5].

Table 2: Modified ILU^{EN}: multigrid convergence factors ρ_h, $h = \frac{1}{128}$

a)	$\mathcal{L}^{\varepsilon}$		
$\varepsilon \backslash \delta$	0	0.5	1
10000	0.233	0.105	0.190
1000	0.645	0.251	0.400
100	0.582	0.252	0.394
10	0.267	0.157	0.261
2	0.126	0.102	0.144
1	0.127	0.096	0.130
0.5	0.141	0.096	0.126
0.1	0.161	0.102	0.126
0.01	0.161	0.096	0.120
0.001	0.104	0.071	0.088
0.0001	0.011	0.013	0.017

b)	$\mathcal{L}^{\tau}$		
$\tau \backslash \delta$	0	0.5	1
1.99	0.651	0.263	0.407
1.95	0.447	0.221	0.350
1.9	0.352	0.188	0.307
1.7	0.202	0.150	0.221
1.0	0.168	0.105	0.131
0.0	0.126	0.097	0.131
−1.0	0.099	0.135	0.164
−1.7	0.144	0.168	0.184
−1.9	0.145	0.146	0.246
−1.95	0.467	0.165	0.294
−1.99	>1	0.354	0.344

c)	$\mathcal{L}^{\alpha,\varepsilon}$, $\varepsilon = 0.001$		
$\alpha \backslash \delta$	0	0.5	1
0.0	0.104	0.071	0.088
7.5	0.122	0.177	0.236
15.0	0.226	0.217	0.303
22.5	0.675	0.241	0.320
30.0	>1	0.259	0.342
37.5	0.410	0.255	0.360
45.0	0.704	0.272	0.419
52.5	>1	0.883	0.906
60.0	0.902	0.915	0.926
67.5	0.914	0.925	0.931
75.0	>1	0.919	0.933
82.5	>1	0.850	0.886
90.0	0.645	0.251	0.400
97.5	0.557	0.222	0.363
105.0	>1	0.289	0.358
112.5	>1	0.518	0.364
120.0	>1	0.814	0.366
127.5	>1	0.776	0.369
135.0	>1	0.515	0.368
142.5	>1	0.355	0.365
150.0	>1	0.247	0.351
157.5	0.860	0.223	0.349
165.0	0.263	0.178	0.322
172.5	0.142	0.155	0.249

present conference it was seen that modifications similar to (13) are currently investigated also by M. Khalil and G. Wittum.

Remark: The semi- and fully-modified decompositions are characterized by special properties. The semi-modified ILU^{EN} has *optimal smoothing factors* (computed by local mode analysis) if applied to $\mathcal{L}^{\varepsilon}$ with $\varepsilon \to \infty$ and $\mathcal{L}^{\tau}$ with $\tau \to 2$. The fully-modified ILU, on the other hand, is guaranteed to *converge* for every positive definite matrix A (as long as the non-zero structure of ILU is symmetric). This can immediately be seen by observing that R_δ ($\delta = 1$) is positive semi-definite and hence the iteration operator of the fully-modified ILU, i.e., $S_\delta = (A + R_\delta)^{-1} R_\delta$, has a spectral radius smaller than 1.

The numerical results given in Table 2 show the influence of the above modification on the convergence of multigrid. One of the most remarkable effects is that, for $\mathcal{L}^\varepsilon$ in the critical range of large ε, the convergence factor ρ_h is now bounded away from 1 *independently* of h and ε. This is predicted by local mode analysis which, for $\varepsilon \to \infty$, gives two-grid convergence factors of $\frac{1}{3}$ and $\frac{1}{2}$ in case of the semi- and fully-modified ILU^{EN}, respectively. This prediction is in perfect agreement with our numerical experiments. For $\varepsilon \to 0$, the modified versions behave as well as the non-modified one. According to the above Remark, it is not surprising that multigrid using the semi-modified ILU^{EN} converges somewhat faster than the fully-modified analog.

For $\mathcal{L}^\tau$ and $\tau \to 2$, the modification has a similiar effect on the multigrid convergence as for $\mathcal{L}^\varepsilon$ and $\varepsilon \to \infty$ (cf. Table 2b): ρ_h is bounded away from 1. Actually, local mode analysis gives even the same two-grid convergence factors, namely, $\frac{1}{3}$ and $\frac{1}{2}$ in the semi- and fully-modified case, respectively. Also for $\tau \to -2$ the multigrid behavior is essentially improved. In particular, in contrast to the non-modified ILU^{EN}, the fully-modified ILU^{EN}-iteration no longer diverges (see the Remark from above). As a matter of fact, the convergence factors ρ_h of the corresponding multigrid method stay bounded away from 1, which can be verified both by local mode analysis and numerical computations. The situation is different for the semi-modified ILU^{EN}, which still diverges for low frequencies. In contrast to the non-modified ILU^{EN}, however, this did not yet lead to multigrid divergence for the examples shown in the table. It *does* lead to divergence, though, if the number of smoothing steps is increased (e.g., if a V-(2,0)-cycle is used instead of a V-(1,0)-cycle).

Finally, for the operator $\mathcal{L}^{\alpha,\varepsilon}$ (see Table 2c), multigrid convergence is essentially improved, except for the range $\alpha \in (52.5°, 82.5°)$ where convergence is still very poor. Everywhere else we have good or at least acceptable convergence. We point out that, for $\alpha \in (112.5°, 135°)$, $\delta = 1$ gives the better multigrid variant, otherwise $\delta = 0.5$ is the better choice. The remaining deficiencies of modified ILU^{EN} will be overcome if alternating ILU is used for smoothing (see below).

5. Alternating modified ILU

In the case of the operator $\mathcal{L}^\varepsilon$, smoothing by row-oriented (non-modified) ILU^{EN} is very efficient for *small* ε (see Table 1a). On the other hand, for reasons of symmetry, column-oriented (non-modifed) ILU^{SW} is very efficient for *large* ε. Actually, ILU^{SW} applied to $\mathcal{L}^\varepsilon$ has the same smoothing properties as ILU^{EN} applied to $\mathcal{L}^{1/\varepsilon}$. Consequently, if these two smoothers are combined into one alternating step, ILU$^{SW\text{-}EN}$, one can expect good smoothing independent of ε. This is confirmed by the results of Table 3a for the non-modified as well as for the modified versions of ILU$^{SW\text{-}EN}$. It clearly shows that the modifications do not pay.

Table 3: Alternating modified $\mathrm{ILU}^{SW\text{-}EN}$: multigrid convergence ρ_h, $h = \frac{1}{128}$

a)	$\mathcal{L}^\varepsilon$		
$\varepsilon \backslash \delta$	0	0.5	1
10000	0.002	0.001	0.001
1000	0.032	0.016	0.027
100	0.039	0.033	0.049
10	0.028	0.041	0.053
2	0.031	0.044	0.057
1	0.033	0.046	0.058
0.5	0.031	0.045	0.057
0.1	0.029	0.041	0.053
0.01	0.039	0.034	0.049
0.001	0.032	0.016	0.027
0.0001	0.002	0.001	0.001

b)	$\mathcal{L}^\tau$		
$\tau \backslash \delta$	0	0.5	1
1.99	0.434	0.130	0.187
1.95	0.211	0.114	0.150
1.9	0.141	0.098	0.129
1.7	0.068	0.069	0.090
1.0	0.024	0.032	0.040
0.0	0.033	0.046	0.058
−1.0	0.055	0.072	0.086
−1.7	0.070	0.083	0.092
−1.9	0.037	0.042	0.047
−1.95	0.022	0.013	0.057
−1.99	0.051	0.029	0.091

c)	$\mathcal{L}^{\alpha,\varepsilon}$			
	$\varepsilon = .001$	$\varepsilon = .0001$		
$\alpha \backslash \delta$	0	0	0.5	1
0.0	0.032	0.002	0.001	0.001
7.5	0.142	0.188	0.155	0.296
15.0	0.173	0.322	0.193	0.335
22.5	0.386	>1	0.238	0.344
30.0	0.582	>1	0.273	0.350
37.5	0.191	>1	0.256	0.336
45.0	0.512	0.262	0.063	0.115
52.5	0.190	>1	0.256	0.337
60.0	0.631	>1	0.282	0.349
67.5	0.430	>1	0.237	0.342
75.0	0.178	0.331	0.197	0.331
82.5	0.143	0.194	0.156	0.291
90.0	0.032	0.002	0.001	0.001
97.5	0.018	0.059	0.034	0.093
105.0	0.043	0.084	0.055	0.105
112.5	0.061	0.076	0.058	0.109
120.0	0.069	0.107	0.061	0.110
127.5	0.076	0.136	0.088	0.111
135.0	0.078	0.170	0.093	0.113
142.5	0.077	0.114	0.089	0.111
150.0	0.071	0.167	0.065	0.110
157.5	0.064	0.084	0.054	0.108
165.0	0.050	0.082	0.054	0.102
172.5	0.017	0.062	0.034	0.092

Table 3b shows corresponding results for $\mathcal{L}^\tau$. Also here, except for values of τ close to 2, the modifications do not pay. Compared to the results of Table 1b, it is interesting to note that, for $\tau \to -2$, multigrid with non-modified $\mathrm{ILU}^{SW\text{-}EN}$ no longer diverges. A heuristic explanation for this has already been given in Section 3. There we stated that the two partial ILU steps should have their origin in *opposite* corner points of the grid, which is satisfied for $\mathrm{ILU}^{SW\text{-}EN}$. That this is really essential can be seen by replacing $\mathrm{ILU}^{SW\text{-}EN}$ with $\mathrm{ILU}^{NE\text{-}EN}$ (both components of which have the *same* origin). The resulting multigrid method diverges even stronger than the one using plain ILU^{EN}.

In order to obtain uniform multigrid convergence for $\mathcal{L}^\varepsilon$ as well as for $\mathcal{L}^\tau$, one does not need alternating ILU. This can already be achieved by the modifications discussed in the previous section. However, taking computational work into account, smoothing by alternating ILU is generally somewhat more efficient [10].

On the other hand, to obtain good and robust multigrid convergence for the operator $\mathcal{L}^{\alpha,\varepsilon}$, one really needs alternating *and* modified ILU for smoothing. Table 3c shows that numerical results may easily mislead if one is not careful. Using ***non-modified*** $\mathrm{ILU}^{SW\text{-}EN}$ as smoother gives reasonable convergence for $\varepsilon = 0.001$. However, if ε is decreased even further to $\varepsilon = 0.0001$, we see that non-modified $\mathrm{ILU}^{SW\text{-}EN}$ is still not robust. This is overcome by using modified versions of $\mathrm{ILU}^{SW\text{-}EN}$.

We want to point out that alternating (non-modified) 9-point ILU-smoothing was first examined theoretically in [12] for both $\mathcal{L}^{\varepsilon}$ (for which 9-point ILU actually degenerates to the 7-point ILU) and a 9-point discretization of $\mathcal{L}^{\tau}$.

6. Different orderings of grid points

According to the 8 possible orderings of grid points by rows or columns, there are 8 types of (7-point) ILU-smoothers. In Section 3 and 4, we considered only the particular case ILU^{EN}. Note, however, that for the model operators $\mathcal{L}^{\varepsilon}$, $\mathcal{L}^{\tau}$ and $\mathcal{L}^{\alpha,\varepsilon}$ one does not have to distinguish among all 8 smoothers. Actually, there are only two groups of smoothers, namely

$$\{\mathrm{ILU}^{EN}, \mathrm{ILU}^{WS}, \mathrm{ILU}^{SW}, \mathrm{ILU}^{NE}\} \quad \text{and} \quad \{\mathrm{ILU}^{WN}, \mathrm{ILU}^{ES}, \mathrm{ILU}^{SE}, \mathrm{ILU}^{NW}\}.$$

For each group, it is, for reasons of symmetry, sufficient to investigate one representative only. For instance, in case of the operator $\mathcal{L}^{\varepsilon}$, ILU^{EN} and ILU^{WS} (and similiarly ILU^{SW} and ILU^{NE}) give identical convergence behavior of the corresponding multigrid methods. On the other hand, as already mentioned, ILU^{SW} behaves for $\mathcal{L}^{\varepsilon}$ in the same way as ILU^{EN} for $\mathcal{L}^{1/\varepsilon}$.

To conclude our numerical investigations, we want to give some results for one representative of the second group of smoothers, namely ILU^{WN}, which is characterized by the following structure

$$L_h \mathrel{\hat=} \begin{bmatrix} \cdot & \cdot & \\ \cdot & * & * \\ & * & * \end{bmatrix}_h, \qquad U_h \mathrel{\hat=} \begin{bmatrix} * & * & \\ * & * & \cdot \\ & \cdot & \cdot \end{bmatrix}_h, \qquad R_h \mathrel{\hat=} \begin{bmatrix} \cdot & \cdot & * \\ \cdot & \cdot & \cdot \\ * & \cdot & \cdot \end{bmatrix}_h. \tag{14}$$

Note that, for $\mathcal{L}^{\varepsilon}$, ILU^{WN} actually degenerates to a 5-point ILU, which for $\delta = 0$ has already in [13] been pointed out to be less efficient than the 7-point ILU. In particular, multigrid convergence is not satisfactory both for $\varepsilon \to \infty$ and $\varepsilon \to 0$ (see Table 4a, $\delta = 0$). On the other hand, our results clearly show that convergence is essentially improved if, in particular, semi-modified ILU^{WN} is used instead of non-modified ILU^{WN}.

Table 4: Modified ILU^{WN}: multigrid convergence factors ρ_h, $h = \frac{1}{128}$

a)	$\mathcal{L}^{\varepsilon}$		
$\varepsilon \backslash \delta$	0	0.5	1
10000	0.240	0.107	0.194
1000	0.699	0.261	0.415
100	0.737	0.284	0.436
10	0.467	0.250	0.356
2	0.248	0.275	0.326
1	0.226	0.280	0.327
0.5	0.245	0.271	0.324
0.1	0.466	0.251	0.355
0.01	0.740	0.284	0.439
0.001	0.697	0.262	0.415
0.0001	0.237	0.107	0.193

b)	$\mathcal{L}^{\tau}$		
$\tau \backslash \delta$	0	0.5	1
1.99	0.162	0.091	0.116
1.95	0.165	0.100	0.124
1.9	0.167	0.101	0.127
1.7	0.168	0.101	0.127
1.0	0.167	0.104	0.132
0.0	0.226	0.280	0.327
−1.0	0.509	0.566	0.610
−1.7	0.789	0.823	0.841
−1.9	0.888	0.904	0.914
−1.95	0.915	0.924	0.930
−1.99	0.936	0.944	0.946

In this case, local mode analysis shows that the two-grid convergence factor is bounded from above by $\frac{1}{3}$, independent of ε and h.

For $\mathcal{L}^{\tau}$, the ILU^{WN}-decomposition is really a 7-point one. A comparison of ILU^{WN} in Table 4b with ILU^{EN} in Table 2b shows two remarkable facts. On the one hand, multigrid with ILU^{WN}-smoothing converges quite well for $\tau > 0$, even for $\tau \to 2$, while, on the other hand, it is worse for $\tau < 0$ but is at least never divergent. Note that modifying ILU^{WN} does not help for $\tau \leq 0$.

7. Conclusions

Having started from the observation that multigrid with standard ILU-smoothing (i.e., non-modified ILU^{EN}) is not really robust, we pointed out that for the operators $\mathcal{L}^{\varepsilon}$ and $\mathcal{L}^{\tau}$ one obtains multigrid convergence factors which are *uniformly* bounded away from 1 if only the ILU is properly modified. Such a modification also helps for the case $\mathcal{L}^{\alpha,\varepsilon}$ but still does not yield satisfactory multigrid convergence for all values of α and ε. A multigrid variant which turns out to be robust for all problems considered and arbitrary values of the parameters is obtained if alternating modified ILU is used for smoothing.

More extensive investigations are found in [10]. In particular, results are given for 9-point ILU-smoothing, some more test problems and different discretizations.

References

[1] Brandt, A.: *Multigrid Techniques: 1984 guide with applications to fluid dynamics.* GMD-Studien Nr. 85. Gesellschaft für Mathematik und Datenverarbeitung, St. Augustin, 1984.

[2] Hackbusch, W.; Trottenberg, U. (eds.): *Multigrid methods. Proceedings of the conference held at Köln-Porz, November 23-27, 1981.* Lect. Notes in Mathematics 960. Springer Verlag, Berlin, 1982.

[3] Hemker, P.W.: *The incomplete LU-decomposition as a relaxation method in multigrid algorithms.* In: Miller, J.J.H. (ed.), Boundary and interior layers – computational and asymptotic methods, Proceedings BAIL II Conference. Boole Press, Dublin, 1980, pp. 306-311.

[4] Hemker, P.W.: *On the comparison of line-Gauss-Seidel and ILU relaxation in multigrid algorithms.* In: Miller, J.J.H. (ed.), Computational and asymptotic methods for boundary and interior layers. Boole Press, Dublin, 1982, pp. 269-277.

[5] Hemker, P.W.: *Multigrid methods for problems with a small parameter in the highest derivative.* In: Watson, G.A. (ed.), Numerical analysis, Proceedings, Dundee 1983. Lect. Notes in Math. 1066, Springer Verlag, Berlin, 1984, pp. 106-121

[6] Hemker, P.W.; Kettler, R.; Wesseling, P.; de Zeeuw, P.M.: *Multigrid methods: development of fast solvers.* In: McCormick, S.F., Trottenberg, U. (eds.), Proceedings of the International Multigrid Conference, April 6-8, 1983, Copper Mountain, CO, Appl. Math. Comp. 13, North Holland, 1983, pp. 311-326.

[7] Jennings, A.; Malik, G.M.: *Partial elimination.* J. Inst. Math. Appl. 20, 1977, pp. 307-316.

[8] Kettler, R.: *Analysis and comparison of relaxation schemes in robust multigrid and preconditioned conjugate gradient methods.* In [2], pp. 502-534.

[9] Meijerink, J.A.; van der Vorst, H.A.: *An iterative solution method for linear systems of which the coefficient matrix is a symmetric M-Matrix.* Math. Comp. 31, 1977, pp. 148-162.

[10] Oertel, K.-D.: *Praktische und theoretische Untersuchungen zur ILU-Glättung bei Mehrgitterverfahren.* Diplomarbeit, Institut für angewandte Mathematik, Universität Bonn, to appear.

[11] Stüben, K.; Trottenberg U.: *Multigrid methods: Fundamental algorithms, model problem analysis and applications.* In [2], pp. 1-176.

[12] Thole, C.A.: *Beiträge zur Fourieranalyse von Mehrgittermethoden: V-Cycle, ILU-Glättung, anisotrope Operatoren.* Diplomarbeit, Institut für angewandte Mathematik, Universität Bonn, 1983.

[13] Wesseling, P.: *Theoretical and practical aspects of a multigrid method.* SIAM J. Sci. Stat. Comp 3, 1982, pp. 387-407.

[14] Wesseling, P.: *A robust and efficient multigrid method.* In [2], pp. 614-630.

[15] Wesseling, P.; Sonneveld, P.: *Numerical experiments with a multiple grid and a preconditioned Lanczos type method.* In: Rautmann, R. (ed.), Approximation methods for Navier-Stokes problems. Proceedings of the IUTAM-Symposium, Paderborn 1979. Lect. Notes in Math. 771, Springer Verlag, Berlin, 1980, pp. 543-562.

MULTILEVEL PRECONDITIONING MATRICES AND MULTIGRID V-CYCLE METHODS

Panayot Vassilevski*)

Institute of Mathematics, Bulgarian Academy of Sciences,
1090 Sofia, P.O. Box 373, Bulgaria

SUMMARY

The relationship between the multilevel preconditioning matrices and a natural multigrid method is outlined. The multilevel preconditioning matrices are derived by an approximate block-factorization of the stiffness matrix computed by piecewise linear nodal basis functions. The block ordering of the nodes corresponds to a certain refining procedure, which starting with a coarse triangulation of the considered plane polygonal region generates successively refined triangulations. At each step of this refining procedure we add a new group of nodes which give the corresponding block of nodes in our multilevel ordering of the final nodal set. Under this multilevel block ordering, starting from the top level our stiffness matrix is factorized approximately, where any successive Schur complement is replaced (approximated) by the stiffness matrix of the current level. Another alternative is to view this method in the framework of the classical multigrid method of V-cycle type. For this natural multigrid method it is enough to use the corresponding two-level ordering of the stiffness matrix at each discretization level. Then the smoothing procedure is naturally derived from the stiffness matrix. The main computational task in performing one smoothing step is solution of problems, on the current level, with a matrix (the first pivot block of the stiffness matrix in the two-level ordering), which has a condition number bounded independently on the number of levels used. The convergence properties of these iterative methods are compared.

AMS Subject Classifications: 65F10, 65N20, 65N30

Key words: multigrid method, smoothing, V-cycle, optimal convergence, elliptic problems, finite element method, preconditioning

1. Preliminaries

Consider the bilinear form, bounded and W_2^1 coercive,

$$a(u,v) = \int_\Omega \sum_{i,j} a_{ij}(x) \frac{\partial}{\partial x_i} u \frac{\partial}{\partial x_j} v \, dx, \quad u,v \in V = \{w \in W_2^1(\Omega),\ w = 0 \text{ on } \Gamma_D\},$$

where Ω is a plane polygonal region, which is triangulated into triangles and with this triangulation is associated the finite element space of piecewise linear and continuous in $\bar{\Omega}$ functions, which vanish on the Dirichlet part of the boundary $\Gamma_D \subset \Gamma = \partial\Omega$.

Let $\{\Phi_1\}$ be the nodal basis functions, i.e.

---

*) This research of the author is supported in part by the Committee of Science under Grant No 55/26.3.87

$$\Phi_i(x_j) = \delta_{i,j},$$

where x_j runs over all nodes of the initial grid, i.e. the vertices of the triangles which do not lie on Γ_D.

Then the following linear algebraic problem will be of our further interest

$$A\,\underline{x} = \underline{b},$$

where

$$A = (a(\Phi_i, \Phi_j))$$

is the stiffness matrix associated with the bilinear form a(.,.).

Let us now have a sequence of discretization spaces

$$V_1 \subset V_2 \subset \ldots \subset V_l,$$

where V_i consists of piecewise linear and continuous functions on the triangulation τ_i of Ω. τ_{i+1} is a direct refinement of τ_i, which means that it T is a triangle in τ_i then the triangles in τ_{i+1} are obtained in one of the following ways, by using - a) regular refinement by pairwise connecting the midpoint of the edges of T; b) irregular refinement by connecting a vertex of T by the midpoint of the opposite edge of T; c) the triangle T remains not changed. We satisfy the following rule: if a triangle is refined in a irregular way then the resulting triangles are not further refined. In this way we preserve the minimal angle condition for every triangle in our triangulations.

Let N_k be the set of nodes (the vertices of the triangles of τ_k that do not lie on Γ_D) on the level k, $1 \leqq k \leqq l$.

Thus at each level k we have a stiffness matrix $A^{(k)}$, $1 \leqq k \leqq l$.

As τ_{k+1} is a direct refinement of τ_k the nodes of N_{k+1} can be partitioned into the following disjoint groups $N_{k+1} \setminus N_k$ and N_k. Thus for any function $v \in V_{k+1}$ we have

$$v = \sum_{x_i \in N_{k+1}} v_i \Phi_i^{(k+1)} = \sum_{x_i \in N_{k+1} \setminus N_k} v_i^* \Phi_i^{(k+1)} + \sum_{x_i \in N_k} v_i \Phi_i^{(k)} .$$

$$/v_i = v(x_i)/.$$

This expression defines a mapping $J = J_{k+1}$, which transforms the coefficient vector $\underline{v}^* = \begin{bmatrix} \underline{v}_1^* \\ v_2^* \end{bmatrix}$ of a function $v \in V_{k+1}$ with respect to the so-called two-level basis

$$\{\Phi_i^{k+1},\ x_i \in N_{k+1} \setminus N_k,\ \Phi_i^{(k)},\ x_i \in N_k\}$$

to the coefficient vector $\underline{v} = \begin{bmatrix} \underline{v}_1 \\ \underline{v}_2 \end{bmatrix}$ of v with respect to the usual nodal basis of V_{k+1}. The coefficient vectors are partitioned into two blocks corresponding to the groups of nodes $N_{k+1} \backslash N_k$ and N_k, respectively.

In matrix notations (using the above ordering) J has the following 2 by 2 block structure

$$J = \begin{pmatrix} I & J_{12} \\ 0 & I \end{pmatrix}.$$

Let us now consider a function $v \in V_k$. Then its coefficient vector with respect to the two-level basis has the form

$$\underline{v}^* = \begin{pmatrix} 0 \\ \underline{v}_2 \end{pmatrix}.$$

Thus its coefficient vector with respect to the nodal basis has the form

$$\underline{v} = J \begin{pmatrix} 0 \\ \underline{v}_2 \end{pmatrix} = \begin{pmatrix} J_{12}\, \underline{v}_2 \\ \underline{v}_2 \end{pmatrix}.$$

I.e. the mapping

$$I_k^{k+1} = \begin{pmatrix} J_{12} \\ I \end{pmatrix} \tag{1.1}$$

is the natural prolongation (or coarse - to - fine transfer) matrix, cf., e.g., [8].

Finally we shall need the so-called smoothing procedure.

Consider the following problem

$$A^{(k+1)} \underline{v} = \underline{rhs}$$

and let $\underline{v}^{(o)}$ be an initial guess. Then a smoothing procedure is defined as follows, cf., e.g., [8]

$$(\underline{v}^{(o)}, A^{(k+1)}, \underline{rhs}) \rightarrow \underline{v}^{(\nu)}, \ \nu \text{ some given integer}$$

and

$$\underline{v}^{(j+1)} = (I - T\, A)\, \underline{v}^{(j)} + T\, \underline{rhs}, \quad j = 0, 1, \ldots, \nu-1 \tag{1.2}$$

for some properly chosen matrix $T = T^{(k+1)}$ /$A = A^{(k+1)}$/.

In what follows the following block-factorization of $A = A^{(k+1)}$ will be used

$$A = \begin{pmatrix} A_{11} & A_{12} \\ A_{21} & A_{22} \end{pmatrix} = \begin{pmatrix} I & O \\ A_{21}A_{11}^{-1} & I \end{pmatrix} \begin{pmatrix} A_{11} & O \\ O & S \end{pmatrix} \begin{pmatrix} I & A_{11}^{-1}A_{12} \\ O & I \end{pmatrix}, \quad S = A_{22} - A_{21}\, A_{11}^{-1}\, A_{12} \tag{1.3}$$

and A is partitioned using the two-level block ordering on the level k+1.

The rest of the paper is organized as follows. In the second section the multilevel approximate factorization of $A = A^{(1)}$ is recalled and is redefined in the framework of the classical MG method by appropriate choice of the matrix T in the definition of the smoothing procedure (1.2). In the third part a natural multigrid method is studied, which generalizes the multilevel preconditioning method. And finally we present some implementation details.

2. Multilevel preconditioning matrices

We recall the definition of the multilevel preconditioning matrix $M = M^{(1)}$ proposed in [10]

$$M^{(1)} = A^{(1)}, \; M^{(k+1)} = \begin{pmatrix} I & O \\ A_{21}^{(k+1)} A_{11}^{(k+1)-1} & I \end{pmatrix} \begin{pmatrix} A_{11}^{(k+1)} & O \\ O & M^{(k)} \end{pmatrix} \begin{pmatrix} I & A_{11}^{(k+1)-1} A_{12}^{(k+1)} \\ O & I \end{pmatrix}$$

which can be given in the following more explicit from

$$M^{(k+1)} = \begin{pmatrix} A_{11}^{(k+1)} & & \\ \hline & A_{11}^{(k)} & O \\ A_{21}^{(k+1)} & \overline{} \; \ddots & \\ & A_{21}^{(k)} & A^{(1)} \end{pmatrix} \begin{pmatrix} I & A_{11}^{(k+1)-1} A_{12}^{(k+1)} & \\ \hline & I & A_{11}^{(k)-1} A_{12}^{(k)} \\ \hline O & & \ddots \\ & & I \end{pmatrix},$$

and this defines our block multilevel approximate (incomplete) factorization of $A^{(k+1)}$.

In [10] the following main result has been proven

<u>Theorem</u> 2.1 For any vector $\underline{v} \in R^{n_k}$ / n_k is the number of nodes in N_k/ the following inequalities are valid

$$\underline{v}^t A^{(k)} \underline{v} \leq \underline{v}^t M^{(k)} \underline{v} \leq (1 + C k^2) \underline{v}^t A^{(k)} \underline{v}$$

with C a positive constant independent of the number of levels used.

Denote now by L and U the lower unit triangular and upper unit triangular factors of the factorization (1.3) of $A = A^{(k+1)}$, respectively. Choose now in the smoothing procedure (1.2) the following matrix

$$T = U^{-1} \begin{pmatrix} A_{11}^{-1} & O \\ O & O \end{pmatrix} L^{-1}. \tag{2.1}$$

We are going to redefine the multilevel approximate factorization matrices in the framework of the classical MG method.

For this let us consider the error propagation matrices

$$\hat{M}^{(k+1)} = I - M^{(k+1)-1} A^{(k+1)}$$

and the iteration matrices in the smoothing procedure

$$G = I - T A^{(k+1)} = \begin{pmatrix} -A_{11}^{-1} A_{12} \\ I \end{pmatrix} (O \quad I).$$

Then one can easily check the following identity

$$\hat{M}^{(k+1)} = G\,(A^{(k+1)-1} - I_k^{k+1} (I - \hat{M}^{(k)})\, A^{(k)-1} I_{k+1}^k) A^{(k+1)} G,$$

where

I_k^{k+1} is the natural coarse - to - fine transfer matrix and $I_{k+1}^{k} = (I_k^{k+1})^t$.

The last identity points out that the iterative method with $\hat{M}^{(k)}$ as an error propagation matrix is a V-cycle MG method with natural coarse - to - fine matrix and the specified smoothing iteration ($\nu = 1$). For this cf., e.g., [9].

In the next section we consider some generalization of the last multigrid method.

3. Generalization to a natural MG method

Instead of (2.1) we may use the following nonsingular smoothing matrix

$$T = U^{-1} \begin{pmatrix} A_{11}^{-1} & 0 \\ 0 & 1/\tau I \end{pmatrix} L^{-1}.$$

Then the multilevel preconditioning matrix can be obtained as a limit case, when $\tau \to \infty$.

Let us denote

$$\hat{S} = \tau I. \tag{3.1}$$

Then for any $\nu \geq 1$ one can prove the following identity (see [11])

$$G = (I - T A^{(k+1)})^{\nu} = \begin{pmatrix} -A_{11}^{-1} A_{12} \\ I \end{pmatrix} (0,\ (I-\hat{S}^{-1}S)^{\nu}),$$

and S is the Schur complement $A_{22} - A_{21}A_{11}^{-1} A_{12}$.

Set

$$\hat{G} = I - \hat{S}^{-1} S. \tag{3.2}$$

Then the following identity is valid (see [11])

$$A^{(k+1)}\hat{M}^{(k+1)} = \begin{pmatrix} 0 \\ I \end{pmatrix} S\, \hat{G}^{\nu}\, (S^{-1} - (I - \hat{M}^{(k)})A^{(k)-1})S\, \hat{G}^{\nu}\ (0,\ I), \tag{3.3}$$

where

$\hat{M}^{(k)}$ are the error propagation matrices of the MG method, defined with parameters

$$T^{-1} = L \begin{pmatrix} A_{11} & 0 \\ 0 & \hat{S} \end{pmatrix} U, \quad I_k^{k+1} = \begin{pmatrix} J_{12} \\ I \end{pmatrix}$$

for the smoothing procedure (1.2) and natural coarse-to-fine transfer matrix (1.1). ν is the number of smoothing iterations performed at each discretization level.

<u>Remark</u> 3.1. J_{12} is not presented in the formula (3.3).

(3.1) - (3.3) define our <u>natural V-cycle MG method.</u> Another candidate for a name is MG method is subspace as at each discretization level we eliminate the first block of the unknowns exactly and iterate over the second block.

In order to prove optimal convergence of this natural MG method one may use the theory of Braess-Hackbusch (V-cycle) [5]. It is based on the following assumptions

(P1) W_2^2 regularity

I.e. for every $f \in L_2(\Omega)$ there exists a solution $u \in V \cap W_2^2(\Omega)$ of the variational problem

$$a(u, v) = (f, v), \text{ all } v \in V.$$

In our case Ω must be convex polygon, cf., e.g., [7].

(P2) uniform refining

This ensures

(P2.1) inverse estimate $|.|_1 \leq C h^{-1} |.|_o$;

(P2.2) $\lambda_{min}(A) = O(h^2)$, $\lambda_{max}(A) = O(1)$.

(P1) ensures the so-called approximation property.

Let Q be the following mapping

$Q : u_k \in V_k \to u_k - u_{k-1}$ orthogonal to V_{k-1} in the $a(.,.)$ - inner product. As a matrix we have

$$Q = A^{(k)-1} - \begin{pmatrix} J_{12} \\ I \end{pmatrix} A^{(k-1)-1} (J_{12}^t \quad I) = \begin{pmatrix} * & * \\ * & S^{-1} - A^{(k-1)-1} \end{pmatrix}.$$

Then the corresponding approximation property reads

$$\|Q u_k\|_{energy} \leq C h_k |||\underline{u}_k|||_2,$$

where

$|||.|||_\theta$ is a scale of norms related to the corresponding generalized eigenvalue problem (for details see [5] and [11]). In our particular case we have to consider the following generalized eigenvalue problem

$$A \underline{q} = \lambda L \begin{pmatrix} A_{11} & O \\ O & \tau I \end{pmatrix} U \underline{q},$$

or

$$(U \underline{q})_1 = \lambda (U \underline{q})_1,$$

$$S \underline{q}_2 = \lambda \tau \underline{q}_2.$$

The last equation gives us the idea to call the smoothing iteration damped Jacobi in subspace. This subspace is obtained for $\lambda < 1$, then

$$(U \underline{q})_1 = 0,$$

i.e. $\underline{q}$ belongs to the subspace

$$A_{11}\,\underline{v}_1 + A_{12}\,\underline{v}_2 = 0,$$

or

$$A\,\underline{v} = \begin{pmatrix} 0 \\ S\,\underline{v}_2 \end{pmatrix}.$$

If we define the measure of smoothness, cf. [5]

$$r(\underline{v}) = \frac{(\hat{G}\,\underline{v}_2)^t\, S\,\underline{v}_2}{\underline{v}_2^t\, S\,\underline{v}_2} \qquad /\hat{G} = I - \hat{S}^{-1} S/$$

for vectors in this subspace, the proof of Braess, Hackbusch [5] gives the following result

Theorem 3.1 The natural MG method has spectral convergence factor

$$\rho(\hat{M}^{(k)}) \leq \frac{C}{C + 2\nu}$$

in the regular case.

In the general case (possibly irregular problems and local refinement) we may see, that algebraically

$$\rho(\hat{M}^{(k)}) < \rho(\hat{M}^{(k)}_{\tau=\infty}) = 1 - O(k^{-2}) \quad \text{/based on Theorem 2.1)}$$

For further details and $W_2^{1+\alpha}$ - regular problems see [11].

4. Remarks on the implementation

The main task in the implementation of the studied natural multigrid method is solution of systems with matrices

$$\begin{pmatrix} A_{11} & 0 \\ A_{21} & \tau I \end{pmatrix} \begin{pmatrix} I & A_{11}^{-1} A_{12} \\ 0 & I \end{pmatrix}$$

i.e. with A_{11} at every discretization level (twice).

Here we use the following main fact

Theorem 4.1 ([4], [2])

$$\mathrm{Cond}(A_{11}^{(k)}) = O(1),$$

independent of the number of levels used.

For such a situation methods based on (block-) incomplete factorization of A_{11} are ideally suited. Another alternative is to use polynomially accelerated iterative methods, as good spectral bounds of A_{11} available on element level exist.

We consider here the former method. For the model Poisson equation in the unit square, discretized by piecewise linear triangular elements on a square grid the structure of A_{11} on every level is block - tridiagonal, using as blocks nodes on the lines parallel to one of the coordinate axes.

For solving systems with A_{11} we use the PCG method with preconditioning matrices, derived by the "inverse - free" blook-incomplete factorization of A_{11}. If $B = \text{block tridiag}(B_{i,i-1}, B_{i,i}, B_{i,i+1})_{i=1}^{n}$ is an s.p.d. block - tridiagonal M-matrix, the "inverse - free" block - incomplete factorization is defined as follows (for details, see [3])

$$(L + Y^{-1})\, Y\, (U + Y^{-1}),$$

where

$$L = \begin{pmatrix} 0 & & & O \\ B_{21} & O & & \\ & \ddots & \ddots & \\ O & & B_{n,n-1} & O \end{pmatrix}, \quad U = \begin{pmatrix} O & B_{12} & & \\ & O & \ddots & O \\ & & \ddots & B_{n-1,n} \\ O & & & O \end{pmatrix}$$

and

$$Y_1 = (B_{11}^{-1})^{(p)},$$

for i = 2 to n

$$Y_i = ((B_{i,1} - B_{i,i-1} Y_{i-1} B_{i-1,i})^{-1})^{(p)}.$$

$$Y = \text{block diag } (Y_i)_{i=1}^{n}.$$

Above for any matrix $C = (c_{ij})_{ij=1}^{m}$ by $C^{(p)}$ we mean the matrix with entries c_{ij} for $|i-j| \leqq p$ and 0 for $|i-j| > p$.

Algorithms for computing $(C^{-1})^{(p)}$ of complexity $O(m\, p^2)$ for C 2p+1 - banded may be found in [1], [12] or for banded approximation to C^{-1} in [6].

On Table 1 we show the number of iterations in the PCG method, used to solve systems of the form

$$A_{11}\, \underline{v}_1 = \underline{b}_1 ,$$

for various r.h.s., gridsize h and halfbandwidth p, used in the "inverse-free" block - incomplete factorization of A_{11}.

Table 1

Solving systems with A_{11} by the "inverse-free" block-incomplete factorization PCG method

p \ h^{-1}	32	64
1	4, $\underline{2-3}$	4 $\underline{2-3}$
2	3, $\underline{1}$	2 $\underline{1}$
4	2, $\underline{1}$	3 $\underline{1}$
8	2, $\underline{1}$	*

*) not computed

The numbers in the first columns stand for a maximal number of iterations achieved in our test. The second column indicates for an average number of iterations achieved in our test. As a general conclusion from

Table 1 we may see that for p sufficiently large (p ≧ 2 for the model problem) we may just replace A_{11} by its block-incomplete factorization matrix in the definition of the smoothing procedure of our natural MG method as well as in the definition of the multilevel preconditioning matrices in section 2.

REFERENCES

[1] Axelsson, O.: "A survey of vectorizable preconditioning methods for large scale finite element matrix problems", Colloquium Topics in Applied Numerical Analysis, J.G. Verwer (ed.), CWI syllabus 4, pp. 21-47, Mathematical Center, Amsterdam, 1984.

[2] Axelsson, O., Gustafsson, I.: "Preconditioning and two-level multigrid methods of arbitrary degree of approximation", Math. Comp., 40 (1983) pp. 219-242.

[3] Axelsson, O., Polman, B.: "On approximate factorization methods for block-matrices suitable for vector and parallel processors, Lin. Alg. Appl., 77 (1986) pp. 3-26.

[4] Bank, R.E., Dupont, T.: "Analysis of a two-level scheme for solving finite element equations", Report CNA-159, Center for Numerical Analysis, University of Texas at Austin, 1980.

[5] Braess, D., Hackbusch, W.; "A new convergence proof for the multigrid method including the V-cycle", SIAM J. Numer. Anal., 20 (1983) pp. 967-975.

[6] Eijkhout, V., Vassilevski, P.: "The positive defineteness requirement for vectorizable preconditioners based on incomplete block - factorizations", Report 8629, Dept. Math., University of Nujmegen, The Netherlands, 1986 (To appear in Parallel Computing, 1989).

[7] Grisvard, P.: "Elliptic problems in nonsmooth domains", Pitman Publishing Inc., Boston-London-Melbourne, 1985.

[8] Hackbusch, W.: "Multigrid convergence theory", in: "Multigrid methods", Proceedings of the Conference on Multigrid methods, held at Cologne-Porz, November 23-27, 1981, W. Hachbusch, U. Trottenberg (Eds.), Lecture Notes in Mathematics, Springer, 960 (1982), pp. 177-219.

[9] Stuben, K., Trottenberg, U.: "Multigrid methods: Fundamental algorithms, model problem analysis and applications", in: W. Hackbusch, U. Trottenberg (Eds.) Multigrid Methods, Lecture Notes in Math., Sprin- Springer, 960 (1982), pp. 1-176.

[10] Vassilevski, P.S.: "Nearly optimal iterative methods for solving finite element elliptic equations based on multilevel splitting of the matrix", submitted to Math. Comp., 1987.

[11] Vassilevski, P.S.: "A natural multigrid method of V-cycle type", submitted to Numer. Math., 1988.

[12] Vassilevski, P.S.: "On some ways of approximating inverses of banded matrices in connection with deriving preconditioners based on incomplete block factorizations", submitted to Computing, 1987.

TWO REMARKS ON MULTIGRID METHODS

P. Wesseling
Department of Mathematics and Informatics, University of Technology
P.O. Box 356, 2600 AJ Delft, The Netherlands

SUMMARY

This paper consists of two parts. In the first part, a partially successful attempt is made to extend multigrid theory to the case of a discontinuous coefficient, and an open problem is formulated. In the second part a non-recursive formulation of the fundamental multigrid algorithm is presented that contains only one goto statement, and is better structured than the usual formulation.

CELL-CENTERED MULTIGRID

In cell-centered multigrid, the unknowns are associated with nodes that are centers of cells, and coarse grids are constructed by taking unions of fine grid cells, cf. Figure 1.

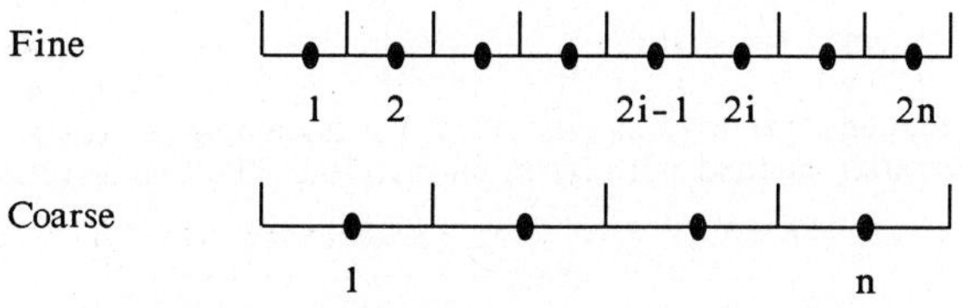

Figure 1 Fine and coarse cell-centered grid.

Coarse cell i is the union of fine cells 2i-1 and 2i. Only the one-dimensional case is considered here.

The following equation is considered:

$$-\frac{d}{dx}\left(a\frac{d\phi}{dx}\right) = f, \quad x \in (0,1), \quad \phi(0) = \phi(1) = 0. \tag{1}$$

Straightforward cell-centered finite volume discretization results in the following system of linear algebraic equations on the fine grid:

$$A\phi = f. \tag{2}$$

From now on, ϕ and f denote grid functions. The stencil of A in cell i, denoted by $[A]_i$, is given by

$$h^2[A]_1 = [0 \quad 2a_1+w_1 \quad -w_1], \quad h^2[A]_{2n} = [-w_{2n-1} \quad w_{2n-1}+2a_{2n} \quad a_{2n}],$$
$$h^2[A]_i = [-w_{i-1} \quad w_{i-1}+w_i \quad -w_i], \quad i = 2,3,...,2n-1. \tag{3}$$

Here the central entry of the stencil refers to cell i; h = 1/2n, and w_i is given by

$$w_i = 2a_i a_{i+1}/(a_i + a_{i+1}), \qquad i = 1,2,...,2n-1, \tag{4}$$

where a_i is (an approximation to) a(x) in the center of cell i.

For a description of a cell-centered multigrid method for the two-dimensional version of (2), see [1,2]. Here only the two grids specified in Figure 1 are required. The restriction operator R is the adjoint of linear interpolation:

$$[R]_1 = [0 \quad 2 \quad 3 \quad 1]/8, \qquad [R]_n = [1 \quad 3 \quad 2 \quad 0]/8,$$
$$[R]_i = [1 \quad 3 \quad 3 \quad 1]/8, \qquad i = 2,3,...,n-1. \tag{5}$$

The entries of $[R]_i$ refer to fine cells 2i-2, 2i-1, 2i, 2i+1, respectively. Prolongation is piecewise constant interpolation:

$$P\bar{\phi}_{2i-1} = P\bar{\phi}_{2i} = \bar{\phi}_i. \tag{6}$$

Overbars refer to the coarse grid. The coarse grid matrix is defined by

$$\bar{A} = RAP. \tag{7}$$

It turns out that this cell-centered multigrid method works for continuous a(x) as well as for discontinuous a(x) ([1,2]). When a(x) is discontinuous, the more common vertex-centered multigrid methods need complicated matrix-dependent R and P ([3,4,5,6,7,8]), which seems to preclude the development of theory. For the present more simple cell-centered multigrid method the prospects for the development of theory seem more bright; here we make a first attempt.

For simplicity we assume

$$a_i = 1, \quad 1 \le i \le m; \quad a_i = \alpha > 1, \quad m+1 \le i \le 2n. \tag{8}$$

We will try to prove h-independent two-grid convergence. We assume no pre-smoothing and p post-smoothings with an iteration method with iteration matrix S. The two-grid iteration matrix M is given by

$$M = S^p B \tag{9}$$

with the coarse grid correction matrix B given by

$$B = I - P\bar{A}^{-1}RA. \tag{10}$$

THE COARSE GRID CORRECTION MATRIX

We will in this section explicitly express ψ given by

$$\psi = B\phi \tag{11}$$

in terms of ϕ. Let

$$\phi = P\bar{\phi} + \nu \tag{12}$$

with $\bar{\phi}_i = \frac{1}{2}(\phi_{2i-1} + \phi_{2i})$. Obviously,

$$BP\bar{\phi} = 0, \tag{13}$$

$$\nu_{2i-1} = -\mu_{2i}, \quad \nu_{2i} = \mu_{2i}, \quad \mu_{2i} = \frac{h}{2}\nabla\phi_{2i}, \tag{14}$$

with the backward divided difference ∇ defined by

$$\nabla\phi_i = (\phi_i - \phi_{i-1})/h. \tag{15}$$

First we treat the case m = 2k-1. One finds:

$$\begin{aligned} &RA\nu_i = 0, \quad i \neq k\pm 1,\, k; \quad RA\nu_{k-1} = (1 - w)\mu_{2k}/\bar{h}^2; \\ &RA\nu_k = (\alpha - 1)\mu_{2k}/\bar{h}^2; \quad RA\nu_{k+1} = (w - \alpha)\mu_{2k}/\bar{h}^2; \end{aligned} \tag{16}$$

where $\bar{h} = 2h$ and $w = 2\alpha/(1 + \alpha)$. Inspection shows that

$$\bar{A}^{-1}RA\nu_i = (w - 1)\mu_{2k}, \quad i = k; \quad = 0, \quad i \neq k. \tag{17}$$

Hence

$$\begin{aligned} &B\phi_{2i-1} = -\frac{h}{2}\nabla\phi_{2i}, \qquad B\phi_{2i} = \frac{h}{2}\nabla\phi_{2i}, \quad i \neq k; \\ &B\phi_{2k-1} = -\frac{hw}{2}\nabla\phi_{2k}, \quad B\phi_{2k} = \frac{hw}{2}\nabla\phi_{2k}. \end{aligned} \tag{18}$$

Next we treat the case m = 2k. One finds:

$$RA\nu_i = 0, \quad i \neq k,\, k+1; \qquad RA\nu_k = \delta/\bar{h}^2; \qquad RA\nu_{k+1} = -\delta/\bar{h}^2 \tag{19}$$

with

$$\delta = (w - 1)\frac{h}{2}\nabla\phi_m + (w - \alpha)\frac{h}{2}\nabla\phi_{m+2} = \frac{h^2}{2}(w - 1)A(\phi_{m+1} + \phi_m). \tag{20}$$

Inspection shows that

$$\bar{A}^{-1}RA\nu = \bar{\eta}; \quad \bar{\eta}_i = a(i - 1/2), \quad i \leq k; \quad \bar{\eta}_i = b(n + 1/2 - i), \quad i \geq k+1, \tag{21}$$

with

$$a = \frac{\alpha + 1}{2}\delta/(n + \alpha k - k), \quad b = -a/\alpha. \tag{22}$$

Finally, we obtain:

$$B\phi_{2i-1} = -\frac{h}{2}\nabla\phi_{2i} - \bar{\eta}_i, \quad B\phi_{2i} = \frac{h}{2}\nabla\phi_{2i} + \bar{\eta}_i. \tag{23}$$

TWO-GRID CONVERGENCE

Let us define the following scale of norms:

$$|\phi|_s = |A^{s/2}\phi|_0, \quad |\phi|_0^2 = (\phi,\phi) = h\sum_{i=1}^{2n}\phi_i^2. \tag{24}$$

Define

$$v_1 = 2; \quad v_i = 1,\ 2 \leq i \leq m; \quad v_{m+1} = w; \quad v_i = \alpha,\ m+2 \leq i \leq 2n; \quad v_{2n+1} = 2\alpha. \tag{25}$$

Then we can write

$$A\phi = -\Delta(v\nabla\phi) \tag{26}$$

with the forward divided difference Δ defined by

$$\Delta\phi_i = (\phi_{i+1} - \phi_i)/h, \tag{27}$$

and defining $\phi_i = 0$ for $i < 1$ or $i > 2n$. Using the partial summation formula

$$(\nabla\phi,\psi) = -(\phi,\Delta\psi) \tag{28}$$

we find that

$$|\phi|_1^2 = (A^{1/2}\phi,A^{1/2}\phi) = (A\phi,\phi) = (v\nabla\phi,\nabla\phi) = |v^{1/2}\nabla\phi|_0^2. \tag{29}$$

From (18) we find that, since $w < \alpha$,

$$|B\phi|_0 \leq \frac{h}{2}\sqrt{\alpha}\;|\phi|_1, \tag{30}$$

so that

$$|B|_{0\leftarrow 1} \leq \frac{h}{2}\sqrt{\alpha} \tag{31}$$

for the case $m = 2k-1$. For the case $m = 2k$, (23) gives

$$|B\phi|_0 \leq \frac{h}{2}|\phi|_1 + |\bar{\eta}|_0 \tag{32}$$

with $|\bar{\eta}|_0 = h\sum_1^n \bar{\eta}_i^2$. We find that

$$|\bar{\eta}|_0 < c\alpha|\delta| \tag{33}$$

with c a numerical constant of about 1/3. We have

$$|\delta| \leq h|2w-1-\alpha||\nabla\phi|_\infty \leq h^{1/2}\frac{(\alpha-1)^2}{\alpha+1}|\nabla\phi|_0 \leq h^{1/2}\frac{(\alpha-1)^2}{\alpha+1}|\phi|_1, \tag{34}$$

where $|\nabla\phi|_\infty = \max_i |\nabla\phi_i|$. Hence,

$$|B|_{0\leftarrow 1} = O(h^{1/2}). \tag{35}$$

Our two-grid convergence proof follows the lines laid down in [10]. Equations (31) and (35) establish an approximation property. We proceed to derive a smoothing property. As smoothing method we choose Jacobi with parameter 1/2, so that S is given by:

$$S = I - C, \quad C = \frac{1}{2}D^{-1}A, \quad D = \mathrm{diag}\,(A). \tag{36}$$

We have

$$|S^p|_{1\leftarrow 0} = |A^{1/2}S^p|_{0\leftarrow 0} = \sqrt{2}\;|D^{1/2}C^{1/2}(I-C)^p|_{0\leftarrow 0}, \tag{37}$$

$$|D^{1/2}|_{0\leftarrow 0} = \sqrt{2\alpha}/h. \tag{38}$$

Furthermore, corollary 6.2.3 in [10] gives

$$|C^{1/2}(I-C)^p|_{0\leftarrow 0} \leq (\eta_0(2p))^{1/2} \tag{39}$$

with $\eta_0(q) = q^q/(q+1)^{q+1}$, so that the following smoothing property is obtained:

$$|S^p|_{1\leftarrow 0} \le 2\sqrt{\alpha\eta_0(2p)}\ h^{-1}. \tag{40}$$

Combining (31) and (40) gives h-independent two-grid convergence for the case $m = 2k-1$:

$$|S^pB| < 1, \quad \text{p large enough.} \tag{41}$$

But for the case $m = 2k$, (35) and (41) leave us with $|S^pB| = O(h^{1/2})$, so that we have not succeeded. The improvement of the proof is left as a challenge for other researchers. During the conference, Wolfgang Hackbusch suggested to incorporate the supremum norm in the smoothing property, which is perhaps possible by generalizing exercise 6.6.6 in [10] to equation (1) with a(x) given by (8).

Numerical evidence suggests that we have h-independent two-grid convergence irrespective of the value of m. In [1] and [2] good convergence of the multigrid method discussed here is found in a variety of interface problems. Note that one would expect the case $m = 2k-1$, for which the above proof works, to be more critical in practice than the case $m = 2k$, because in the first case the discontinuity is not resolved on the coarse grid.

NON-RECURSIVE FORMULATION OF THE FUNDAMENTAL MULTIGRID ALGORITHM WITH ONLY ONE GOTO STATEMENT

The fundamental multigrid algorithm is defined as follows (cf. [10] for example):

procedure MG(k,ϕ,f);	k = grid index
if $k = 1$ **then** $\phi^1 := (A^1)^{-1}f^1$ **else**	k = 1 is coarsest grid
begin $\phi^k := S^1(\phi^k,f^k)$;	pre-smoothing
$f^{k-1} := R^{k-1}(f^k - A^k\phi^k)$;	defect computation
$\phi^{k-1} := 0$;	initialization of correction
for i := 1 **step** 1 **until** γ **do** MG(k-1,ϕ,f);	recursive call of MG
$\phi^k := \phi^k + P^k\phi^{k-1}$;	coarse grid correction
$\phi^k := S^2(\phi^k,f^k)$	post-smoothing
end of MG;	

This is the correction variant. This may be called the fundamental multigrid algorithm, because the other variants, such as the full approximation and nonlinear variants, have the same structure.

In FORTRAN recursivity is not allowed. A well-known flow diagram ([10],[11],[12]) for a non-recursive algorithm is given in Figure 2. The finest grid has index $k = \ell$, the coarsest $k = 1$. S^1 represents pre-smoothing, S^2 post-smoothing. This flow diagram leads to code with four **goto** statements. Generally speaking, fewer **goto** statements mean better structure. Better structure can be obtained by designing the algorithm with a structure diagram instead of a flow diagram. We found that we could represent the fundamental multigrid algorithm by a structure diagram in one way only. Therefore one might call this the *canonical form* of the fundamental multigrid algorithm. This structure diagram is presented in Figure 3. The parts A,B,C of the algorithm are defined in Figure 2. This structure diagram leads to the following FORTRAN code. Because FORTRAN does not know **while** statements, still one **goto** is required.

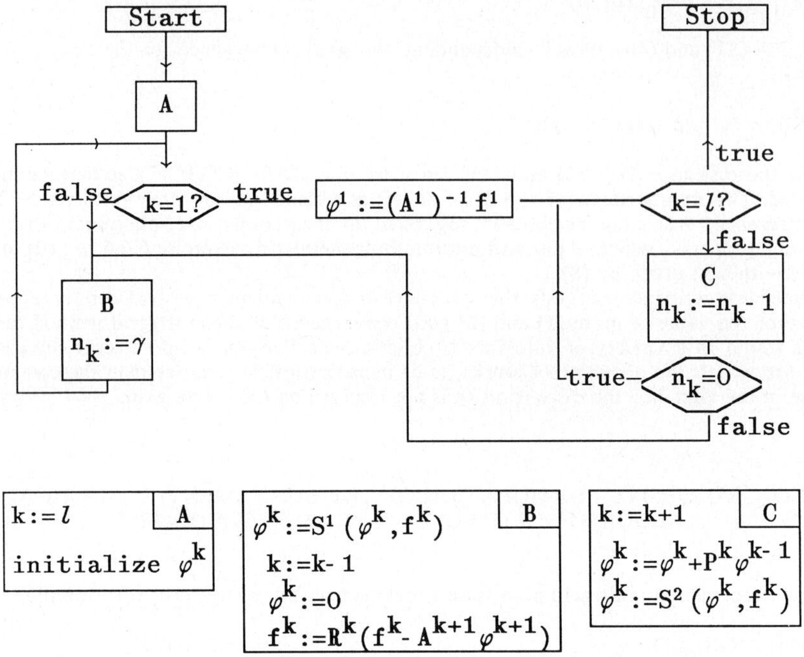

Figure 2 Flow diagram of fundamental multigrid algorithm.

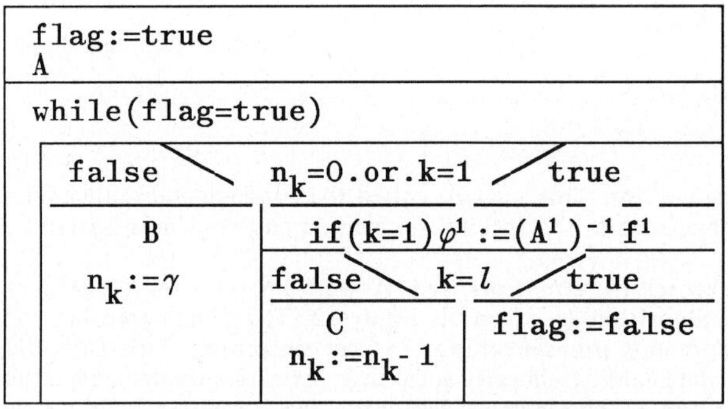

Figure 3 Structure diagram of fundamental multigrid algorithm.

```
     subroutine MG(k,φ,f)
     flag = true
     A
10   if flag then
        if (n_k.eq.0.or.k.eq.1) then
          if (k.eq.1) φ^1 = (A^1)^-1 f^1
          if (k.eq.ℓ) then
             flag = false
        else
           C
           n_k = n_k - 1
        endif
        else
           B
           n_k = γ
        endif
        goto 10
        return
     end
```

The flow diagram associated with the structure diagram of Figure 3 is given in Figure 4.

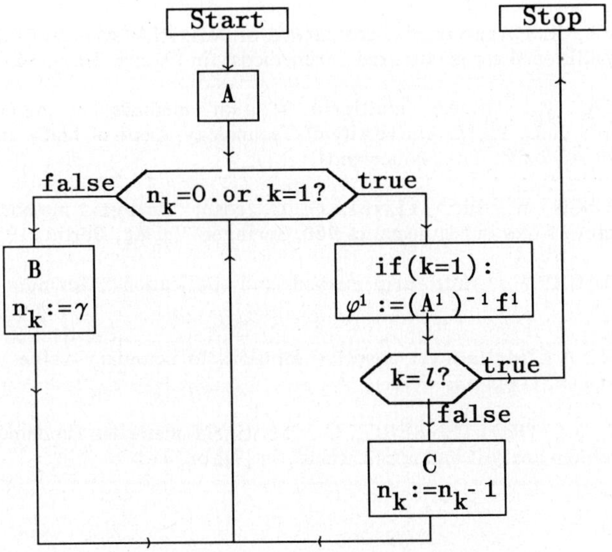

Figure 4 Canonical flow diagram of fundamental multigrid algorithm.

During the conference, Peter Bastian and Graham Horton from Erlangen University, West-Germany showed the author how the **while** statement can be replaced by a **for** statement, thus eliminating the remaining **goto**. But the resulting code was less transparent, and, unlike the code above, left no freedom in choosing other (e.g. adaptive) criteria for the decision to go to a coarser or finer grid. Which shows that not all **goto**'s are bad.

REFERENCES

[1] WESSELING, P.: "Finite volume multigrid". In: S.F. MacCormick (ed.), Proceedings, Third Copper Mountain Conference on Multigrid Methods, April 1987. To appear, 1988.

[2] WESSELING, P.: "Cell-centered multigrid for interface problems". To appear in J. Comp. Phys.

[3] ALCOUFFE, R.E., BRANDT, A., DENDY Jr., J.E., PAINTER, J.W.: "The multigrid method for the diffusion equation with strongly discontinuous coefficients", SIAM J. Sci. Stat. Comp., **2** (1981) pp. 430-454.

[4] DENDY Jr., J.E.: "Black box multigrid", J. Comp. Phys., **48** (1982) pp. 366-386.

[5] DENDY Jr., J.E.: "Black box multigrid for nonsymmetric problems", Appl. Math. Comp., **13** (1983) pp. 261-283.

[6] KETTLER, R., MEIJERINK, J.A.: "A multigrid method and a combined multigrid-conjugate gradient method for elliptic problems with strongly discontinuous coefficients in general domains", Shell Publ. 604, 1981, KSEPL, Rijswijk, The Netherlands.

[7] KETTLER, R.: "Analysis and comparison of relaxation schemes in robust multigrid and preconditioned conjugate gradient methods", in [9], pp. 502-534.

[8] KETTLER, R.: "Linear multigrid solution methods in numerical reservoir simulation", Ph.D. Thesis, University of Technology, Dept. of Math. and Inf., P.O. Box 356, 2600 AJ Delft, The Netherlands.

[9] HACKBUSCH, W., TROTTENBERG, U. (eds.): "Multigrid methods", Proc., Köln-Porz, Lecture Notes in Mathematics **960**, Springer-Verlag, Berlin, 1982.

[10] HACKBUSCH, W.: "Multi-grid methods and applications", Springer-Verlag, Berlin, 1985.

[11] BRANDT, A.: "Multi-level adaptive solutions to boundary-value problems", Math. Comp., **31** (1977) pp. 333-390.

[12] STÜBEN, K., TROTTENBERG, U.: "Multigrid methods: fundamental algorithms, model problem analysis and applications", in [9], pp. 1-176.

On the Robustness of ILU-Smoothing

Gabriel Wittum
SFB 123, Universität Heidelberg
Im Neuenheimer Feld 294
D-6900 Heidelberg

Summary: In the present paper we give a detailed analysis of a multi-grid method with ILU-smoother applied to singularly perturbed problems. Based on the analysis of a simple anisotropic model problem we introduce a variant of the usual incomplete LU-factorization, which is especially suited as robust smoother. For this variant we give a detailed analysis and a proof of robustness. Further, we explain some contradictions between the smoothing rates predicted by local Fourier analysis and the practically observed convergence factors,(see [7], [13], [19]). The theoretical results are confirmed by numerical tests.

1. Introduction

The treatment of singular perturbations is a severe problem in solving partial differential equations. The discretization of a singularly perturbed operator contains characteristic difficulties which may affect multi-grid convergence, as the discretization process is hidden in the hierarchy of grids Ω_l, $l=1,\ldots,l_{max}$, used in multi-grid technique. To handle such problems "robust" multi-grid methods are required. In the present paper we introduce a robust multi-grid method based on a variant of incomplete LU-smoothing. This method is proven to be robust for the anisotropic model-problem and related ones. The theoretical results are confirmed by numerical tests.

1.1. Preliminaries and Notations

An operator $K = K(\varepsilon)$ depending on a parameter ε is called singularly perturbed, if the limiting operator $K(0) = \lim_{\varepsilon \to 0} K(\varepsilon)$ is of a type other than $K(\varepsilon)$ for $\varepsilon > 0$. We are especially interested in operators of the form

(1.1.1) $$K(\varepsilon) = \varepsilon K^{I} + K^{II} \quad ,$$

where $K(\varepsilon)$ is elliptic for $\varepsilon > 0$ and K^{II} is non-elliptic or elliptic of lower order. A simple model-problem for this situation is

(1.1.2a) $$K(\varepsilon)\, u = f \text{ in } \Omega, \qquad u = g \quad \text{on } \partial\Omega$$

with

(1.1.2b) $$K(\varepsilon) = - \left(\varepsilon \frac{\partial^2}{\partial x^2} + \frac{\partial^2}{\partial y^2} \right)$$

in a domain $\Omega \subset \mathbf{R}^2$. For the moment let $\Omega = (0,1)\times(0,1)$ and (1.1.2) be discretized by

(1.1.3) $$K_l(\varepsilon) = h_l^{-2} \begin{bmatrix} & -1 & \\ -\varepsilon & 2(1+\varepsilon) & -\varepsilon \\ & -1 & \end{bmatrix}$$

on a rectangular equidistant grid with stepsize h_l, ordering the grid-points lexicographically. Stencil (1.1.3) applies to grid-points lying totally inside Ω. For points having a boundary-point as neighbour, the corresponding entry of the stencil has to be skipped (cf. [8]).

This problem is very simple, however, the typical difficulties arising from singular perturbations show up there and can be analyzed thoroughly. In addition, the insight obtained by the analysis of the model-problem gives rise to certain generalizations as presented in sec. 2.2.For ease of notation we used the unit square in (1.1.3), the results, however, apply to a general region (see theorem 2.1.6).

Concerning multi-grid technique, we refer to Hackbusch's book, [7]. The two-grid operator, corresponding to a two-grid method on levels l and l-1 is given by

(1.1.4) $$T_{2,l}(\nu,0) = (I - p\, K_{l-1}^{-1}\, r\, K_l)\, S_l^{\nu} \quad ,$$

using ν steps of a smoother S_l, a prolongation p and a restriction r. We restrict our analysis to this two-grid operator, as from there the convergence of the corresponding multi-grid method can be concluded easily (see [7], §7).

Given two matrices A and B, we denote by

(1.1.5) $$A \underset{(=)}{>} B$$

the positive (semi-)definiteness of the matrix A-B.

1.2. Background

There are several approaches treating problems of type (1.1.1) by multi-grid methods. On the one hand methods based on special coarsenig techniques and sophisticated extensions of Gauß-Seidel smoothers (see [2], [18]), on the other hand methods based on incomplete LU-smoothing. Further, Hackbusch introduced the frequency-decomposition multi-grid method in [9]. This method looks very promising towards a general-purpose robust and efficient multi-grid method being applicable even to 3D problems. In the 2D-case, however, ILU will still be competitive, at least for non-parallel computers. The ILU-smoothers were introduced by P. Wesseling and R. Kettler (cf. [21], [22], [23], [13]) on account of their "robustness". By robustness we denote the following:

Definition 1:
Denote by $\zeta(\varepsilon)$ the contraction-number of a multi-grid method for the singularly perturbed problem (1.1.1). We call the method $K_1(\varepsilon)$ - robust iff

(1.2.1) $\qquad \zeta(\varepsilon) \leq \zeta < 1 \qquad \forall \varepsilon \geq 0.$

holds.

Though being widely used, there was no proof nor rigorous theoretical treatment of robustness up to now. The only analysis existing was given by W. Hackbusch, establishing robustness for a special line-Jacobi smoother applied to problem (1.1.2) (see [7], lemma 10.1.2).

Wesseling and Kettler found the robustness of ILU for model-problem (1.1.2) to be due to the fact that ILU solves $K_1(0)$ exactly. This fundamental observation led Hackbusch to the criterion that a smoothing-iteration suitable for a singularly-perturbed problem $K_1(\varepsilon)$ has to be a very fast or even exact solver for the limit case $K_1(0)$ (cf. [7], crit. 10.1.1). Acutally, this is the main tool to achieve robustness and it is equivalent to assumption (2.1.10), but unfortunately it is not sufficient as results from our analysis. This also becomes visible when doing computations on extremely fine grids using a straightforward ILU-approach. There the convergence rate deteriorates for certain values of ε. These difficulties led to a very different esteem of ILU-smoothing among several groups. For us this is the reason to introduce some variant called ILU_β controlled by a parameter β, similar to the modifications of incomplete LU-decompositions as suggested by Gustafsson, [5], and Jennings-Malik, [12]. In section 2 we prove robustness of ILU_β for problem (1.1.2) and a generalized problem for certain values of β.

Section 3 is concerned with the rôle of local Fourier-analysis to predict convergence rates. There usually influences of the finite dimension of the discrete problems are neglected, thus providing unrealistic results for ILU- smoothing applied to problem (1.1.2) (see [19], [13]). These discrepancies are explained and give rise to a slight modification of local mode-analysis.

In section 4 we optimize the parameter β, using the estimates deduced in section 2. We check this prediction of the optimal paramter by comparing the results of a large bunch of practical computations performed on problem (1.1.2).

We were led to the ILU_β-variant by a careful analysis of robustness and the behaviour of incomplete LU-smoothing, based on the convergence theory by W. Hackbusch (see [7]) and some extensions by the author (see [24]). Using local Fourier-techniques Khalil, [14], and Oertel, [17], suggested the same modification in lectures during the 4. GAMM-Seminar on Robust Multi-Grid methods in Kiel, 22.-24.1.88. On just that conference we presented the main results of the present paper.

2. A Robust Smoother

2.1 The Model Problem

Let K_l from (1.1.3) be split into

(2.1.1) $\quad K_l = M - N$

with

(2.1.2) $\quad M = (L+D)\, D^{-1}\, (L+D)^T$

and

(2.1.3a) $\quad D = h^{-2} \operatorname{diag}\{d_{i,j}\}_{i=1,\dots,m,\, j=1,\dots,m}$,

$d_{i,j}$ corresponding to the grid-point $x=(jh,ih)$, $h = 1/(m+1)$, and

(2.1.3b) $\quad L = h^{-2} \begin{bmatrix} & \cdot & \\ -\varepsilon & 0 & \cdot \\ & -1 & \end{bmatrix}$.

The entries of D are defined by

$$(2.1.4)\quad d_{i,j} = \begin{cases} 2(1+\varepsilon) & i=j=1, \\ 2(1+\varepsilon) - \varepsilon(\varepsilon-\beta)/d_{i,j-1} & i=1,\, j>1, \\ 2(1+\varepsilon) - (1-\beta\varepsilon)/d_{i-1,j} & i>1,\, j=1 \\ 2(1+\varepsilon) - \varepsilon(\varepsilon-\beta)/d_{i,j-1} - (1-\varepsilon\beta)/d_{i-1,j} & i,j>1 \end{cases}$$

with $\beta \in \mathbf{R}$. We are especially interested in $\beta \in [0,1]$.

The corresponding iterative method is given by

(2.1.5) $\quad$ Let $x^{(0)}$ be given. Compute $x^{(i+1)}$ from $x^{(i)}$ by

$$x^{(i+1)} = x^{(i)} - M^{-1} (K_l\, x^{(i)} - b) .$$

We call this iteration ILU_β.

For certain values of β we regain well-known ILU-methods. So ILU_0 is identical with the usual 5-Point ILU-factorization, as described in [16]. ILU_{-1} is identical with the so-called modified incomplete LU-decomposition by Gustafsson [5] introduced to improve the asymptotic condition number of the preconditioned system, while ILU_1 corresponds to a modification suggested by Jennings and Malik [12] yielding a positive definite rest-matrix. These modifications were applied as preconditioners for cg-algorithms. In multi-grid framework only Hemker in [10] investigated about some modification of the standard ILU-algorithm for $\varepsilon=1$.

To prove robustness of a two-grid method with this smoother, we consider the splitting

(2.1.6) $\quad \|T_{2,l}(\nu,0)\| \le \| K_l^{-1} - p\, K_{l-1}^{-1} r \|\, \| K_l\, S_l^{\nu} \|$.

of the two-grid operator $T_{2,l}(\nu,0)$ from (1.1.4) which is the basic tool of Hackbusch's multi-grid convergence theory (cf. [7]). The factors are estimated separately by the "approximation property"

(2.1.7) $\quad \| K_l^{-1} - p\, K_{l-1}^{-1} r \| \le C_A\, h_l^2$,

and the "smoothing property"

(2.1.8a) $\| K_l S_l^{\nu} \| \leq C_S h_l^{-2} \eta(\nu)$

with

(2.1.8b) $\eta(\nu) \to 0, \;\text{ for } \nu \to \infty.$

In [7] Hackbusch uses a more general definition of smoothing property (2.1.8) in order to incorporate also diverging iterations as smoothers. For our purpose, however, definition (2.1.8) is sufficient.

This construction applies to the case of ε=const. Considering $K_l(\varepsilon)$ with ε as parameter, we realize that now the approximation property behaves like

(2.1.7ε) $\| K_l^{-1}(\varepsilon) - p\, K_{l-1}^{-1}(\varepsilon)\, r \| \leq C_A \frac{1}{\varepsilon} h_l^2$

(cf [7], pg. 202).

There are in general two possibilities to get rid of the factor $1/\varepsilon$. Firstly, the coarse-grid correction may be improved, say by semi-coarsening and related, very specialized techniques (cf [18]), or as a general purpose approach by the frequency-decomposition multi-grid method, introduced by Hackbusch in [9]. Then the factor $1/\varepsilon$ drops out of (7ε). Secondly, we may use a smoother satisfying

(2.1.8aε) $\| K_l(\varepsilon)\, S_l^{\nu} \| \leq C_S h_l^{-2} \varepsilon\, \eta(\nu)$

with $\eta(\nu)$ from (2.1.8b). This was the way followed by Wesseling and Kettler when introducing their robust multi-grid method (cf. [21], [13]).The following criterion gives sufficient conditions for (8aε).

<u>Criterion 1:</u>

Let $K_l(\varepsilon)$ be a symmetric and positive definite matrix. Let further $K_l(\varepsilon)$ be split according to (2.1.1) with symmetric and positive definite M. Further suppose

(2.1.9) $\exists\, \alpha > 2:\; K_l(\varepsilon) + \alpha N \geq 0 \quad, \forall\, \varepsilon > 0,$

(2.1.10a) $\| N \|_{\infty} \leq \varepsilon\, h^{-2} C_N \quad,$

and

(2.1.10b) $\| S_l \| \leq C \quad.$

Then (2.1.8aε) is valid with $\eta(\nu) = \eta_0(\nu-2, 1/(\alpha-1))$, η_0 from (2.1.12b), and $C_S = CC_N$.

<u>Proof:</u>

As $M=M^T > 0$, there exists $M^{1/2} = {M^{1/2}}^T > 0$. Let $X := M^{-1/2} N M^{-1/2}$. Since $K_l(\varepsilon) > 0$, we get $X = I - M^{-1/2} K_l(\varepsilon) M^{-1/2} \leq I$. On account of (2.1.9) $(1+\vartheta)/\vartheta\, N + K_l(\varepsilon) \geq 0$ holds with $\vartheta = 1/(\alpha-1)$. Using this, we obtain $0 \leq (1+\vartheta)M - K_l(\varepsilon)$. By virtue of that and by the symmetry of the matrices involved we end up with $(1+\vartheta)I - M^{-1/2} K_l(\varepsilon) M^{-1/2} \geq 0$ and conclude that $X \geq -\vartheta I$. Thus $-\vartheta I \leq X \leq I$ holds. Using (2.1.2), the symmetry of the problem, the following lemma, and $\|S_l\| \leq 1$, we obtain

$\|K_l(\varepsilon) S_l^{\nu}\| = \|(M-N)(M^{-1}N)^{\nu}\| = \|M^{1/2} X (I-X) X^{\nu-2} X M^{1/2}\| \leq \|(I-X)X^{\nu-2}\|\, \|S_l\|\, \|N\|_{\infty} \leq$

$\leq CC_N \varepsilon h^{-2} \eta_0(\nu-2, 1/(\alpha-1)).$ ❑

Criterion 1 is a generalization of the criterion used in [24] to prove the smoothing-property for incomplete LU-decompostitions. The technique is based on Hackbusch's work on the smoothing-property (see [7], §6.2).

Lemma 2:

Let A be symmetric. Let further

(2.1.11) $$-\vartheta I \le A \le I$$

Then the spectral-norm of $A^\nu(I-A)$ is bounded by

(2.1.12a) $$\|A^\nu(I-A)\| \le \eta_0(\nu,\vartheta)$$

with

(2.1.12b) $$\eta_0(\nu,\vartheta) = \max\{\nu^\nu/(\nu+1)^{(\nu+1)}, (1+\vartheta)\vartheta^\nu\}$$

Proof:

According to [7], Lemma 1.3.5 (iii), we have

$$\| A^\nu(I-A)\| \le \max_{-\vartheta\le x\le 1} |x^\nu(1-x)| = \max\{\nu^\nu/(\nu+1)^{(\nu+1)},(1+\vartheta)\vartheta^\nu\}.$$ □

To apply criterion 1 to ILU_β, we prove the following lemma.

Lemma 3

Let $K_l(\varepsilon)$ from (1.1.3) be split as given in (2.1.2) - (2.1.4). Then (2.1.9) and (2.1.10) hold for $\beta \in (0,1)$.

Proof:

ad(2.1.10a): N from (2.1.1) is given by

(2.1.13) $$N = h^{-2}\begin{bmatrix} \gamma^r_{ij} & & \cdot & \\ \cdot & & \beta(\gamma^r_{ij}+\gamma^l_{ij}) & \cdot \\ & \cdot & & \gamma^l_{ij} \end{bmatrix}$$

with

(2.1.14) $$\begin{cases} \gamma^r_{ij} = \dfrac{\varepsilon}{d_{i,j-1}} & \text{for } i \ge 1 \text{ and } j > 1 \\ \gamma^l_{ij} = \dfrac{\varepsilon}{d_{i-1,j}} & \text{for } i > 1 \text{ and } j \ge 1 \end{cases}$$.

As $d \ge d_{i,j} \ge \delta$ (d, δ from (2.1.27)), (2.1.10) holds.

ad (2.1.10b): Using (2.1.3), (2.1.26c), and (2.1.27c) assumption (2.1.10b) is readily proven with C=1 by $\|S_l\| \le \|M^{-1}\| \, \|N\|_\infty \le 1$.

ad(2.1.9): We split $(K_l(\varepsilon) + \alpha N)$ in the following way. Let $i,j \le m$ then

$$h^2(K_1(\varepsilon)+\alpha N) = \begin{bmatrix} & \alpha\gamma_{ij}^r & -1 & \\ & -\varepsilon & 2(1+\varepsilon)+\alpha\beta(\gamma_{ij}^r+\gamma_{ij}^l) & -\varepsilon \\ & & -1 & \alpha\gamma_{ij}^l \end{bmatrix} =: A =$$

$= B_1 + B_2$ with

(2.1.15)
$$B_1 = \begin{bmatrix} & \frac{q\alpha\varepsilon}{\delta} & -1 & \\ & -\varepsilon & 2(1+\varepsilon+\frac{\alpha\beta\varepsilon}{\delta}) & -\varepsilon \\ & & -1 & \frac{q\alpha\varepsilon}{\delta} \end{bmatrix} \text{ and }$$

$$B_2 = \begin{bmatrix} \alpha\varepsilon\left(\frac{1}{d_{i,j-1}}-\frac{q}{\delta}\right) & & & \\ & \alpha\beta\varepsilon\left(\frac{1}{d_{i,j-1}}+\frac{1}{d_{i-1,j}}-\frac{2}{\delta}\right) & & \\ & & & \alpha\varepsilon\left(\frac{1}{d_{i-1,j}}-\frac{q}{\delta}\right) \end{bmatrix}$$

where
(2.1.16) $\quad q = \frac{1-\beta}{2}$

and δ from (2.1.27). Denote by $\lambda_{min}(C)$ the smallest eigenvalue of the symmetric matrix C. Then

(2.1.17) $\quad \lambda_{min}(A) \geq \lambda_{min}(B_1) - \|B_2\|_\infty$

holds according to usual perturbation theory. Now we show the estimate

(2.1.18) $\quad \|B_2\|_\infty \leq 2(1-q)\frac{\alpha\varepsilon}{\delta}$

With $d = \min(d_{i,j-1}, d_{i-1,j})$ and by Lemma 4 we have

$$\|B_2\|_\infty = \alpha\varepsilon\left(\left|\frac{1}{d_{i,j-1}}+\frac{1}{d_{i-1,j}}-2\frac{q}{\delta}\right|+\beta\left|\frac{1}{d_{i,j-1}}+\frac{1}{d_{i-1,j}}-\frac{2}{\delta}\right|\right) \leq$$

$$\leq \frac{\alpha\varepsilon}{\delta}\frac{2(1-\beta)\delta + (3\beta-1)d}{d}$$

Since $d \geq \delta$ (cf. (2.1.27)), $\frac{2(1-\beta)\delta + (3\beta-1)d}{d} \leq 1+\beta = 2(1-q)$ holds.

Thus (2.1.18) is proved. So it is sufficient to show that
(2.1.19) $\quad \lambda_{min}(B_1) - 2(1-q)\frac{\alpha\varepsilon}{\delta} \geq 0$.

We prove (2.1.19) by constructing a diagonal matrix Φ with
(2.1.20) $\quad B_1 - 2(1-q)\frac{\alpha\varepsilon}{\delta}I \geq (\Lambda+\Phi)\Phi^{-1}(\Lambda+\Phi)^T$

and $\Lambda := h^2 L$ with L from (2.1.3b). The left-hand side of (2.1.20) reads

(2.1.21) $B_1 - 2(1-q)\frac{\alpha\varepsilon}{\delta}I =$

$$\begin{bmatrix} q\frac{\alpha\varepsilon}{\delta} & -1 & \\ -\varepsilon & 2(1+\varepsilon) + \frac{2\alpha\varepsilon\beta}{\delta} - 2(1-q)\frac{\alpha\varepsilon}{\delta} & -\varepsilon \\ & -1 & q\frac{\alpha\varepsilon}{\delta} \end{bmatrix}$$

while the right-hand side is

$$(2.1.22) \qquad (\Lambda+\Phi)\Phi^{-1}(\Lambda+\Phi)^T = \begin{bmatrix} \frac{\varepsilon}{\varphi} & -1 & \\ -\varepsilon & \varphi + \frac{1+\varepsilon^2}{\varphi} & -\varepsilon \\ & -1 & \frac{\varepsilon}{\varphi} \end{bmatrix}$$

with $\Phi = \operatorname{diag}\{\varphi,...,\varphi\}$. Comparing both, we obtain

$$(2.1.23) \qquad \varphi = \frac{2\delta}{(1-\beta)\alpha}$$

and

$$(2.1.24) \qquad 2(1+\varepsilon) + \frac{\alpha\varepsilon}{\delta}(\beta-1) \geq \varphi + \frac{1+\varepsilon^2}{\varphi},$$

with

$$(2.1.25) \qquad \alpha = \frac{2\delta}{(1-\beta)(1+\varepsilon)},$$

the value of α still satisfying (2.1.24). As $\beta > 0$ is a fixed Parameter and $\delta = 1+\varepsilon+\sqrt{2\varepsilon(1+\beta)} > 1+\varepsilon$, $\alpha > 2$ holds. □

Some properties of recursion (2.1.4), used in the above proof, are given in

Lemma 4:

Let $d_{i,j}$ be computed by (2.1.4). Then for $\varepsilon \geq \beta$

$$(2.1.26a) \qquad d_{i,j} \leq d_{i-1,j}, \; i>1, j\geq 1,$$

$$(2.1.26b) \qquad d_{i,j} \leq d_{i,j-1}, \; i\geq 1, j>1,$$

hold and $d_{i,i}$ converges:

$$(2.1.26c) \qquad d_{i,i} \searrow \delta_1 = 1+\varepsilon+\sqrt{2\varepsilon(1+\beta)},$$

whereas in general

$$(2.1.27a) \qquad d \geq d_{i,j} \geq \delta$$

with

$$(2.1.27b) \qquad d = \max\{d_{11}, d_{12}\},$$

and

$$(2.1.27c) \qquad \delta = \min\{\delta_1, \delta_2 = 1+\varepsilon+\sqrt{(2+\beta+\varepsilon)\,\varepsilon}\}$$

is satisfied.

Proof:

(2.1.26a,b) are readily proved by induction, using the monotonicity property of M-matrices. So these propositions apply to incomplete decompositions of a general M-matrix. By virtue of (2.1.26a,b) $d_{i,i}$ is monotonically decreasing. By the stability theorem of Meijerink - van der Vorst , [16], it is also bounded below by 0. Hence $d_{i,i}$ converges and the limit $\delta=1+\varepsilon+\sqrt{2\varepsilon(1+\beta)}$ can be computed from $\delta = 2(1+\varepsilon) - \frac{\varepsilon(\varepsilon-\beta)}{\delta} - \frac{1-\varepsilon\beta}{\delta}$. (2.1.27) follows from (2.1.26) for $\varepsilon\geq\beta$ and for $\varepsilon<\beta$ from the inequalities $d_{12} > d_{i2}$ for $i>1$, $d_{i1} < d_{ik} < d_{i2}$, for $k > 2$ and $d_{i,1} < d_{i+1,1}$ for $i\geq 1$ which is still valid. □

Remark 5:

Lemma 3 carries over to a general domain, provided an equidistant rectangular grid is used, treating boundaries by linear interpolation. Then the corresponding stencils are of the form

$$\begin{bmatrix} & -1 & \\ \cdot & g_{1j} & -\varepsilon \\ & -1 & \end{bmatrix}$$ at the left-hand boundary. The other ones are given analogously.

g_{ij} satisfies $g_{ij} \geq 2(1+\varepsilon)$. Thus it is sufficient to replace the splitting of $h^2 (K_l(\varepsilon)+\alpha N) = B_1+B_2$ in the proof of lemma 3 by $h^2 (K_l(\varepsilon)+\alpha N) = B_1+B_2+G$ with a diagonal matrix $G = \mathrm{diag}\{g_{ij}- 2(1+\varepsilon)\} \geq 0$ and B_1, B_2 from (2.1.15).

Combining criterion 1, lemma 3, and remark 5 and regarding that (2.1.9) and (2.1.10) are satisfied for $\beta\geq 1$ we obtain

Theorem 6:

Let $K_l(\varepsilon)$ be the usual 5-point discretization of the anisotropic problem from (1.1.2) on an arbitrary connected domain $\Omega \subset \mathbf{R}^2$. Then the ILU_β-smoother, defined by (1) - (5) is $K_l(\varepsilon)$-robust, provided $\beta > 0$. More precisely: The smoothing property

(2.1.28a) $$\|K_l(\varepsilon)S_l^\nu\| \leq \varepsilon\, h^{-2}\, C_S\, \eta_0(\nu-2,1/(\alpha-1))$$

with

(2.1.28b) $$C_S = \frac{2(1+\beta)}{\delta},$$

is satisfied, and

(2.1.29) $$\|T_{2,1}(\nu,0)\| \leq C\, \eta_0(\nu-2,1/(\alpha-1))$$

with

(2.1.30) $$\alpha = \begin{cases} \frac{2\delta}{(1-\beta)(1+\varepsilon)} & \text{for } 0<\beta<1 \\ \infty & \text{for } \beta\geq 1 \end{cases}$$

holds independently of the stepsize h_l and the anisotropy-constant ε.

2.2 The General Case

ILU_β can easily be extended to a general problem.

Let

(2.2.1) $$K u = f$$

be a system of linear equations with a sparse mxm-matrix K. Suppose that the usual incomplete decomposition using some prescribed sparse-pattern exists for K (cf. [16], [1]). I.e. there exist strictly lower and upper triangular matrices L and U and a diagonal matrix D > 0, so that

(2.2.2a) $$K = M - N$$

with

(2.2.2b) $$M = (L+D)D^{-1}(U+D)$$

and

(2.2.2c) $$N = \begin{pmatrix} n_{11} & \cdots & n_{1m} \\ \vdots & & \vdots \\ n_{m1} & \cdots & n_{mm} \end{pmatrix}$$

being some rest-matrix. Then for $\beta \in \mathbf{R}$ we define

(2.2.3) $$D_\beta = \operatorname{diag}\{d_i + \beta \sum_{j=1}^{m} |n_{ij}|\}$$

where d_i denote the elements of D. Replacing D in (2.2.2b) by D_β, we obtain ILU_β for (2.2.1):

(2.2.4a) $$K = M_\beta - N_\beta$$

with

(2.2.4b) $$M_\beta = (L+D_\beta)\, D_\beta^{-1}(U+D_\beta) \quad .$$

As mentioned above, this corresponds to some known modifications for certain values of β. All these modifications are designed to shift the spectrum of the rest-matrix N.

Remark 1:

Let K be positive definite and the sparse-pattern on which the incomplete decomposition is based include the one of K. Then for $\beta = 0$ the rest-matrix is indefinite, for $\beta = -1$, it is negative definite, while for $\beta \geq 1$ it is positive (semi-)definite. In addition, for $\beta = -1$ M can again be regarded as discretization of some differential operator, so that the largest eigenvalue of M^{-1} is bounded below. This allows MILU(=ILU_{-1}) to improve the asymptotic condition number (cf. [5]). As smoother, however, it is not suited at all, as the eigenvalues of $M^{-1}N$ are unbounded. So we concentrate here on $\beta \geq 0$.

Remark 2:

Since $D_\beta \geq D$ holds elementwise, the stability of ILU_β is ensured for M-matrices. Further, ILU_β is at least as stable as the usual ILU_0.

Proof:

Follows immediately from [16], theorem 2.3. □

Remark 3:
Let K be an M-matrix and $\beta \geq 0$. Then ILU_β from (2.2.4) is a regular splitting of K.

Proof:
By virtue of Remark 2 the ILU_β-decomposition is stable. According to [16], theorem 2.3 and its proof, ILU_0 from (2.2.2) is a regular splitting of K. As $D_\beta \geq D$ elementwise, $(U+D_\beta)^{-1} \geq 0$, $(L+D_\beta)^{-1} \geq 0$, $D_\beta > 0$ hold elementwise. So we obtain $M_\beta^{-1} \geq 0$ and $N_\beta \geq 0$, elementwise, completing the proof. □

Suppose $K_1(\varepsilon)$ is a discretization of a differential operator, depending on some parameter ε. Let $K_1(\varepsilon)$ be positive definite and symmetric. Let further (2.1.7ε) be the corresponding approximation-property. To compensate for the factor $1/\varepsilon$ we again need smoothing property (2.1.8ε). Using criterion 1.1, we obtain the following theorem.

Theorem 4:
Let $K_1(\varepsilon)$ be positive definite and symmetric for all $\varepsilon \geq 0$. Suppose ILU_1 to exist for $K_1(\varepsilon)$. Further let (2.1.10) hold.
Then ILU_β satisfies (2.1.8aε) for $\beta \geq 1$ and $\varepsilon \geq 0$.

Proof:
By virtue of remark 1 (2.1.9) is satisfied, yielding the proposition. □

Thus ILU_β is $K_1(\varepsilon)$-robust for $\beta \geq 1$.

The above construction also covers factorizations of higher order as 7-point and nine point stencils (see [24]) and incomplete block decompositions (see [1], [13]). There the same modification can be applied.

Now we turn to the practical aspects.

3. The Value of Local Fourier-Analysis to Predict the Practical Behaviour of a Multi-Grid Method

The first convergence proofs for multi-grid methods were given by analyzing model problems with Fourier-techniques(see [4], [6], [2]). Though being no rigorous proof for more complicated situations, local Fourier-analysis has been established as a tool to predict the practical performance of a multi-grid method (cf. [7], [3]). In many cases the results are rather reliable and offer a nice possiblity to check the correctness of a program (cf. [3]). Not so with ILU. The practical computations for model problem (1.1.3) using a 5-point ILU-smoother corresponding to ILU_0 from section 2, show the convergence factor κ_{10} to approach 0 for $\varepsilon \to 0$, as can be seen from fig. 1. κ_i is defined by

$$(3.1) \qquad \kappa_i = \sqrt[i]{\frac{\|r_i\|}{\|r_0\|}}$$

where r_i denotes the residual after step i.

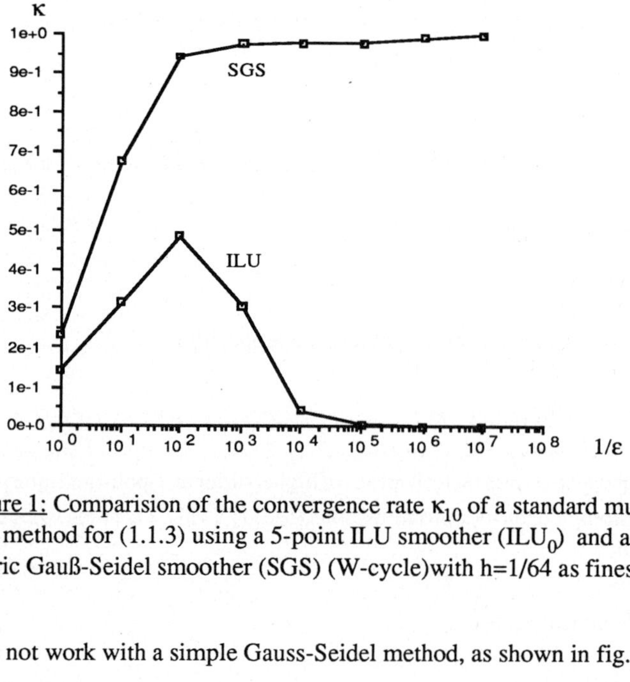

Figure 1: Comparision of the convergence rate κ_{10} of a standard multi-grid method for (1.1.3) using a 5-point ILU smoother (ILU_0) and a symmetric Gauß-Seidel smoother (SGS) (W-cycle)with h=1/64 as finest step-size.

This does not work with a simple Gauss-Seidel method, as shown in fig. 1.

The local Fourier-analysis, however, yields as prediction for the smoothing rate

$$(3.2) \qquad \rho_B(\varepsilon) = \frac{1}{1+2(\varepsilon+\sqrt{2\varepsilon})},$$

tending to 1, $\varepsilon \to 0$ (cf. [13], [19]). Up to now, this difference was thought to arise from boundary effects and/or rounding errors (cf [19], [17]). Interested in the reason for this discrepancy and also in the reason for the peak, showing up in fig.1, we computed the convergence rates for the same multi-grid method as in fig. 1 gradually refining the stepsize of the finest grid. Using $h_f = 1/2^i$, i=5,...,10, as finest stepsize we obtained the results shown in fig.3.

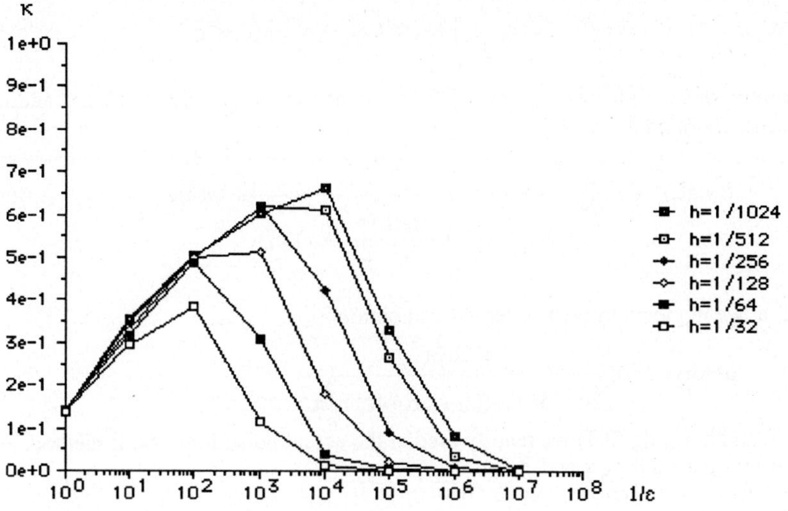

Figure 2: Convergence rate for the multi-grid method from fig.1 (ILU_0) with varying finest stepsize.

Obviously, the peak depends on ε and h. So we cannot expect a realistic prediction of the convergence-rate by usual local Fourier-analysis, as there $h \to 0$ is incorporated (cf. [2]). This limit, however, will never be reached on a real computer, as firstly computing time and storage are limited and secondly the condition number of $K_l(\varepsilon)$ is proportional to $1/h^2$, making computations with too small stepsizes and a finite number of digits senseless. It is mainly used, since multi-grid methods, involving rather fine grids, quickly approach the limit rates for usual problems (cf. [7]). Model problem (1.1.3), however, consists of a set of problems, controlled by the parameter ε which is ranging in a much larger interval than h. Hence it is reasonable to assume

(3.3) $\quad h \geq h_0 > 0$

for practical purposes. Usually, $h_0 \approx 10^{-3}$. Looking for a reasonable prediction of the practical behaviour of our method, we simply include the finite stepsize h into the local Fourier-analysis.

To predict the smoothing-rate, we have to compute

$$(3.4) \qquad \rho(\varepsilon, h_l) = \max_{\lambda \in \Lambda_{high}(M_l^{-1}(\varepsilon) K_l(\varepsilon))} \left| 1 - \lambda(M_l^{-1}(\varepsilon) K_l(\varepsilon)) \right|$$

with Λ_{high} being the set of eigenvalues corresponding to the high-frequency eigenfunctions (cf. [7], [11], [18]) and $M_l(\varepsilon)=M$ from (2.1.2). Computing these eigenvalues by local Fourier-analysis results in

$$\text{(3.5)} \qquad \rho(\varepsilon,h_l) = \max_{\mu,\nu \in J} \frac{\varepsilon}{\delta} \frac{|\cos(\mu-\nu)h_l\pi|}{\left|1+\varepsilon-\varepsilon\cos\mu h_l\pi - \cos\nu h_l\pi + \frac{\varepsilon}{\delta}\cos(\mu-\nu)h_l\pi\right|}$$

with $J = [1, n_l=1/h_l-1]^2 \setminus [1,n_{l-1}=1/h_{l-1}-1]^2 \cap \mathbb{N}^2$ and $\delta = 1+\varepsilon+\sqrt{2\varepsilon}$.

The maximum of the right-hand side of (3.5) is attained at $\mu = (1-h_l)/h_l$, $\nu = 1$. Expanding the cosine terms, we obtain

$$\text{(3.6)} \qquad \rho(\varepsilon,h_l) = \frac{\varepsilon}{\delta} \frac{1-2h_l^2\pi^2}{\varepsilon\left(2-\frac{1}{\delta}\right) + \left(1-\varepsilon+\frac{4\varepsilon}{\delta}\right)\frac{h_l^2\pi^2}{2} + O(h_l^4)} + O(h_l^4)$$

For small h we neglect terms of order ≥ 4 and obtain

$$\text{(3.7)} \qquad \rho(\varepsilon,h_l) \leq 2\varepsilon \frac{1-2h_l^2\pi^2}{2\varepsilon\,(2\delta-1)+((1-\varepsilon)\,\delta+4\varepsilon)\,h_l^2\pi^2}$$

for small h and ε. Using (3.7) we readily predict the actual behaviour of our method, as can be seen by comparing figs. 2 and 3.

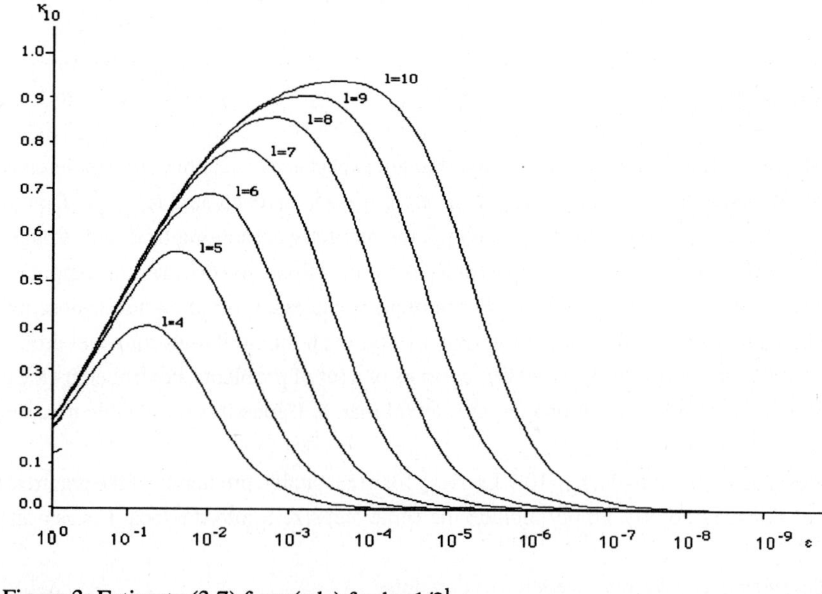

Figure 3: Estimate (3.7) for $\rho(\varepsilon,h_l)$ for $h_l=1/2^l$.

Thus we conclude with the following remark

Remark 1:

Suppose (3.3). Then the usual 5-point ILU scheme (ILU_0) is robust for $K_l(\varepsilon)$. This confirmes rather well the early assertion by P. Wesseling, [21], and R. Kettler, [13], about the robustness of incomplete LU-smoothing.

Admitting $h \to 0$, however, we are in the same situation as in section 2, where we had to require $\beta > 0$ in order to obtain robustness.

4. The Optimial Choice of β

4.1 Optimizing the Smoothing-Property Estimate

For practical purposes we are interested in the dependence of the convergence factor of a multigrid method with ILU_β-smoothing on β. To investigate about the influence of β in (2.1.28), we define

$$(4.1.1) \qquad \chi(\beta) = \frac{1+\beta}{\delta}\,\eta_0\left(\nu-2,\frac{1}{\alpha(\beta)-1}\right).$$

First we compute for $\nu\geq 3$

$$(4.1.2) \qquad \eta_1(\nu) := \min_{\beta\in(0,1]} \eta_0\left(\nu-2,\frac{1}{\alpha(\beta)-1}\right) = \min_{\beta\in(0,1]} \max\left\{\frac{(\nu-2)^{\nu-2}}{(\nu-1)^{\nu-1}},\ \frac{\alpha(\beta)}{(\alpha(\beta)-1)^{\nu-1}}\right\}.$$

As $(\nu-2)^{(\nu-2)}/(\nu-1)^{\nu-1}$ does not depend on β, while $\alpha/(\alpha-1)^{\nu-1}$ is monotonously decreasing with $\beta\in[0,1]$, we look for β_2:

$$(4.1.3) \qquad \frac{\alpha(\beta_2)}{(\alpha(\beta_2)-1)^{\nu-1}} = \frac{(\nu-2)^{\nu-2}}{(\nu-1)^{\nu-1}}.$$

Let $\xi = \xi(\nu)$ denote a solution of

$$(4.1.3') \qquad \xi = \frac{(\nu-2)^{\nu-2}}{(\nu-1)^{\nu-1}}(\xi-1)^{\nu-1}$$

for fixed $\nu\in\mathbb{N}$. For $3\leq\nu\leq 11$ the maximum values of ξ are given in table 1.

Table 1: $\xi(\nu)$ from (4.1.3') for $3\leq\nu\leq 11$.

ν	3	4	5	6	7	8	9	10	11
$\xi(\nu)$	5.828	4.0	3.379	3.063	2.879	2.740	2.646	2.574	2.518

Equating $\alpha(\beta_2) = \xi(\nu)$ with $\alpha(\beta)$ from (2.1.30), we obtain for $\beta < 1$

(4.1.4) $$\frac{2\delta}{(1-\beta_2)(1+\varepsilon)} = \xi .$$

Using (2.1.27) and performing some elementary calculations, we end up with

(4.1.5) $$\beta_2 = \frac{\xi(\xi-2)(1+\varepsilon)^2 - 4\varepsilon - 4\sqrt{\varepsilon(\varepsilon+\xi(1+\varepsilon)^2)}}{\xi^2(1+\varepsilon)^2} .$$

β_2 satisfies $\beta_2 \le 1$, and $\beta_2 \ge 0$ holds, provided

(4.1.6) $$\xi \ge 2 + \frac{\sqrt{8\varepsilon}}{1+\varepsilon} .$$

For $2 \le \nu \le 10$ the values of ε for which (4.1.6) is valid are given in table 2.

<u>Table 2:</u> ε_0, so that (4.1.6) holds for ε: $0 \le \varepsilon \le \varepsilon_0$.

ν	3	4	5	6	7	8	9	10	11
ε_0	1	1	0.638	0.205	0.121	0.079	0.058	0.045	0.036

Hence

(4.1.7) $$\beta_{opt} = \begin{cases} \beta_2 & \text{for } \xi \ge 2+\frac{\sqrt{8\varepsilon}}{1+\varepsilon} \\ 0 & \text{else} \end{cases}$$

and

(4.1.8) $$\eta_1(\nu) = \begin{cases} \left\{ \begin{array}{ll} \frac{\alpha(\beta)}{(\alpha(\beta)-1)^{\nu-1}} & \text{for } 0 \le \beta < \beta_{opt} \\ \frac{(\nu-2)^{\nu-2}}{(\nu-1)^{\nu-1}} & \text{for } \beta_{opt} \le \beta \le 1 \end{array} \right\} & \text{for } \xi \ge 2+\frac{\sqrt{8\varepsilon}}{1+\varepsilon} \\ \frac{(\nu-2)^{\nu-2}}{(\nu-1)^{\nu-1}} & \text{else} \end{cases}$$

$\chi(\beta)$ will attain its minimum in β_{opt}, since $\alpha(1+\beta)/((\alpha-1)^\nu\delta)$ decreases monotonously, while $(1+\beta)/\delta$ increases with β. So for $\beta=\beta_{opt}$ estimate (2.1.29) will be optimal and we may expect this choice to yield a reasonable prediction of the practically best parameter.

Figure 1 shows β_{opt} from (4.1.7) for $\nu = 3,\ldots,7$.

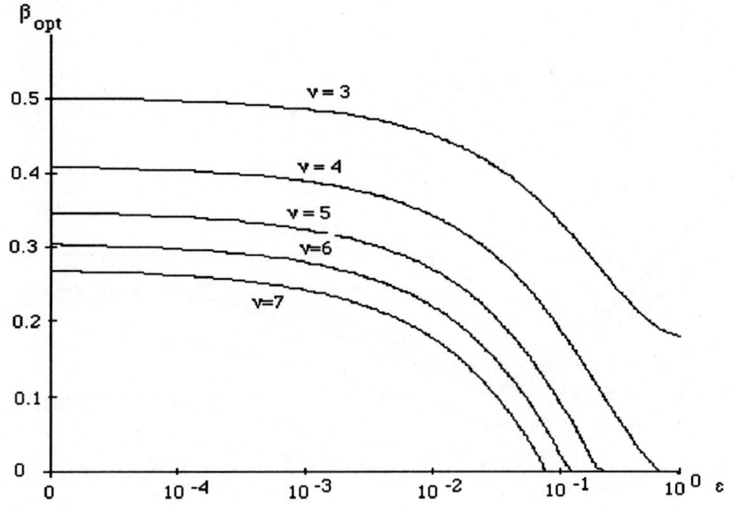

Figure 1: β_{opt} from (4.1.7) for $\nu = 3,\ldots, 7$.

4.2. Numerical Results

To check the validity of our theoretical considerations from sec. 4.1 we implemented ILU_β for problem (1.1.3) and tested the convergence-rate for several parameters h, ε and β. The results are given in the following figures.

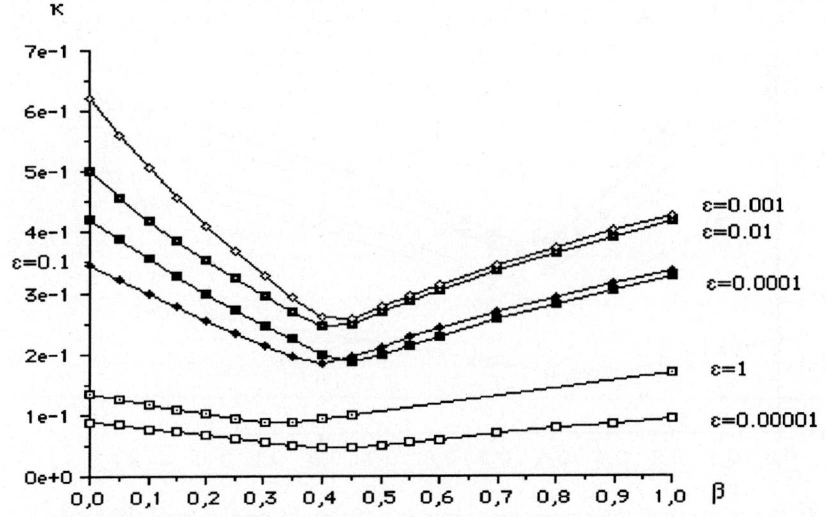

Figure 1: Dependence of the convergence-rate κ_{10} of a multi-grid method (W-cycle) with ILU_β-smoothing on β with finest stepsize h=1/256 and ν=1.

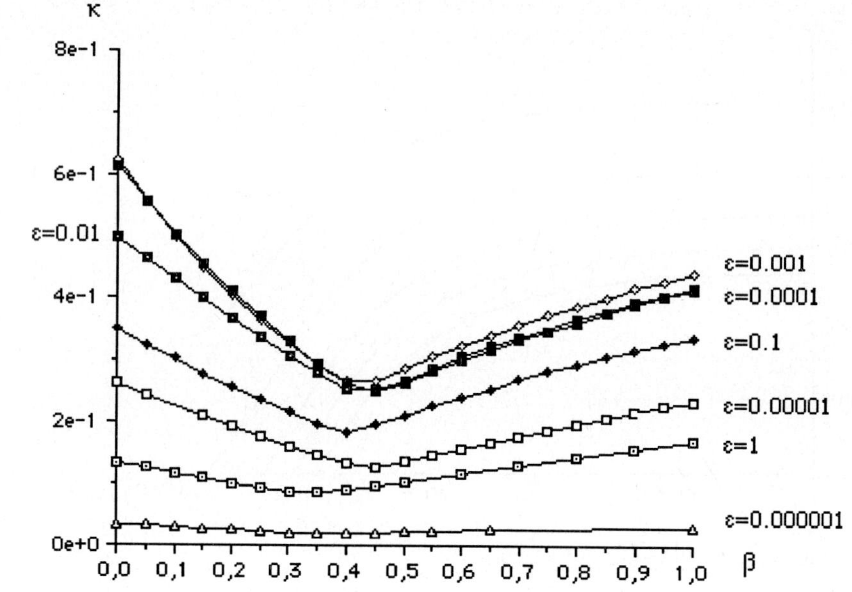

Figure 2: Dependence of the convergence-rate κ_{10} of a multi-grid method (W-cycle) with ILU_β-smoothing on β with finest stepsize h=1/512 and ν=1.

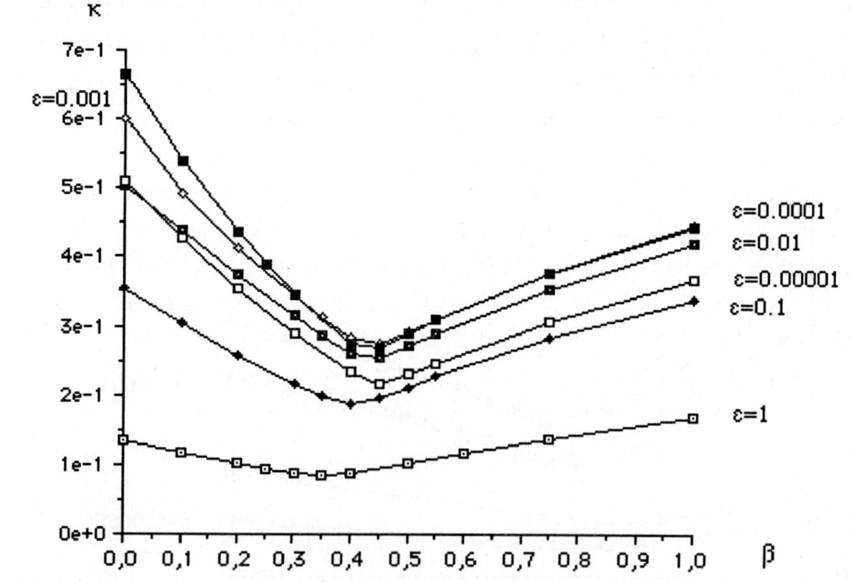

Figure 3: Dependence of the convergence-rate κ_{10} of a multi-grid method (W-cycle) with ILU_β-smoothing on β with finest stepsize h=1/1024 and ν=1.

The curves in figs. 1-3 show the behaviour predicted in sec. 4.1. So $\kappa(\beta)$ attaines a well defined minimum in (0,1) depending on ε.

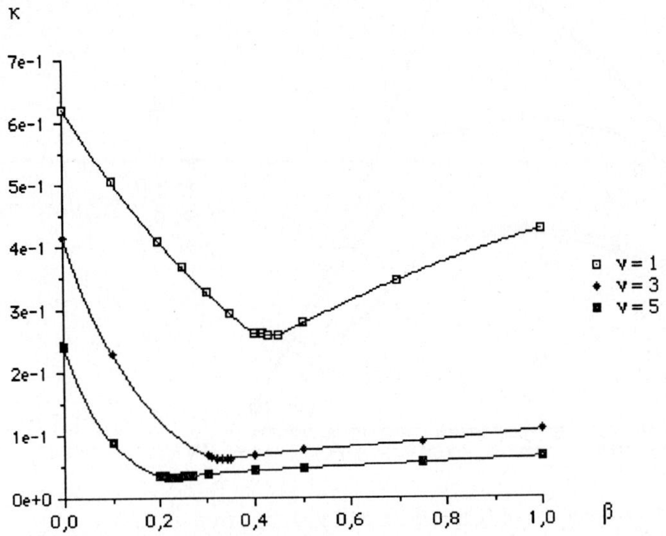

Figure 4: $\kappa_{10}(\beta)$ for $h=1/512$, $\varepsilon=10^{-3}$ and variable ν.

Figure 4 shows the dependence of $\kappa(\beta)$ on ν for $\varepsilon=10^{-3}$. As predicted by (4.1.7), β_{opt} tends to 0 with increasing ν. Comparing $\beta_{opt}(\varepsilon,\nu)$ from (4.1.7) with the optimal value of β from figures 1-4, however, we have to be careful about the correct choice of ν. It is well-known that splitting (2.1.6) of $T_{2,1}(\nu,0)$ is primarily constructed to allow a short and well-structured qualitative convergence-proof for sufficiently large ν and tends to give an overestimate for the number of smoothing steps required, which can be avoided only by constructing special problem-dependent norms (cf. [7]). In addition, we had to pay for estimate (2.1.28) by taking $\eta(\nu-2)$ instead of $\eta(\nu)$. Hence, we cannot expect ν in our theoretical considerations to be identical with the number of smoothing steps used in practice. A comparison of the practical results as given in figs. 1-4 with the values of β_{opt} from (4.1.7), shows that $\nu=3$ in (4.1.7) corresponds to the results for one smoothing-step and a W-cycle. Once this number is fixed, our predictions using (4.1.7) for different values of ε and ν turn out to be good approximations to the optimal values of β, as shown in figures 5 and 6.

From figs. 5 and 6 we see that β_{opt} from (4.1.7) yields a resonable prediction for the practically best parameter $\beta_{opt,em}$, as the convergence rates obtained for the empirically determined $\beta_{opt,em}$ differ only approx. 0.02 from the ones obtained with $\beta=\beta_{opt}$. For higher ν $\beta_{opt,em}$ is not so easy to determine, as the convergence rates are extremely fast and differ only very little in a neighbourhood of β_{opt}.

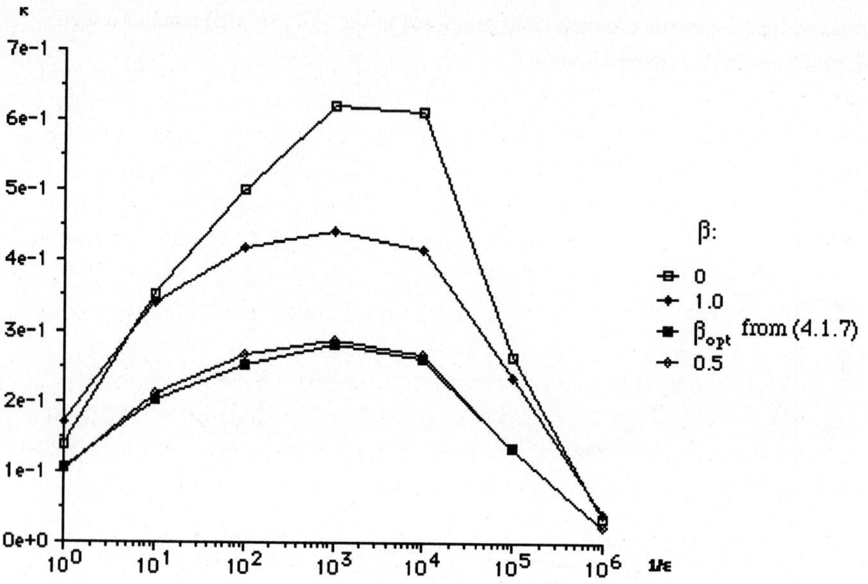

Figure 5: $\kappa(\varepsilon)$ for different values of β, h=1/512, and ν=1. β_{opt} from (4.1.7)

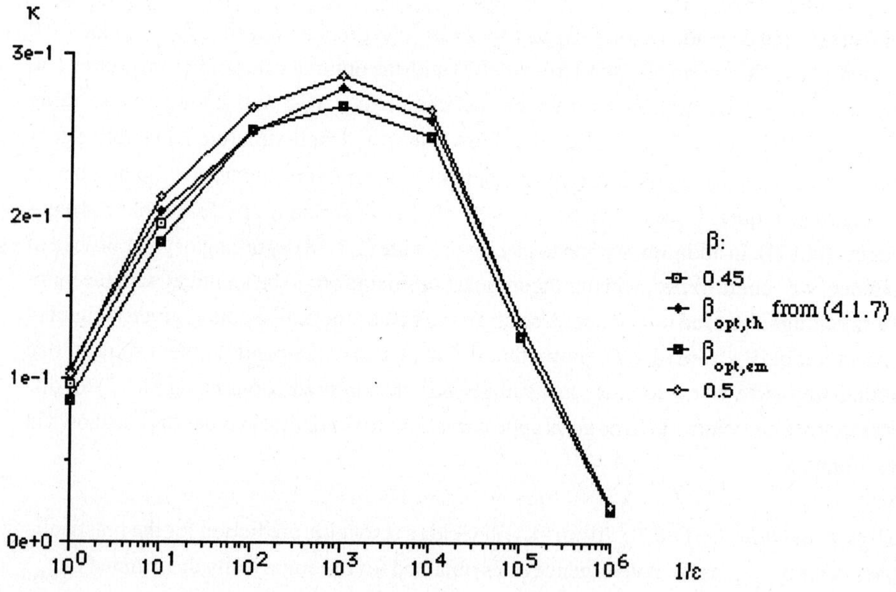

Figure 6: Comparison of convergence factors for different values of β for ν=1 and h=1/512.

Of course, β can also be optimized by local Fourier-techniques which has actually been done for $\varepsilon=0$ by Khalil, [14] and Oertel, [17], and for $\varepsilon=1$ by Hemker, [10]. These techniques, however, suffer from certain drawbacks, since they do not include effects arising from the finite dimension of the problem and from the boundary, which are rather important for ILU-factorizations as pointed out in sec. 3. In addition, it seems impossible to obtain an analytical formulation for β_{opt} depending on ε and ν by local Fourier-analysis (cf. [15]). So Oertel, [17] and Khalil, [14], recommend to apply in general $\beta = 0.5$. This is the correct choice for small ε and $\nu = 1$. There the practice, local Fourier-analysis, and (4.1.7) agree. For $\varepsilon=1$, $\nu=1$, and $h=1/512$ we get $\beta_{opt,em} \approx 0.35$ from fig.2, while $\beta_{opt} = 0.18$ from (4.1.7), and the result of Hemker from [10] obtained with local Fourier-analysis corresponds to $\beta_{opt,F} = 0.14$, showing that Hackbusch's approach may yield results which are even closer to the empirical ones than those produced by local-Fourier analysis. However, this is no problem, since the corresponding convergence rates do not differ much, as can be seen from fig. 4. The same holds for the dependence on ν. Hence it is reasonable to use $\beta = 0.5$ or better $\beta = 0.45$ for all ε and ν.

We tested ILU_β also in connection with our transforming approach for the Navier-Stokes equations (cf. [25]). There the results are very similar.

Thus we conclude stating that ILU_β-smoothing is robust and should be applied with $\beta = \beta_{opt}(\varepsilon,\nu)$ from (4.1.7) or $\beta = 0.45$. Thus the original ILU-factorization developped by Meijerink and van der Vorst, [16], shows to be an ingenious guess, comprising nice properties as preconditioner as well as smoother and providing a basis for many variants optimized for special purposes.

References:

[1] Axelsson,O., Barker,V.A.: Finite element solutions of boundary value problems. Theory and computation. Academic Press, New York, 1985.

[2] Brandt,A.: Multi-level adaptive solutions to boundary-value problems. Math. Comp. 31 (1977), 333-390.

[3] Brandt,A.: Guide to multigrid development. In: Hackbusch,W.,Trottenberg,U.(eds.): Multi-grid methods. Proceedings, Lecture Notes in Math. 960, Springer, Berlin (1982).

[4] Fedorenko,R.P.: A relaxation method for solving elliptic difference equations. USSR Comput. Math. and Math. Phys. 1,5 (1961), 1092-1096.

[5] Gustafsson,I.: A class of 1st order factorization methods. Report 77.04, Department of Computer Sciences, Chalmers University Göteborg (1977).

[6] Hackbusch,W.: Ein iteratives Verfahren zur schnellen Auflösung elliptischer Randwertprobleme. Report 76-12, Universität Köln (1976).

[7] Hackbusch,W.: Multi-grid methods and applications.
Springer, Berlin, Heidelberg (1985).

[8] Hackbusch,W.: Theorie und Numerik elliptischer Differentialgleichungen.
Teubner, Stuttgart (1986).

[9] Hackbusch,W.: A new approach to robust multi-grid solvers.
Preprint, Institut für Informatik und praktische Mathematik, CAU, Kiel (1987).

[10] Hemker,P.W.: The incomplete LU-decomposition as a relaxation method in multigrid algorithms.In: Miller,J.J.H.(ed.): Boundary and interior layers - computational and asymptotic methods . Proceedings,, Boole Press Dublin (1980).

[11] Hemker,P.W.: Fourier analysis of grid functions, prolongations and restrictions.
NW 93/80, Mathematisch Centrum, Amsterdam (1980).

[12] Jennings,A., Malik,G.M.: Partial elimination.
J. Inst. Maths. Applics. 20 (1977), 307-316.

[13] Kettler,R.: Analysis and comparison of relaxation schemes in robust multi-grid and preconditioned conujugate gradient methods. In:Hackbusch,W.,Trottenberg,U.(eds.): Multi-grid methods. Proceedings, Lecture Notes in Math. 960, Springer, Berlin (1982).

[14] Khalil,M.: On some modifications of ILU-smoothers.
Lecture given at the 4.GAMM-Seminar Kiel, 22.-24.1.1988.

[15] Liebau,F.: Unvollständige LU-Zerlegungen mit angenäherten Koeffizienten als Glätter in Mehrgitterverfahren. Diplomarbeit, Institut für Informatik und praktische Mathematik, Kiel (1987).

[16] Meijerink,J.A, Van der Vorst,H.A.: An iterative solution method for linear systems of which the coefficient matrix is a symmetric M-matrix.
Math. Comp. 31 (1977), 148-162.

[17] Oertel,K.-D.: ILU-smoothing: Systematic tests and improvements. Lecture given at the 4. GAMM-Seminar Kiel, 22.-24.1.1988.

[18] Stüben,K., Trottenberg,U.: Multigrid methods: fundamental algorithms, model problem analysis and applications.In: Hackbusch,W., Trottenberg,U.(eds.): Multi-grid methods. Proceedings, Lecture Notes in Math. 960, Springer, Berlin (1982).

[19] Thole,C.A.: Beiträge zur Fourieranalyse von Mehrgitterverfahren. Diplomarbeit, Universität Bonn (1983).

[20] Varga,R.S.: Matrix iterative analysis. Prentice Hall (1962).

[21] Wesseling,P.: A robust and efficient multigrid method. In: Hackbusch,W., Trottenberg,U.(eds.): Multi-grid methods. Proceedings, Lecture Notes in Math. 960, Springer, Berlin (1982).

[22] Wesseling,P.: Theoretical and practical aspects of a multigrid method. SIAM J. Sci. Statist. Comp. 3 (1982), 387-407.

[23] Wesseling,P., Sonneveld,P.: Numerical experiments with a multiple grid and a preconditioned Lanczos type method.In: Rautmann,R.(ed.): Approximation methods for Navier-Stokes Problems.Lecture Notes in Math. 771. Springer Berlin (1980), 543-562.

[24] Wittum,G.: Distributive Iterationen für indefinite Systeme. Thesis, Universität Kiel, 1986.

[25] Wittum,G.: Multi-grid methods for Stokes and Navier-Stokes equations. Transforming smoothers – algorithms and numerical results. Submitted to Numer. Math.

List of Participants:

Prof. Dr. J. Albrecht, Institut für Mathematik,
Technische Universität Clausthal, Erzstraße 1, 3392 Clausthal-Zellerfeld 1

Manfred Alef, KfK-IDT, Postfach 3640, 7500 Karlsruhe 1

Prof. O. Axelsson, Department of Mathematics, University of Nijmegen,
Toernooiveld, NL - 6525 ED Nijmegen

P. Bastian, IMMD III, Universität Erlangen, Martensstraße 3, 8520 Erlangen

C. Becker, Strömungsmechanik, Universität Erlangen,
Egerlandstraße 13, 8520 Erlangen

V. Bertram, Institut für Schiffbau, Universität Hamburg,
Lämmersieth 90, 2000 Hamburg 60

Frau M. Bloß, Antonstraße 25, 1000 Berlin 65

Dr. H. Blum, Fachbereich Angewandte Mathematik und Informatik,
Universität des Saarlandes, 6000 Saarbrücken

A. von Borzyskowski, Institut für Angewandte Mathematik, Universität Hamburg,
Bundesstraße 55, 2000 Hamburg 13

Chwalinski, IABGmbH, Abteilung TFF, Einsteinstr. 20, 8012 Ottobrunn

Prof. Dr. L. Collatz, Institut für Angewandte Mathematik, Universität Hamburg,
Bundesstraße 55, 2000 Hamburg 13

P. Conradi, Technische Elektronik, Technische Universität Hamburg-Harburg,
Eissendorfer Straße 38, 2100 Hamburg 90

Prof. Dr. W. Dahmen, Institut für Mathematik III,
Freie Universität Berlin, 1000 Berlin 33

Dr. R. Damm, DFVLR, Institut für Physikalische Chemie,
Pfaffenwaldring 38/40, 7000 Stuttgart 80

J. Dehnhardt, Institut für Angewandte Mathematik, Universität Hannover,
Welfengarten 1, 3000 Hannover

Prof. Dr. P. Deuflhard, Konrad-Zuse-Zentrum,
Heilbronnerstraße 10, 1000 Berlin 31

Dr. E. Dick, Department of Mach., University of Gent,
Sint Pietersnieuwstraat 41, B - 9000 Gent

J. Dörfer, Institut für Angewandte Mathematik, Universität Düsseldorf,
Universitätsstraße 1, 4000 Düsseldorf

Thomas Dreyer, SFB 123, Universität Heidelberg,
Im Neuenheimer Feld 294, 6900 Heidelberg

Dr. U. Gärtel, Gesellschaft für Mathematik und Datenverarbeitung F1/T, Postfach 1240, 5205 Sankt Augustin 1

C. Gebhardt, Institut für Angewandte Mathematik, Universität Hannover, Welfengarten 1, 3000 Hannover

Dr. D. Hänel, Aerodynamisches Institut, RWTH Aachen, Templergraben 55, 5100 Aachen

Priv.Doz. Dr. F.K. Hebeker, Fachbereich Mathematik THD, Schloßgartenstraße 7, 6100 Darmstadt

E. Heineck, Friedelstraße 37, 1000 Berlin 44

Dr. Wilhelm Heinrichs, Angewandte Mathematik, Universitätsstraße 1, 4000 Düsseldorf 1

Priv. Doz. Dr. Ronald Hoppe, Fachbereich 3, Technische Universität Berlin, Straße des 17.Juni 135, 1000 Berlin 12

Graham Horton, IMMD III, Universität Erlangen, Martensstraße 3, 8520 Erlangen

Miloš Huněk, CKD - Prague, Compressors, Klecakova 1947, Prague 9, 192 00 CSSR

G. Jensen, Institut für Schiffbau, Universität Hamburg, Lämmersieth 90, 2000 Hamburg 60

Mohammed Khalil, Laboratoire d'Analyse Numerique, Batiment 1R1, 118, Route de Narbonne, F - 31062 Toulouse Cedex

B. Koren, Centre for Mathematics, P O Box 4079, NL - 1009 AB Amsterdam

Dr. Ralf Kornhuber, Konrad-Zuse-Zentrum, Heilbronner Straße 10, 1000 Berlin 31

Dr. K. Kozel, Katedra vypocetni techniky a informatiky, FSI CVUT Praha, Suchbatarova 4, CS - 16607 Praha

A. Krechel, GMD - F1/T, Postfach 1240, 5205 Sankt Augustin 1

Dr. Jürgen Kux, Institut für Schiffbau, Universität Hamburg, Lämmersieth 90, 2000 Hamburg 60

P. Leinen, Boltestr. 34, 4630 Bochum 7

G. Lill, Fachbereich Mathematik, Universität Darmstadt, Schlossgartenstraße 7, 6100 Darmstadt

Frau Y. Luh, GMD F1/T, Postfach 1240, 5205 Sankt Augustin 1

R. Maehr, Am Roten Stein 9 G, 1000 Berlin 22

Dr. B. Mantel, AMD-BA, F - 92214 St.Cloud Cedex

Y. Marx, CFD Group ENSM, 1, rue de la Noe, F - 44072 Nantes Cedex

Dipl.-Ing. J. Mayer, FORD-Werke Köln, Produktentwicklung MD/PN-220,
John-Andrews-Entwicklungszentrum, Postfach 606002, 5000 Köln

D. F. Mayers, Oxford University, Computing Laboratory,
Numerical Analysis Group, 8 - 11 Keble Road, GB Oxford OX1 3QD

Dr. B. Müller, DFVLR-AVA, SM-TS, Bunsenstraße 10, 3400 Göttingen

Rafael Munoz-Sola, I.N.R.I.A., Domaine de Voluceau, BP 105,
78153 Le Chesnay Cedex, France

Prof. Dr. Frank Natterer, Institut für Numerische und Instrumentelle Mathematik,
Westf. Wilhelms-Universität, Einstein Straße 62, 4400 Münster

K.-D. Oertel, GMD - F1/T, Postfach 1240, 5205 Sankt Augustin 1

Prof. Dr. H. Pfau, Ingenieurhochschule, Abteilung Mathematik,
Philipp-Müller-Straße, DDR - 24 Wismar

J. Piquet, CFD Group ENSM, 1, rue de la Noe, F - 44072 Nantes Cedex

C. Pöppe, SFB 123, Universität Heidelberg,
Im Neuenheimer Feld 294, 6900 Heidelberg

P. Rem, Shell Research BV, Badhuisweg 3, NL - 1031 CM Amsterdam

Arnold Reusken, Mathematisch Instituut, Rijksuniversiteit Utrecht,
Budapestlaan 6, NL - 3508 TA Utrecht

Michael Riedel, Seydlitzstraße 47, 1000 Berlin 46

Dipl.-Ing. W. Rust, Institut für Baumechanik und Numerische Mechanik,
Universität Hannover, Appelstraße 9A, 3000 Hannover

Dr. R. Sawatzky, Institut für Angewandte Mathematik,
Universität Hamburg, Bundesstraße 55, 2000 Hamburg 13

Dr. Georg Scheuerer, Lehrstuhl für Strömungsmechanik,
Universität Erlangen-Nürnberg, Egerlandstraße 13, 8520 Erlangen

Dr. H. Schippers, National Lucht- en Ruimtevaartlaboratorium,
Postbus 153, NL - 8300 AD Emmeloord

Dr. W. Schmidt, Dornier GmbH, BF 30, 7990 Friedrichshafen

Dr.-Ing. Dietmar Schröder, TU Hamburg-Harburg - Technische Elektronik -,
Eißendorfer Straße 38, 2100 Hamburg 90

M. Schulte, FU Berlin, Fachbereich Mathematik III,
Arnimallee 2-6, 1000 Berlin 33

G. J. Shaw, Computing Laboratory, Oxford University,
8-11 Keble Road, B Oxford, OX1 3QD, England

M. Smoch, Institut für Numerische und Instrumentelle Mathematik,
Westf. Wilhelms-Universität, Einstein Straße 62, 4400 Münster

Th. Sonar, DFVLR Braunschweig, Institut für Entwurfsaerodynamik,
Am Flughafen, 3200 Braunschweig

Prof. Dr. E. Spedicato, Dipartimento di Matematica, Universita di Bergamo,
Via Salvecchio, 19, I - 24100 Bergamo

Clemens-August Thole, SUPRENUM, Hohe Straße 73, 5300 Bonn 1

S. Turek, Fachbereich Angewandte Mathematik und Informatik,
Universität des Saarlandes, 6000 Saarbrücken

Dr. V. Vasanta Ram, Institut für Thermo- und Fluiddynamik,
Ruhr-Universität Bochum, Postfach 102148, 4630 Bochum 1

Dr. P. Vassilevski, Institute of Mathematics, Bulgarian Academy of Sciences,
1090 Sofia, P.O.Box 373, Bulgaria

Prof. Dr. Rüdiger Verfürth, Institut für Angewandte Mathematik,
Universität Heidelberg, Im Neuenheimer Feld 294, 6900 Heidelberg

Dr. G. Warnecke, Mathematisches Institut A, Universität Stuttgart,
Pfaffenwaldring 57, 7000 Stuttgart 80

Prof. Erich Wehrhahn, Philips Kommunikations Industrie AG,
Kommunikationssysteme, Postfach 4943, 8500 Nürnberg 10

Prof. Dr. Pieter Wesseling, Onderafdeling der Wiskunde en Informatica,
Technische Hogeschool Delft, Julianalaan 132, NL - 2628 BL Delft

Frau D. Wessels, Institut für Numerische und Instrumentelle Mathematik,
Westf. Wilhelms-Universität, Einstein-Straße 62, 4400 Münster

C. Weyand, Mathematisches Institut, Universität Düsseldorf,
Universitätsstraße 1, 4000 Düsseldorf

Dr. P. Wild, Forschingszentrum BBC, CH - 5405 Baden - Dättwil

Prof. Dr. K. Witsch, Mathematisches Institut, Universität Düsseldorf,
Universitätsstraße 1, 4000 Düsseldorf

Dr. Gabriel Wittum, SFB 123, Universität Heidelberg,
Im Neuenheimer Feld 294, 6900 Heidelberg

Zhao, Institut für Schiffbau, Universität Hamburg,
Lämmersieth 90, 2000 Hamburg 60

Addresses of the editors of the series „Notes on Numerical Fluid Mechanics":

Prof. Dr. *Ernst Heinrich Hirschel* (general editor)
Herzog-Heinrich-Weg 6
D-8011 Zorneding
FRG

Prof. Dr. *Kozo Fujii*
High-Speed Aerodynamics Div.
The ISAS
Yoshinodai 3-1-1, Sagamihara
Kanagawa 229
Japan

Prof. Dr. *Keith William Morton*
Oxford University Computing Laboratory
Numerical Analysis Group
8-11 Keble Road
Oxford OX1 3QD
Great Britain

Prof. Dr. *Earll M. Murman*
Department of Aeronautics and Astronautics
Massachusetts Institut of Technology (M.I.T.)
Cambridge, Ma 02139
U.S.A.

Prof. Dr. *Maurizio Pandolfi*
Dipartimento di Ingegneria Aeronautica e Spaziale
Politecnico di Torino
Corso Duca Degli Abruzzi, 24
I-10129 Torino
Italy

Prof. Dr. *Arthur Rizzi*
FFA Stockholm
Box 11021
S-16111 Bromma 11
Sweden

Dr. *Bernard Roux*
Institut de Mećanique des Fluides
Laboratoire Associé au C.R.N.S. LA 03
1, Rue Honnorat
F-13003 Marseille
France